# Exploring the Impact of OTT Media on Global Societies

Nithin Kalorth
*Mahindra University, India*

IGI Global
PUBLISHER of TIMELY KNOWLEDGE

A volume in the Advances in Media, Entertainment, and the Arts (AMEA) Book Series

Published in the United States of America by
IGI Global
Information Science Reference (an imprint of IGI Global)
701 E. Chocolate Avenue
Hershey PA, USA 17033
Tel: 717-533-8845
Fax: 717-533-8661
E-mail: cust@igi-global.com
Web site: http://www.igi-global.com



Library of Congress Cataloging-in-Publication Data

Names: Kalorth, Nithin, 1988- editor.
Title: Exploring the impact of OTT media on global societies / edited by Nithin Kalorth.
Description: Hershey, PA : Information Science Reference, [2024] | Includes bibliographical references and index. | Summary: "This book aims to discuss influence, challenges, and implications of over the top media on the global societies"-- Provided by publisher.
Identifiers: LCCN 2024008898 (print) | LCCN 2024008899 (ebook) | ISBN 9798369335260 (hardcover) | ISBN 9798369335277 (ebook)
Subjects: LCSH: Mass media--Social aspects. | Mass media--Economic aspects.
Classification: LCC HM1206 .E995 2024 (print) | LCC HM1206 (ebook) | DDC 302.23--dc23/eng/20240312
LC record available at https://lccn.loc.gov/2024008898
LC ebook record available at https://lccn.loc.gov/2024008899

This book is published in the IGI Global book series Advances in Media, Entertainment, and the Arts (AMEA) (ISSN: 2475-6814; eISSN: 2475-6830)

British Cataloguing in Publication Data
A Cataloguing in Publication record for this book is available from the British Library.

# Advances in Media, Entertainment, and the Arts (AMEA) Book Series

Giuseppe Amoruso
Politecnico di Milano, Italy

ISSN:2475-6814
EISSN:2475-6830

## Mission

Throughout time, technical and artistic cultures have integrated creative expression and innovation into industrial and craft processes. Art, entertainment and the media have provided means for societal self-expression and for economic and technical growth through creative processes.

The **Advances in Media, Entertainment, and the Arts (AMEA)** book series aims to explore current academic research in the field of artistic and design methodologies, applied arts, music, film, television, and news industries, as well as popular culture. Encompassing titles which focus on the latest research surrounding different design areas, services and strategies for communication and social innovation, cultural heritage, digital and print media, journalism, data visualization, gaming, design representation, television and film, as well as both the fine applied and performing arts, the AMEA book series is ideally suited for researchers, students, cultural theorists, and media professionals.

## Coverage

- Fabrication and prototyping
- Gaming
- Drawing
- Cross-Media Studies
- Film & Television
- Blogging & Journalism
- Popular Culture
- Design of Interiors
- Visual Computing
- Applied Arts

IGI Global is currently accepting manuscripts for publication within this series. To submit a proposal for a volume in this series, please contact our Acquisition Editors at Acquisitions@igi-global.com or visit: http://www.igi-global.com/publish/.

The Advances in Media, Entertainment, and the Arts (AMEA) Book Series (ISSN 2475-6814) is published by IGI Global, 701 E. Chocolate Avenue, Hershey, PA 17033-1240, USA, www.igi-global.com. This series is composed of titles available for purchase individually; each title is edited to be contextually exclusive from any other title within the series. For pricing and ordering information please visit http://www.igi-global.com/book-series/advances-media-entertainment-arts/102257. Postmaster: Send all address changes to above address. 

## Titles in this Series

*For a list of additional titles in this series, please visit:*
*http://www.igi-global.com/book-series/advances-media-entertainment-arts/102257*

***Advancements in Socialized and Digital Media Commnications***
Gülbuğ Erol (Iğdır University, Turkey) and Michael Kuyucu (Alanya University, Turkey)
Information Science Reference • © 2024 • 364pp • H/C (ISBN: 9798369308554) • US $230.00

***Using Traditional Design Methods to Enhance AI-Driven Decision Making***
Tien V. T. Nguyen (Industrial University of Ho Chi Minh City, Vietnam) and Nhut T. M. Vo (National Kaohsiung University of Science and Technology, Taiwan)
Information Science Reference • © 2024 • 503pp • H/C (ISBN: 9798369306390) • US $245.00

***Survival Strategies and Transformations of the Global Media Industry Mitigating the Impact of the COVID-19 Pandemic***
Floribert Patrick C. Endong (University of Dschang, Cameroon)
Information Science Reference • © 2024 • 300pp • H/C (ISBN: 9798369305737) • US $250.00

***Digital Space in the Wake of Shrinking Traditional Media***
Alfred Okoth Akwala (Technical University of Kenya, Kenya) Denish Ouko Otieno (Kenya Institute of Mass Communication, Kenya) and Amos Moraro Marube (Technical University of Kenya, Kenya)
Information Science Reference • © 2024 • 300pp • H/C (ISBN: 9781668493564) • US $225.00

***News Media and Hate Speech Promotion in Mediterranean Countries***
Elias Said Hung (Universidad Internacional de la Rioja, Spain) and Julio Montero Diaz (Universidad Internacional de la Rioja, Spain)
Information Science Reference • © 2023 • 364pp • H/C (ISBN: 9781668484272) • US $215.00

***Examinations and Analysis of Sequels and Serials in the Film Industry***
Emre Ahmet Seçmen (Beykoz University, Turkey)
Information Science Reference • © 2023 • 389pp • H/C (ISBN: 9781668478646) • US $215.00

***Using Innovative Literacies to Develop Leadership and Agency Inspiring Transformation and Hope***
Limor Pinhasi-Vittorio (Lehman College, CUNY, USA) and Elite Ben-Yosef (The BYEZ Foundation, USA)
Information Science Reference • © 2023 • 287pp • H/C (ISBN: 9781668456149) • US $215.00

***Music and Engagement in the Asian Political Space***
Uche Titus Onyebadi (Texas Christian University, USA) and Delaware Arif (University of South Alabama, USA)
Information Science Reference • © 2023 • 254pp • H/C (ISBN: 9781799858171) • US $215.00

IGI Global
PUBLISHER of TIMELY KNOWLEDGE

701 East Chocolate Avenue, Hershey, PA 17033, USA
Tel: 717-533-8845 x100 • Fax: 717-533-8661
E-Mail: cust@igi-global.com • www.igi-global.com

# Table of Contents

# Detailed Table of Contents



Popular culture reinforces dominant ideologies, maintaining existing power structures by normalizing specific beliefs and values. This results in the marginalization of certain social groups and distorted media representations. However, resistance movements and counter-narratives challenge these norms, particularly in postcolonial contexts where the concept of the 'subaltern' is prominent. The expansion of OTT platforms in India and Pakistan has stimulated conversations on sensitive topics and disrupted conventional entertainment. OTT series employ subaltern characters to shed light on the challenges, fostering a sense of empathy. India and Pakistan, despite their differences, share cultural similarities. This study compares the portrayal of 'subalterns' in both nations, analyzing the highly rated Indian web series, 'Dahaad,' and Pakistani series, 'Churails.' Guided by Gayatri Spivak's subaltern theory, the study assesses how these OTT series, whether fully or partially embracing this theory, contribute to equitable depictions of marginalized communities and enduring social change.



This chapter delves into the dynamic landscape of over-the-top platforms, focusing on Bengali content, exploring their growth trajectory, unique initiatives, and content diversification. It highlights the emergence of ventures like Hoichoi, which has garnered widespread acclaim through its original and diverse subscription strategies. It underscores the role of quality content in attracting audiences and emphasizes the significance of stories. It discusses how regional content, including adaptations from literature, has succeeded, illustrating this with examples from Hoichoi's innovative projects. Furthermore, the chapter sheds light on the growing demand for women-centric narratives and how OTTs provide a platform for progressive portrayals of women's roles and issues. Despite the progress, the chapter acknowledges challenges, including the need for sustained originality and the expanding competition in the OTT sphere. In conclusion, it foresees a promising future for Bengali content on OTT platforms, driven by creative storytelling, audience preferences, and an evolving landscape.



This research tries to review the growth of OTT platforms throughout the lockdown. It's vital to understand the extent of increase within the popularity of OTT platforms throughout lockdown to understand their future scope. It's evident that, since their launch, OTT platforms have solely discovered an upward curve in their usage, and because of the pandemic and lockdown there has been exponential increase in its popularity because of the modification in consumption patterns of individuals for diversion through numerous media platforms. This analysis conducted a survey and analyzed the opinions of individuals relating to OTT platforms, their consumption patterns, and its comparison with cinema to examine if OTT platforms were slowly taking on the foremost fashionable standard medium of diversion. It had been found that individuals used OTT over the other platforms outside of TV and YouTube to pass their time or for diversion.



This article studies celebrity status in digital media, focusing on OTT platforms and social media. New fame paths have changed famous dynamics. This has changed content development, audience engagement, personal branding, and professional-personal boundaries. This study compares digital celebrities to film and television stars, and finds a complex world where renown is achieved by competence, deliberate online presence, and active audience connection. This study examines digital celebrities' professional trajectories, their capacity to migrate into traditional media, and OTT platform fame's future. The results show that digital fame requires constant ingenuity, flexibility, and digital pattern understanding. With more material and specialist personalities, over-the-top (OTT) platforms have a promising stardom potential. This study analyses modern celebrity culture and its potential effects on digital stardom.



This chapter examines the growth and influence of over-the-top (OTT) media within China's unique economic and legal context. It examines the business models, revenue sources, and revolutionary shift in content consumption and creation patterns brought about by major platforms like iQiyi, Tencent Video, and Youku. The regulatory practices in China are highlighted, along with the potential and difficulties that they provide to the OTT scene. Questions concerning media pluralism, content quality, and algorithmic recommendation systems are highlighted as the authors look at the broader societal ramifications, including the effect on cultural trends, celebrity culture, and the film and television industry.



This chapter is bent at, through audience perspectives, examining the role of over-the-top media; precisely, streamed South African soap operas in the creation of hybrid cultures and the impact of these cultures on the nation building exercise in Zimbabwe. This qualitative chapter utilises the uses and gratifications theory and the reception theory as theoretical toolkits. Its empirical data is drawn from in-depth interviews with sampled Zimbabwean journalists who stream South African soap operas. The chapter fleshes out that South African soap operas have cultivated new and hybrid cultures in Zimbabwe—among them acceptance of the LGBT community, acceptance of ATR, the use of different languages during communication processes and the appreciation of traditional dressing among others. Empirical data from the participants further exhibits that these hybrid cultures have subsequently become an ideological engineering hub for nation building motives as they carry discourses that can be used for unifying the Zimbabwean people.



This chapter explores uses and gratifications of watching "The Last Dance," a sports documentary series that premiered in 2020 on ESPN and Netflix about Michael Jordan and the 1990s Chicago Bulls basketball team. The study examines the influence of uses and gratifications on brand-user imagery fit and the intention to watch a Chicago Bulls game after having seen the show. Primary data was collected through a self-administered online questionnaire from people that had been watching "The Last Dance." Results were computed through exploratory and confirmatory factor analysis and structural equation modeling. Results suggest that the realistic nature of the story and characters portrayed in "The Last Dance," the relaxation experienced from watching the show, and social interaction with other viewers influence the intention to watch a Chicago Bulls game when mediated through brand-user imagery fit. These are new findings to the respective field of study and contribute towards the understanding for producing similar documentaries.

## Section 3
## Psychological and Social Implications



Exploring the prosperous union of New Malayalam Cinema and OTT platforms, this chapter scrutinises the circumstances under which the former thrived on OTTs during the COVID-19 pandemic. It delves into Malayalee market trends, the content diversification incited by OTTs, and the expansion of New

The battle between JioCinema and Star Sports for live sports viewership reached its peak with Indian Premier League (IPL) 23. In stage one, this chapter uses the case study method to present the story of IPL 23 viewership in the light of free subscriptions. Data was collected through secondary sources like newspaper articles, online newsfeeds, syndicated reports, and expert comments over a period of six months. This data was then analysed on the basis of audience size, advertiser dynamics, and advertising revenues for both JioCinema and Star Sports. Focusing on free subscriptions as a tactic and the sustainability of such promotions on the long-term standing of an over-the-top platform (OTT), an OTT platform performance model (PPM) is proposed in stage two of this chapter. The model offers a five-pronged approach to long term sustainability of an OTT platform.



The way we consume video content has undergone a radical transformation. Gone are the days of VCRs and brick-and-mortar video lending stores. Today, digital streaming platforms using the internet have become a global business. The seeds of streaming were sown in the 1990s when audio compression technologies like MP3 came into being and services like RealPlayer and Microsoft's Windows Media Player allowed users to listen to music of their choice. This laid the groundwork for video streaming. The 2000s saw the launch of digital video streaming, changing the entertainment landscape forever. YouTube, founded in 2005, paved the way for user-generated content. Netflix, initially a DVD rental service, switched to an unlimited viewing streaming service in 2007, based on a monthly model. These platforms faced challenges like bandwidth and piracy, but fueled by technological advancements and an insatiable consumer demand for on-demand entertainment, they thrived.

# Foreword

I am happy to write the foreword and introduce the book, *Exploring the Impact of OTT Media on Global Societies*, edited by Dr Nithin Kalorth. This book comes at a time when OTT platforms across the world have changed the way audiences are finding and consuming entertainment content.

They are now consuming more content of their choice thanks to the reach and penetration of OTT platforms. This has had four major effects. One, it has ended the dominance of television and cinema; two, it has expanded the audience; three, it has led to the emergence of new content creators who explore themes that were earlier considered unconventional; and four, it is has started impacting social and cultural values globally.

In this book, leading academicians and researchers from across the world have shared their findings as to how OTT content is impacting societies in their respective countries. There is a detailed chapter that explores the role of OTT platforms in advancing digital inclusion in the Global South.

Equally fascinating are two studies of cross-border content. One study explores the influence of South African soap operas on Zimbabwean society as they stream across the border. Another provides a riveting comparison of how characters are portrayed in OTT serials in Inda and Pakistan. The book also provides a glimpse of the rise of OTT platforms in China.

An important feature of OTT content is its regional penetration. OTT platforms have opened new markets and brought in inclusivity through content in regional languages. The book examines the impact of OTT content on regional language audiences in India, with two case studies – one on Bengali markets and another on Malayali audiences.

The book also explores the challenges that this new platform is posing to the deeply entrenched film industry in India. Besides this, there are two interesting studies on the impact of sports broadcasting – one on Indian Premier League and another on Chicago Bulls.

I must congratulate the Editor for visualising such a book. It analyses important trends related to OTT content in different parts of the world, especially its impact on social and cultural norms, and fills an important gap in the knowledge about this emerging platform for film and media students, scholars and OTT researchers.

*Sunil Saxena*
*Independent Researcher, India*

**Sunil Saxena** *is an award-winning media professional with four and a half decades of experience in Digital Media, Social Media, Mobile Journalism, Print Journalism, Media Education and Research. He piloted the award-winning Gaon Ki Awaaz, India's first voice-based news and information service for rural India. He is the author of four media books; Breaking News, Headline Writing, Web Journalism 2.0 and The Power of Content Marketing.*

# Preface

The edited volume *Exploring the Impact of OTT Media on Global Societies* is an anthology that aims to discuss influence, challenges, and implications of OTT media on the global societies. The rise OTT media is understood as a technological shift in distribution and consumption and also an opportunity for content producers to place their narrative with a new visual grammar and language. The OTT platform are appreciated by all stakeholders in the entertainment business and the race to the top is also intense.

The OTT platforms and content has impacted both business and audiences. The shift in the concept of audience to user is an important turn in understanding in digital medium. The current volume is an attempt to explore OTT not only as conduits of content but catalysts for societal transformation. The chapters of this books are contributed by scholars from different parts of the world. The topics covered in the chapters are of interdisciplinary nature with transnational understanding of impact of OTT in global societies. The book is divided into four thematic sections such as: (i) Shaping the Future of OTT Media, (ii) User Experience, (iii) Psychological and Social Implications; and (iv) Societal Impact and Inclusion. The chapters within this volume span a wide spectrum of topics, covering the technological infrastructure underpinning OTT services, the intricacies of film and series production, distribution strategies, regulatory and policy considerations, and the profound social and cultural implications of the OTT media revolution. It is our belief that this book fills a crucial gap in the literature, offering fresh insights into the visual media landscape from the vantage point of OTT media.

This book will be useful for media and entertainment professionals, media students and researchers, user experience designers, technology enthusiasts, policy makers and societal impact analysts.

The first chapter of the book and the section Shaping the Future of OTT Media is *The 'Subaltern' Will Speak: Investigating Portrayals in Acclaimed OTT Series of India & Pakistan* by Dr. Saad Ullah Khan and others is the first chapter of the section Shaping the Future of OTT Media. The chapter conducted an in-depth analysis of critical dimensions related to the portrayal of subaltern characters within the context of selected OTT series of India and Pakistan. The authors aims to illuminate the underlying power dynamics that shape the depiction of these characters, providing a nuanced understanding of their representation. The chapter goes beyond surface-level analysis by undertaking a comprehensive examination of storytelling techniques employed in crafting subaltern characters. By doing so, the chapter aims to uncover the intricate nuances inherent in the narrative construction of these characters, shedding light on the intentional choices made by creators in shaping their identities within the series.

Next chapter, *Rise of OTT Bengali Originals: Niche Content for the Diverse Audiences* by Lokesh Chakma and Dr. Mausumi Bhattacharyya. This chapter explores the transformative impact of the Covid-19 pandemic on the entertainment landscape, specifically the surge in popularity of Over-The-Top (OTT) platforms in India. Focusing on the rise of Bengali original content, the chapter outlines the changing viewer preferences, accelerated adoption of OTT platforms, and the strategic shift of major players like Disney+Hotstar, Netflix, Amazon Prime Video, and regional platforms like Hoichoi and Addatimes. The chapter emphasize the significance of regional content, the chapter highlights the demand for niche content among OTT subscribers, reflecting the platforms' ability to offer diverse, authentic, and hyper-localized storytelling. The chapter discusses major trends in the OTT market, the rise in demand for original content, and the potential prospects for major production companies in the evolving digital era. The chapter also covers path-breaking initiatives, such as Hoichoi's efforts to expand its reach globally, launch Hoichoi Studios, and curate remarkable content, underscoring the dynamic and evolving nature of the OTT landscape in the Bengali content market.

In chapter *Study on Growth of OTT Platform during Lockdown and its Future Scope: Study on Growth of OTT* by Sakshi Raj provides a comprehensive understanding of the factors influencing the growth of OTT platforms during the Covid 19 lockdown. The chapter is result of systematic quantitative study in order to understand the patterns of consumer behaviour related to entertainment consumption during and after the lockdown. The chapter concludes that there has been a significant increase in OTT consumption during the pandemic, supporting the hypothesis. The research also validates the prediction that OTT platforms may surpass cinema in terms of popularity, with a noticeable rise in user preference for OTT over traditional media. The findings suggest that OTT platforms could pose a potential threat to cinema halls in the future, emphasizing the changing landscape of entertainment consumption.

The last chapter of the section *The Rise of Digital Celebrities Analyzing the Impact and Sustainability of Stardom on OTT Platforms* by Dr. Archan Mitra and Sayani Das studies the impact of the digital revolution on the concept of fame. The chapter tried to explain the distinctive characteristics that differentiate digital celebrities from traditional ones, emphasizing their proficiency in leveraging OTT platforms and social media for global audience connections. Digital celebrities excel in content creation, brand representation, and influence-building, employing strategies centered on audience engagement, authenticity, and adaptability. The chapter details the complex relation between digital and traditional media, emphasizing the importance of managing personal and professional boundaries in the digital celebrity's career. Despite facing challenges marked by swift transformations and the need for constant innovation, successful digital celebrities showcase exceptional adaptability, transcending into conventional media platforms. Looking towards the future of stardom on OTT platforms, the chapter suggest that the outlook is promising, marked by a diverse content landscape and the emergence of specialized digital stars.

The second section of the book User Experience starts with the chapter *Riding the Dragon: The Emergence and Impact of Over-The-Top Media in China*, by Mohamad Zreik and Runping Zhu, examines the growth and influence of OTT media in China, including business models, regulatory practices, and societal ramifications, offering insights into cultural trends, celebrity culture, and the film and television industry. The chapter discusses important questions concerning media pluralism, content quality, and algorithmic recommendation systems are highlighted as the chapter look at the broader societal ramifications, including the effect on cultural trends, celebrity culture, and the film and television industry.

The chapter titled *The Role of South African Streamed Soap Operas in Contemporary Cultural Imperialism and Implications on the Project of Nation Building in Zimbabwe*, by Jennings Chibike, explores the impact of South African streamed soap operas on the creation of hybrid cultures in Zimbabwe, using audience perspectives and theoretical toolkits to analyze the role of OTT media in nation-building. The qualitative chapter points that South African soap operas have cultivated new and hybrid cultures in

Zimbabwe among them acceptance of the LGBV community, acceptance of ATR, the use of different languages during communication processes and the appreciation of traditional dressing among others.

The last chapter of this section *Uses and Gratifications of 'The Last Dance' and the Viewing Intention for a Chicago Bulls Game*, by Sebastiano Mereu, is an exceptional chapter which looks into the uses and gratifications of watching the sports documentary series "The Last Dance," exploring its influence on brand-user imagery fit and the intention to watch a Chicago Bulls game. The chapter uses exploratory and confirmatory factor analysis and structural equation modeling techniques to bring new results in the field of research in documentaries.

Next section is dedicated to Psychological and Social Implications of OTT media and its starts with the chapter *Streaming Success: Examining the Impact of OTT Media on Content Production and Consumption of New Malayalam Cinema*, by Sanjna T., scrutinizes the successful union of New Malayalam Cinema and OTT platforms during the COVID-19 pandemic, exploring market trends, content diversification, and changing audience consumption patterns.

The chapter titled *Analyzing Narrative Structure of a Tent-Pole Bollywood Film Versus Exclusive Indian OTT Cinema*, authored by Kanika Arya, Manish Verma, and Anisha Jain, compares the narrative structure of a traditional Indian cinema film and an exclusive Indian OTT release, offering insights into changing storytelling techniques. The chapter is built upon studying two popular Bollywood films Jab We Met and Meenakshi Sundareshwar which released in traditional method and OTT representatively.

The chapter *The Impact of OTT Media on Comedy in India: Changing Trends and New Opportunities*, by Sonali Jha and Tamanna Shah, explores the impact of OTT media on the comedy industry, analyzing case studies of comedians and examining trends and opportunities for content creation and audience engagement. The chapter adopts content analysis method which provides analysis of the trends and opportunities of digital platforms for comedians to experiment with content creation, reach varied audiences, and build personal brands.

*A Review of OTT Platforms in India: Challenges and Current Status*, by Nitish Ojha and Abhishek Vaish, critically reviews the regulations, content quality, and accessibility of OTT platforms in India, addressing the challenges and proposing considerations for future legislation. This chapter reviews previously passed regulations, their current status matching the current situation, and the need for a new mechanism if any. This chapter also discusses the pros and cons of censorship hinges, its challenges, and the future scope for new legislation matching the ethics in the shadow of globalization.

The last section of the book Societal Impact and Inclusion starts with the chapter *Bridging the Digital Divide: Exploring the Role of OTT Platforms in Advancing Digital Inclusion in the Global South*, authored by Nitesh Behare, Suraj Sharma, Varada Inamdar, Swapnali Bhosale, and Shrikant Waghulkar, explores the role of OTT platforms in advancing digital inclusion in the Global South, analyzing initiatives, challenges, and the implications for socio-economic development. The chapter analyzes initiatives and strategies employed by OTT platforms to reach remote populations, impact of affordable pricing models, localized content and innovative approaches to overcome connectivity challenges. The challenges in the Global South are also examined in this chapter including infrastructure limitations, connectivity barriers, privacy concerns, regulatory frameworks and ensuring equitable access for all populations.

The chapter written by Antoine Trad titled *Business Transformation Projects: The Impact of the Over-the-Top*, presents the Business Transformation Project (BTP) and the impact of Over-the-Top (OTT) based Platforms and Group Work (OTTBGRW), offering insights into their application across various business domains. This innovative chapter provides a novel insight in understanding OTT media from a business project management framework which is often neglected in mainstream media and communication domain.

The chapter *OTT and Live Sports: IPL'23 and its Impact on the Future of Sports Broadcasting in India*, by Josraj Arakkal and Preetha Menon, examines the unique case of free live sports telecasts of IPL 2023 in the era of subscription-based models, proposing a conceptual framework for its existence and sustenance.

The last chapter of the section and book *Digital Streaming Content: Changing Social Perceptions*, by Vidya Deshpande and Nithin Kalorth, concludes the collection by tracing the evolution of digital streaming platforms, emphasizing the radical transformation in the consumption of video content globally.

Together, these chapters form a comprehensive exploration of the multifaceted impact of OTT media on global societies, covering diverse regions, cultures, and thematic dimensions. The insights and analyses presented in this edited reference book will contribute significantly to the field of media and communication studies, offering valuable perspectives for scholars, researchers, and media professionals navigating the evolving landscape of Over-the-Top media.

The aim is to bring together a collection that not only contributes to the academic discourse in media and communication studies but also serves as a valuable resource for media professionals, policymakers, and anyone intrigued by the rapid evolution of Over-the-Top media. The comprehensive insights presented in this book will inspire further research, spark discussions, and guide future developments in the field. In a world where digital media continues to be a driving force, the exploration of OTT media in this collection is not merely an academic exercise but a reflection of the transformative power these platforms hold in shaping narratives, influencing cultures, and creating new paradigms for media production and consumption. This book hopes to serve as a cornerstone for understanding the complex interplay between technology, media, and society, and inspire continued exploration into the fascinating world of Over-the-Top media.

*Nithin Kalorth*
*Mahindra University, India*

## Section 1

# Shaping the Future of OTT Media

# Chapter 1
# The 'Subaltern' Will Speak:
## Investigating Portrayals in the Acclaimed OTT Series of India and Pakistan

**Saad Ullah Khan**
https://orcid.org/0000-0002-3392-7844
*Amity University Rajasthan, Jaipur, India*

**Sadaf Khan**
https://orcid.org/0009-0005-1348-0386
*IEC College of Engineering and Technology, Greater Noida, India*

**Tanushri Mukherjee**
https://orcid.org/0000-0001-5120-7982
*Amity University Rajasthan, Jaipur, India*

## ABSTRACT

*Popular culture reinforces dominant ideologies, maintaining existing power structures by normalizing specific beliefs and values. This results in the marginalization of certain social groups and distorted media representations. However, resistance movements and counter-narratives challenge these norms, particularly in postcolonial contexts where the concept of the 'subaltern' is prominent. The expansion of OTT platforms in India and Pakistan has stimulated conversations on sensitive topics and disrupted conventional entertainment. OTT series employ subaltern characters to shed light on the challenges, fostering a sense of empathy. India and Pakistan, despite their differences, share cultural similarities. This study compares the portrayal of 'subalterns' in both nations, analyzing the highly rated Indian web series, 'Dahaad,' and Pakistani series, 'Churails.' Guided by Gayatri Spivak's subaltern theory, the study assesses how these OTT series, whether fully or partially embracing this theory, contribute to equitable depictions of marginalized communities and enduring social change.*

DOI: 10.4018/979-8-3693-3526-0.ch001

## INTRODUCTION

Digital era has ushered a significant revolution, reshaping how stories are created and experienced. The emergence of Over-The-Top (OTT) platforms, offering a wide range of content, has not only made the entertainment industry more accessible but has also provided subtle ground for exploring intricate social and cultural narratives (Santhanam, 2020). Primarily in case of India and Pakistan, it has been largely witnessed that the digital realm has given voice to stories that were previously sidelined or neglected by mainstream media (Roy, 2014). India and Pakistan, neighbouring nations with intertwined histories, have produced a wealth of cultural content over the years, reflecting their diverse societies. While traditional cinema and television have played significant roles in shaping cultural narratives, the advent of OTT platforms has broken away from conventional constraints. This newfound freedom not only fosters creative experimentation but also opens doors to discussions that were previously considered taboo or politically sensitive (Bohra, 2021).

Globalization has fostered the proliferation of new media technologies, transforming the way information and entertainment are disseminated worldwide. Over-the-top (OTT) platforms epitomize this correlation by leveraging digital technology and the internet to transcend geographical boundaries. They enable a global audience to access a diverse array of content, reshaping traditional media consumption patterns and illustrating how the convergence of globalization and new media technology has revolutionized the media landscape (Tamandehrou & Khan, 2015). Further, a significant surge in internet usage and smartphone adoption in both the countries have made digital platforms more accessible than ever before (Shehzad, 2017). OTT platforms has brought about a revolution in how people consume entertainment, offering the convenience of accessing content while transcending geographical and temporal boundaries. In India, platforms such as Netflix, Amazon Prime Video, Disney+ Hotstar, among others, have not only made international content readily available but have also fostered the growth of regional content in numerous languages (Tiwari, 2023). This diversification of narratives has allowed for the exploration of themes deeply rooted in specific cultural contexts. Similarly, Pakistan has witnessed the rise of platforms like iFlix and ZEE5, tailored to local audiences and showcasing Pakistani content (Sharma, 2021). This digital transformation has provided storytellers with a platform to address a broader spectrum of issues, including those that impact marginalized communities.

Amidst this ever-evolving landscape, a thought-provoking inquiry emerges: how do highly regarded OTT series depict the experiences of marginalized communities? In many ways, the portrayal of marginalized voices in OTT series mirrors the broader discourse on representation, inclusion, and diversity. It reflects how societies perceive themselves and others, carrying profound implications for social change and mutual understanding (Ghosh, 2023). In this research, the researchers embark on a journey through the digital narratives of India and Pakistan, aiming to uncover the subtleties, challenges, and achievements within the stories of the marginalized.

Narratives possess a unique ability to engage, provoke, and resonate with audiences. In the realm of OTT series, storytelling takes centre stage, and the way marginalized voices are portrayed can significantly influence viewers' perceptions and empathy toward these communities. It is within this narrative power that researchers begin their exploration. The portrayal of marginalized voices in OTT series encompasses a wide range of themes and topics, including narratives related to caste discrimination, gender inequality, religious tensions, socioeconomic disparities, gender rights, and more. These narratives may draw inspiration from real-life events or be entirely fictional, but their resonance with the lived experiences of marginalized communities is what makes them compelling and important. These stories provide a platform for marginalized voices to be heard, challenge stereotypes, and contribute to social awareness and change.

Despite the heavily militarized political borders and seven-and-a-half decades of separation, India and Pakistan exhibit striking similarities in their cultural fabric, societal norms, languages, artistic expressions, psychology, and ideological underpinnings (Hasan, 2005). Across these shared cultural dimensions, elites are frequently portrayed as influential and captivating figures, prompting an intriguing examination of the depictions of 'subalterns' and the identity of these marginalized groups in both countries. Therefore, a significant aim of this study is to uncover any potential parallels or distinctions in the portrayal of marginalized communities within the context of these two nations, with the goal of offering insights into this significant inquiry.

Henceforth, this study employs the Indian web series 'Dahaad' as a representative case study, with 'Churails' serving as its Pakistani counterpart. Both web series have garnered critical acclaim and have achieved exceptional IMDb ratings. The theoretical framework underpinning this study draws upon Gayatri Spivak's 'subaltern theory,' a theoretical construct of notable significance (Morris, 2010). This theory underscores the imperative of achieving a more comprehensive and equitable representation of marginalized groups within various media contexts. The implications of this theoretical perspective are readily observable in ongoing initiatives aimed at amplifying the voices of marginalized populations, challenging prevailing dominant narratives, and fostering the incorporation of intersectional perspectives in the portrayal of marginalized communities within the realm of OTT series and other forms of media. Consequently, this research embarks on a rigorous scholarly inquiry, driven by the aspiration to gain a profound understanding of how the adoption of the 'subaltern theory,' whether in its entirety or in a modified form, holds the potential to contribute to a more equitable and just representation of marginalized communities. This endeavour seeks to assess the extent and multifaceted ways through which such contributions can foster sustainable social change within the broader context of media, and in this case, OTT centric representations.

## SIGNIFICANCE OF THE STUDY

The analysis of how the subaltern is depicted in OTT web series provides an opportunity for the researchers to gauge the degree to which marginalized communities are showcased and made visible within the realm of popular culture. This examination serves to discern whether these series serve as agents of either challenging or reinforcing prevailing power structures and stereotypes, as well as whether they serve as platforms through which subaltern voices can find both auditory and visual expression. OTT web series possess the capacity to present counter-narratives and alternative viewpoints that can disrupt and question the dominance of established ideologies and representations (Miller, Liu, & Ball, 2020). Consequently, the exploration of how these series portray the subaltern allows researchers to focus into the ways in which they destabilize or subvert conventional narratives and, in doing so, contribute to the cultivation of a more inclusive and diverse cultural milieu.

## RELEVANCE OF THE STUDY

OTT web series frequently serve as mirrors and commentators on prevalent social and political concerns. The examination of how the subaltern is depicted within these series offers researchers an avenue to scrutinize how they engage with issues related to societal disparities, power structures, and the experiences of marginalized communities. This analysis provides a means for researchers and readers to grasp the nuanced socio-political commentary embedded within these portrayals and their potential to enhance awareness and catalyse social change. Furthermore, it facilitates an exploration of the decisions made by creators, production teams, and platforms in their representation of the subaltern. Additionally, it affords an opportunity to assess audience reactions to these portrayals and their influence on public perspectives and attitudes.

## OBJECTIVES OF THE STUDY

This study is dedicated to the in-depth exploration of critical dimensions related to the representation of subaltern characters in OTT web series. Its central objective is to closely examine the way these characters are depicted within the context of these series, thereby shedding light on the underlying power dynamics that permeate their portrayal. Furthermore, the study undertakes a comprehensive analysis of the storytelling techniques strategically employed in the crafting of subaltern characters, with the goal of uncovering the intricate nuances inherent in their narrative construction. Additionally, the researchers delve into the interplay of social and political commentary intricately woven into the representation of these subaltern characters. This multifaceted analysis provides valuable insights into the broader societal and political implications that arise from their portrayal.

## DECODING THE 'SUBALTERN': A THEORETICAL OVERVIEW

The theoretical foundation underpinning the concept of the 'subaltern' finds its origins in the scholarship of Antonio Gramsci, an influential Italian Marxist thinker (Thomas, 2018). Over time, this framework has undergone further refinement and expansion through the contributions of scholars from diverse academic fields. The term 'subaltern' serves as a descriptor for individuals or groups who find themselves situated within disadvantaged or marginalized positions in society, often confronting pervasive systems of oppression and societal exclusion. Within the framework of subaltern theory, it is crucial to perceive the subaltern not as a fixed, unchanging identity, but rather as a dynamic concept that underscores the complex interaction of power dynamics, social hierarchies, and historical contexts that perpetuate the marginalization of specific groups (Mallon, 1994). This perspective underscores the imperative of amplifying the voices and experiences of the subaltern, as well as the necessity of challenging prevailing dominant narratives while simultaneously acknowledging the agency and resilience of marginalized communities. Furthermore, subaltern theory embraces the principle of intersectionality, recognizing that individuals may simultaneously belong to multiple marginalized groups, thus shaping their distinctive experiences. This theoretical framework holds particular significance within the realms of postcolonial studies, cultural studies, and social justice movements, providing an analytical lens through which issues of inequality, representation, and transformative social change can be comprehensively examined and addressed.

Gayatri Spivak's interpretation of the subaltern extends beyond Gramsci's original conception, encompassing various dimensions of subordination within society, including class, caste, age, gender, and office. Her examination of the subaltern, as exemplified in her seminal essay 'Can the Subaltern Speak?' (1988), is deeply rooted in the concepts of representation and agency. She argues that the subaltern, frequently marginalized and silenced by prevailing discourses, faces significant hurdles in expressing their experiences and hardships. Spivak contends that intellectuals or elites often speak on behalf of the subaltern, resulting in the suppression of their authentic voices (Prakash, 1994). Her inquiry raises fundamental queries regarding the limitations of representation and highlights the pivotal role of intellectuals in advocating for the subaltern's right to self-expression.

Ranajit Guha, a key figure in the Subaltern Studies collective, approaches the concept from a historical perspective. Guha's definition is anchored in the historical experiences of subaltern groups, notably within the context of colonial India. He places significant emphasis on the agency of subaltern classes in resisting colonial domination and in shaping their own historical narratives. Guha's work underscores the vital importance of uncovering the hidden histories of subaltern communities and recognizing their substantial contributions to resistance movements and broader social transformations (Macwan, 2023). Dipesh Chakrabarty extends the subaltern notion to encompass broader considerations related to temporality and historiography. His scholarship zero in on into the intricate challenges associated with constructing historical narratives from the subaltern perspective. Chakrabarty grapples with the complexities of representing the diverse voices and experiences inherent in subaltern groups within historical accounts. His definition underscores the necessity of adopting a more inclusive and polyphonic approach to historical writing, one that acknowledges the existence of multiple temporalities and subjectivities within subaltern communities (Chakrabarty, 1998).

Partha Chatterjee's scholarly endeavours have played a seminal role in elucidating the poststructuralist foundations of the Subaltern Studies movement. His work signifies a notable departure from the traditional, class-centric interpretation of the 'subaltern'. Instead, Chatterjee's contributions have paved the path toward a more comprehensive, decentralized, and dynamically flexible comprehension of oppression and resistance. This exquisite understanding emanates from a thorough examination of the silences, voids, discrepancies, and contradictions inherent within elite sources and historiography, as underscored by Dworkin ((Radhakrishnan, 1994).

The evolution of Subaltern Studies along this trajectory has been significantly shaped by deconstructive critiques from luminaries such as Gayatri Spivak and Edward Said. A pivotal moment in this journey was marked by the publication of 'The Selected Subaltern Studies' in 1988, scrupulously edited by Ranajit Guha and Gayatri Chakrabarty, featuring a preface by Edward Said. Edward Said's intellectual framework, inspired notably by Michel Foucault's seminal works, particularly 'Archaeology of Knowledge' and 'Discipline and Punish', exerted a profound influence on the Subaltern Studies project. Under Said's guidance, the focus of Subaltern Studies underwent a noticeable shift, increasingly directed towards the critical examination of colonialism and neo-colonialism (Noiriel, 1994).

In its nascent phase, Subaltern Studies primarily aimed to amplify the voices of marginalized groups, with less emphasis on critiquing prevailing historiographical paradigms. However, as the movement expanded and matured, the latter objective gained ascendancy, casting a larger shadow over the former. G Prakash aptly observes that 'The Subaltern Studies' repositioning of subalternity within the dominant discourses inherently demands a critical analysis of the modern West (Prakash, 1994). Sumit Sarkar engages in a rigorous dialogue with Dipesh Chakrabarty's standpoint, as articulated in 'Provincializing Europe', advocating for a judicious assessment of Eurocentrism in Marxist criticism within the Subaltern Studies framework (Lal, 2001).

## WHO ARE THE 'SUBALTERN'? INDIAN AND PAKISTANI PERSPECTIVES

Interestingly, within the contextual framework of India and Pakistan, the term 'subaltern' has been frequently associated with a wide range of individuals. The term is not merely limited to those who are associated with colonized cultures or lack a voice within their broader society but also with various marginalized groups: women, laborers, religious minority groups, and lingual, gendered minority communities etc. dwelling in developing nations. In Indian context, the term primarily comprises of lower castes individuals and groups denied access to hegemonic power [(Pandey, 2008) and (Chakraborty, 2022)]. In Pakistan, the predominant portion of the 'subaltern' comprises of linguistic and ethno-racial minorities, along with children, women, and transgender individuals (Liaqat, 2023).

Theoretically, the truly subaltern individuals are voiceless, as they exist entirely outside the spheres of hegemonic discourse. They remain oblivious to the extent and underlying causes of their oppression and lack the means to fully comprehend or communicate their discontent to a broader audience. Residing in isolated regions, using localized languages often incomprehensible to outsiders, and having minimal interaction with hegemonic forces, they remain profoundly impacted by the actions of the dominant culture (Dhaktode, 2023).

Subaltern individuals encounter numerous barriers preventing them from reaching their oppressors, including language differences and fundamentally distinct worldviews held by diverse cultures. To communicate with a wider audience, they must compromise their unique perspectives to align with the oppressor's understanding, rendering any effort to authentically convey the subaltern voice inherently impractical. Those capable of articulating the subaltern's plight and advocating for greater representation of these marginalized and powerless individuals within society inevitably find themselves positioned within the framework of cultural hegemony. Consequently, in response to Spivak's fundamental query, the subaltern, in essence, cannot articulate their own experiences; instead, their voices can only be represented by others (Sethi & Nayak, 2020).

Contrary to the notion propounded by Gayatri Spivak, the authors of this chapter are of the opinion that new media especially OTT has provided direct platform to the subaltern voices where they can speak to a wider range of audience instead of being represented by some other agency. OTT web series have evolved into a formidable channel for representing the subaltern, providing a platform that gives a voice to the marginalized and underrepresented segments of society. These digital platforms have broken free from the conventional constraints of mainstream media, allowing creators from diverse backgrounds to share their unique narratives and perspectives with a global audience. By doing so, OTT web series have become a powerful tool for challenging established power structures, addressing social issues, and championing inclusivity. They offer a space where a wide range of voices and experiences can flourish, ushering in a new era of storytelling that authentically reflects the subaltern's diverse and often overlooked stories (Agarwal & Das, 2022). In this way, OTT web series have not only expanded the range of narratives available to viewers but have also opened opportunities for subaltern communities to reclaim their own narratives, thereby contributing significantly to a more equitable and representative media landscape.

## RESEARCH GAP AND PROBLEM STATEMENT

In the contemporary media landscape, the portrayal of subaltern identities has arisen as a central concern within the domains of cultural and social studies. The advent of the digital age has ushered in a new era of content consumption, with over-the-top (OTT) series originating from India and Pakistan achieving substantial acclaim and attracting a global viewership. However, a conspicuous void exists in the realm of scholarly inquiry pertaining to the intricate depictions of subaltern voices, experiences, and identities within these lauded OTT series. This research study discerns a critical lacuna within the existing scholarly literature, necessitating a comprehensive and diligent exploration of the multifaceted representations of subaltern narratives embedded within these series. The overarching issue centres on the limited comprehension of the complex dynamics and socio-cultural ramifications inherent in the portrayal of subaltern identities within the context of South Asian OTT content. This underscores the imperative requirement for a scholarly investigation into this hitherto underexplored dimension.

## RESEARCH METHODOLOGY

The present study serves as a convergent viewpoint between sociology of web based visual text, visual aesthetics, and subaltern studies. Not primarily a comparative or correlational study, it is an earnest attempt at interpretative understanding of the notion of 'being a marginalized being' (characterisation) in the selected Web series: Dahaad (India) and Churails (Pakistan). Methodologically speaking, this is primarily a work of interpretative understanding of marginalized characters relying (but not limiting) on assessment of significant subaltern elements/parameters. At the onset and culmination of this study lies the inherent aspect of interpretation, a facet acknowledged by the researchers themselves, fostering a fundamental approach to inquiry. The researchers assertively acknowledge the susceptibility of visual texts to subjective biases, emphasizing the necessity of awareness in this regard. The interaction between the researchers and the subject under scrutiny is notably deep and immersive. The rationale behind the employment of an interpretative approach is rooted in its emphasis on subjective analysis and human reflexivity. Visual media, being regarded as a form of text, comprises a realm of meanings conveyed through narratives, music, imagery, and character portrayal. Henceforth it is strongly felt to understand these meanings and note how marginalised/subaltern are being projected which later fulfils the ulterior framework of a creative process of perpetual exploration of meanings. Not necessarily limited to a subjective interpretative methodology, the attempt is to explore OTT contents as corrective foundations of knowledge building and creations.

## REVIEW OF LITERATURE

To conduct an examination of the subaltern interpretation of visual content, encompassing films and web series, a comprehensive understanding of the concept of the 'subaltern' is indispensable. The inception of the term 'subaltern' as a subject of discourse and scholarly investigation can be attributed to the insights of Italian communist philosopher Antonio Gramsci. Gramsci introduced the concept of subaltern within the context of his analyses of fascist hegemony in Italy, a period during which significant segments of society endured subjugation without always possessing a conscious awareness of their predicament. It is noteworthy that Gramsci himself experienced such subordination (Kroonenberg, 2023). His proposition posits the existence of a subaltern class within a society where dominant powers exert their hegemonic influence. In the Indian context, many scholars engaged in Subaltern Studies have adopted a multitude of ideas they find pertinent to their specific context (Hardiman, 2023).

Ranajit Guha contends that while colonial rule encompassed the systematic subjugation of the Indian masses through a combination of direct force and subversive tactics, the situation underwent minimal transformation following the attainment of independence. He argues that the post-independence era witnessed a perpetuation of the colonial legacy, with the ruling elites utilizing state machinery to impose hegemonic control over the citizenry (Chatterjee, 2023). This situation bred widespread discontent, eventually culminating in the Naxalbari movement in the late 1960s (Srinivasulu, 2017). Gayatri Spivak asserts that the subaltern category comprises individuals who lack the capacity to articulate their experiences and, paradoxically, face further jeopardy when attempting to do so. She places particular emphasis on the historical subjugation of women and offers a critique of early Indian subaltern studies for their oversight of gender-related issues. Spivak insists that the subjugation of women represents an integral facet of subaltern studies (Piu, 2023).

In Gayatri Spivak's seminal critique, 'Can the Subaltern speak?' the concept of the subaltern has emerged as a pivotal and prevalent concern within the sphere of postcolonial studies. Spivak's critique particularly targets the fundamentally essentialist assumptions inherent in Guha's perspectives. She contends that any attempt to establish the composition or identity of the subaltern group inevitably falls prey to essentialism. Within her influential work, Spivak problematizes the creation and retrieval of subaltern discourse, emphasizing its entanglement with controlling and dominant discursive practices. These practices not only determine the modes of expression for subaltern subjects but also shape the context from which they speak or are heard (Spivak, 1982).

Throughout her essay, Spivak undertakes a deeply genealogical exploration to ascertain whether the subaltern can articulate their experiences or is merely reduced to being represented, recognized, and heard in a fundamentally distorted manner. Her primary objective lies in learning to communicate with historically silenced non-elite subjects rather than assuming the role of speaking on behalf of these muted voices. To clarify, Spivak does not seek to problematize the authority of colonial discourse; instead, she focuses on the ways in which imperialism has constructed historical, gender, and class narratives for the subjugated subjects within a singular axis of differentiation—the centrality of Western history and Man (Spivak, 1985). From her standpoint, the actions of the subaltern are interpreted and understood within the dominant frameworks of colonial imperialism and nationalistic elitism. Consequently, she provocatively concludes that 'the subaltern cannot speak.'

Dipesh Chakrabarty articulated that the primary objective of Subaltern enquiry was to establish historical deconstruction portraying subaltern groups as active participants shaping history, not merely passive recipients of it. The subaltern class, viewed through this lens, was not a homogenous entity or a passive subject dictated by historical events, but a dynamic, engaged group actively contributing to societal dynamics. Dipesh Chakrabarty delved into the initial constructs of the subject within subaltern studies, which were rooted in an understanding of subjecthood influenced by concepts of social justice and political resistance (The Subaltern in South Asian History and Society: Report of a Conference, 1983).

Several of these scholars contend that the prospect of subaltern voices gaining recognition remains a distant possibility. Even when subaltern individuals endeavour to voice their experiences or rise against their circumstances, these efforts often result in further subjugation at the hands of the state machinery. Guha suggests that the subaltern masses remain under the pervasive influence of the ruling classes, even when engaging in acts of rebellion or protest (Fracchia, 2004). Nevertheless, this assertion appears to contradict his stance that subaltern individuals possess agency in shaping their own destinies.

Historically, the subaltern's subjugation was maintained using force. However, in the post-colonial era, the predominant approach has shifted towards hegemony as a means of control. Hegemony, in this context, signifies a condition of dominance in which the structure of power enables persuasion to prevail over coercion. In other words, the state now leverages its authority to cajole the citizenry into compliance with its policies, often employing forceful persuasion under the guise of pursuing the 'national interest' (Prakash, 2012).

The discourse presented above underscores the significant impact of societal dynamics on the evolution of Indian cinema. Notably, accomplished directors like Bimal Roy, whose work includes the acclaimed 'Do Bigha Zamin', and Shyam Benegal, recognized for films such as 'Ankur', concentrated their cinematic efforts on portraying the subaltern as a marginalized segment of society confronted with the inability to articulate their concerns, frequently resulting in self-inflicted adversity (Kamble & Biswal, 2023). These films conveyed the disillusionment and perceived betrayal experienced by individuals who believed in their participation within a democratic state. The emergence of the angry young man archetype in the 1970s echoed the frustration and anger of the subaltern (Mubarki, 2018). Characters played by Amitabh Bachchan represented unbridled anger, frequently culminating in a point of no return.

Although cinema production and distribution predominantly emerged from urban settings, the audience extended far beyond urban boundaries. Bollywood cinema, while technologically and cast-wise projecting an elitist image, owed its survival and success to the substantial support from the essential mass audience. Chidananda Das Gupta highlighted the crucial role played by the plebeians and the lumpen proletariat in propelling stars like Amitabh Bachchan, Raj Kapoor, Rajesh Khanna, and others to stardom (Gupta, 2008).

The subaltern's endorsement of cinema played a pivotal role in the growth of the South Indian film industry (Gopalkrishnan,2019). This audience, present across towns and villages, fervently supported regional stars, many of whom ventured into successful political careers. South Indian cinema's extensive reach was facilitated by mini theatres in remote areas, serving as a vital backbone for the triumph of 'masala films,' particularly in Tamil Nadu and Andhra Pradesh.

As the film industry transformed into an economic entity, profit became its primary motivator (Sarkar, Shah & Dutta, 2020). The state recognized cinema's persuasive potential among the masses, prompting initiatives like issue-based feature films and non-feature films aimed at social awareness. The establishment of institutions like the National Film Development Corporation, the Film and Television Institute of India, and the National Film Archives of India facilitated the rise of parallel or art cinema (Kumar, 2013). State patronage was instrumental in promoting parallel cinema, which emerged as a distinct genre from commercial cinema. The Indian parallel cinema became a platform to display the sensibilities of ordinary citizens, particularly focusing on rural themes. While filmmakers like Satyajit Ray, Rithwik Ghatak, and Mrinal Sen showcased rural-centric narratives, the recent trends in Marathi, Malayalam, Tamil, and other regional cinemas present an amalgamation of city-rural, middle class-poor dichotomies, thereby offering a unique narrative distinct from conventional art cinema (Sengupta, 2021)

Streaming platforms (OTT) have granted filmmakers and storytellers newfound artistic freedom, enabling them to present narratives that were previously restricted and censored. This has resulted in increased visibility and representation for minority groups, encompassing aspects like sexuality, gender, religion, and caste. Historically, mainstream cinema was dominated by powerful production companies that perpetuated narratives. However, this conventional approach is now being questioned and reevaluated (Eishita, 2022).

In the realm of cinema, representation plays a pivotal role, extending beyond the struggles of marginalized communities to encompass individuals residing on the fringes of society. Notably, there has been a scarcity of positive portrayals of queer characters in Indian films and television shows. Following the decriminalization of Section 377, filmmakers have started to explore queer representation more openly. Significantly portrayed in principal roles, leading characters from queer backgrounds have been prominently showcased in several OTT series, including Amazon Prime's 'Four More Shots' and 'Made in Heaven' OTT Web series as well as 'Ajeeb Daastaans,', 'Call My Agent,' and 'Feels Like Ishq,' accessible on Netflix (Dhar, 2021).

Furthermore, women have secured a more prominent position in the OTT content landscape. Traditionally, Bollywood predominantly centred its narratives on male protagonists to achieve financial success, often sidelining female characters in storytelling and promotional efforts. However, OTT platforms have been at the forefront of challenging traditional gender roles, as evident in shows like 'Delhi Crime' and 'Aarya.' Notably, 'Delhi Crime,' led by a predominantly female cast, received the prestigious title of 'Best Drama Series' at the 2020 International Emmy Awards (Lal, 2023).

OTT platforms have not only expanded opportunities for actors but have also shattered conventional stereotypes, such as the concept of 'shelf life' within the industry. This deviation from the norm has been warmly embraced by audiences. The inclusion of narratives depicting the dreams, aspirations, and struggles of urban women, exemplified in series like 'Bombay Begums' and the story of a woman taking charge of her corrupt family business in 'Aarya,' is a groundbreaking development. These roles offer a fresh perspective and deviate from the conventional portrayal of the 'good virgin wife' that has often dominated Bollywood (Vyavahare, 2022).

Previously, narratives like these were relegated to the 'independent' category and excluded from mainstream discussions. However, OTT platforms have demonstrated that stories featuring women from diverse backgrounds can resonate with a broad audience. Series like 'Dolly Kitty Aur Woh Chamakte Sitare' and 'Maharani' have portrayed multifaceted and diverse female characters who exhibit strength in various forms. 'Dolly Kitty Aur Woh Chamakte Sitare' adeptly navigates between personal and political themes, portraying realistically flawed individuals dealing with challenging circumstances. Conversely, 'Maharani' narrates the journey of a simple village maid who rises to become the Chief Minister of Bihar, defying powerful male figures in the state's political landscape (Ramnath, 2021).

Furthermore, over-the-top (OTT) content demonstrates an exceptional capability in depicting women in atypical and distinctive roles (Mehta, 2022). Instances include 'Code M', where Jennifer Winget's essays the role of Major Monica. Sanya Malhotra's portrayal of a young widow seeking self-actualization in 'Pagglait,' successfully breaks several stereotypes. In 'Four More Shots Please Season 2', Kirti Kulhari's depiction as a single middle-aged mother win several accolades. These examples serve to exemplify the expansive range of opportunities that OTT platforms offer to women, enabling them to explore diverse roles and portray a broad spectrum of characters.

## DECIPHERING THE NARRATIVES: AN EXAMINATION OF WEB SERIES

**Dahaad (Web Series):** The narrative unfolds in Mandawa (Rajasthan) as Sub-Inspector Anjali Bhaati investigates the mysterious disappearance of twenty-seven women and discovers a pattern linking these cases. This realization suggests the involvement of a serial killer, redirecting Anjali's focus to a systematic criminal plan. Anjali's nonconformity with the traditional village norms and her struggle against caste discrimination pose personal challenges while she remains resolute in her policing.

The narrative introduces two parallel cases with intertwined fates. In one instance, a brother from a scheduled caste reports his sister's disappearance, while another case centres on an inter-caste romance between a prominent village figure's daughter and a Muslim man. Anjali, alongside her colleague Devi Singh, assists the Muslim man in escaping, ultimately shifting their investigative focus to the former case. The investigation discloses that the vanished women share characteristics—belonging to lower castes, financial limitations, and emotional vulnerabilities. Anjali's detailed inquiry exposes a chilling fact: the women were poisoned with cyanide after eloping with their partners. While some suspect a criminal gang's involvement, Anjali insists a single psychopath executed these murders due to the precision and lack of errors.

The narrative also introduces Anand Swarnakar, a college lecturer who secretly teaches underprivileged children. His wife, Vandana, engages in an extramarital affair with a coworker named Jai, leading to a complex web of deceit and infidelity. As the police investigation unfolds, Anand's son, Kapish, discovers a phone in his father's vehicle. The ensuing police inquiry initially points towards Anand, but a lack of concrete evidence forces them to explore other leads. Eventually, suspicion turns to Anand's younger brother, Shiv, who is subsequently arrested. Upon realizing Anand's involvement in the murders, Shiv cooperates with the police to apprehend the elusive killer. Anand, fleeing Rajasthan, starts a new life with Mariam in Goa, only to be captured by Anjali and her team before committing another murder.

In a climactic confrontation, Anjali grills Anand about his motives for taking the lives of innocent women. He justifies his actions by asserting that these women were not as innocent as they seemed and deserved to be punished. Despite her position as a police officer, Anand criticizes Anjali's caste background and questions her relationship with her superior. In the story's denouement, Anjali decides to revert to her original surname, Meghwal, symbolizing her defiance against societal expectations and prejudices. The narrative thus navigates the complexities of caste, crime, and societal expectations in a rural Indian setting, culminating in a poignant exploration of identity and defiance (Kumar, 2023).

**Churails** (Web Series): The web series 'Churails' stands as a poignant commentary on the deep-seated challenges ingrained within Pakistani society, particularly its struggles with misogyny and corruption. The narrative orbits around the lives of four diverse women: Sara, who relinquished her legal career to fulfil the role of a 'perfect wife'; Jugnu, a wedding planner; Zubaida, a passionate boxer seeking both independence and love; and Batool, an ex-convict serving a lengthy 20-year sentence for murder. Their destinies intertwine when Sara stumbles upon evidence of her husband's infidelity. This revelation incites the quartet to establish a clandestine detective agency, operating incognito beneath the facade of a Burkha retail outlet named 'Halal Designs'. Their mission is clear—to unearth marital infidelity concealed within the privileged echelons of the city.

Facilitated by the technological prowess of Zubaida's boyfriend, Shams, and the support of Jugnu's assistant, Dilbar, the agency's ranks soon include women from society's margins, encompassing a prostitute, former convicts, and a transgender woman. Although the agency enjoys rapid growth and financial prosperity, the fault lines beneath the surface begin to emerge. Zubaida is the first to question their focus on serving affluent housewives, recognizing the dissonance between their aspirations for women's emancipation and the clients they currently cater to. As the narrative unfolds, the true adversaries gradually come into focus, with privilege emerging as a formidable adversary that can no longer be disregarded.

In hindsight, it becomes undeniably clear that unfaithful husbands are among the larger villains in this complex narrative. The Churails agency broadens its clientele, venturing deep into the well-guarded secrets within the high-walled mansions of Karachi's elite. This expansion propels the series beyond localized vigilantism, evolving into a comprehensive assault on the patriarchal underpinnings of society. The revelations unearthed by the eponymous 'Churails' illuminate a profound societal decay, touching upon issues such as child abuse, domestic violence, human trafficking, and drug abuse. The series, with its compelling characters and searing critique of societal injustices, underscores the critical need for societal introspection and reform to address deeply rooted inequalities and prejudices (Mushtaq, 2020).

## ANALYSIS AND INTERPRETATION

**Dahaad (Web Series):** This research undertakes an in-depth analysis of the transformative trajectory of the crime thriller 'Dahaad', with a specific focus on its groundbreaking influence in establishing novel paradigms and surpassing conventional benchmarks in the domain of contemporary media. The central facet under scrutiny pertains to the representation of subaltern personas within the narrative. 'Dahaad' is distinguished by its courageous exploration of pressing societal issues, such as casteism, sexism, and xenophobia, which few cinematic endeavours have dared to confront with such fervour (Bijolia, 2023). Although caste plays a crucial role in the storyline, it is not explicitly addressed initially in 'Dahaad'. The series commences by highlighting the theme of communal political unrest within the town, stemming from a situation where a girl from an esteemed Thakur family willingly elopes with a Muslim boy she has fallen in love with. This incident becomes a focal point for a local politician who incites a group of impassioned youth against the Muslim community, as well as against a woman's autonomy to make decisions based on her own volition. (Patil, 2023).

Concurrently, amid this turmoil, a Dalit man emerges, inquiring about the whereabouts of his missing sister. He aligns with the right-wing faction after facing neglect at the police station. In an act of desperation, he navigates through the assembly of fervent extremists brandishing flags and joins in their chants. It resembles an unfolding narrative of the radicalization of an individual: he, much like his peers, was not inherently predisposed to such beliefs. However, a more precise interpretation reveals that he has reached a point of no return. The political operatives exploit the man's vulnerability to advance their own agenda of animosity and apprehension. They afford him a sense of significance and recognition, enticing him with numerous vacuous pledges and personalized attention. This juncture, where systemic indifference propels a subaltern toward dubious associations, stands as the central theme of the series.

Whereas the film industry has frequently mirrored, and in turn, perpetuated the established societal norms rather than challenging them, exacerbating the subaltern women's subjugation; the OTT series 'Dahaad' departs from this pattern by offering a powerful voice to subaltern women who have long been silenced (Bharatvaraj & Subramani, 2018). Through systematic investigation, it reveals the disturbing reality underlying the disappearances and subsequent murders of women from marginalized social backgrounds, attributing these tragedies to the harmful intersection of caste-based discrimination and patriarchal authority. This narrative strikes a chord with the audience on a profound level (Khurana, 2023).

Sub-Inspector Anjali, the protagonist, defies not only the conventions of her village but also the societal injustices ingrained within it, be it due to her low caste or her relentless pursuit of justice, including her support for a Muslim boy involved in an inter-religious love affair and falsely accused of perpetrating 'love-jihad'. The series not only highlights the portrayal of women but also delves into the intricate issue of casteism. The director, Reema Kagti, skilfully weaves this complex issue into the fabric of the series, shedding light on the persistent discriminatory practices that continue to subjugate lower-caste women. Anjali's character is emblematic of a contemporary Dalit woman, who confronts caste-based discrimination and sexism in her daily interactions with colleagues. On each occasion she passes by, a constable ritualistically ignites incense sticks to purify his workspace. Further, several poignant scenes within the series effectively convey the deeply ingrained caste centric beliefs and superstitions that continue to shape societal perceptions. Anjali confronts a barrage of challenges, from street harassment and the unwelcome advances of the superintendent of police in his office to matrimonial proposals orchestrated by her mother within the confines of her home. Even though she has relinquished her father's surname, 'Meghwal', a name associated with the dominant backward castes in Rajasthan, she encounters barriers at the entrances of the town's upper-caste residences. An impactful scene in which Anjali and her superior, SHO Devi Lal Singh, visit the house of Rajni Thakur to question her about her elopement with a Muslim boy exemplifies this deeply ingrained prejudice. The scene, in which Rajni's father questions Devi Lal about Anjali's presence, highlights the society's adherence to caste-based norms. Devi Lal's response, asserting that casteism is a crime in contemporary times, offers a resolute counterpoint to the prevailing prejudices. Subsequently, the series powerfully encapsulates the indomitable spirit of modern low-caste women through Anjali's assertion that many doors have been shut in her face, but no one has managed to silence her (Mishra & Deewan, 2019).

In stark contrast to Anjali Bhati, the character Anand, portrayed as the antagonist, represents a malevolent force that exploits a combination of complicity and shame. It becomes evident that his victims are carefully selected as they all belong to marginalised caste or are from minority background having one attribution in common and i.e., poor economy. He discerns that these women lack alternatives and are constrained by their community's norms, which makes their individualism, as exemplified by their willingness to defy their societal constraints, more vulnerable. Anand is aware that their families might disown them if these abused women choose to elope with him, and most will refrain from reporting their disappearances due to the embarrassment associated with their decisions. In essence, he skillfully manipulates them within the limitations and stigmatization of their own environment. Furthermore, his method of murder, a chilling scenario in which the victims are discovered deceased, seemingly by suicide in public restrooms, capitalizes on their anxiety as members of minority groups in a society predisposed to judgment (Bilal, 2023). Anand's actions can be likened to a seduction, whereby he entraps them and relies on the biases and constraints of their surroundings to facilitate the rest of his sinister plan. This is why the initial episode may appear disconnected from the subsequent events. It serves as a perceptive portrayal of the inherent noise and biases that can allow a criminal to evade detection for an extended period. It also underscores how the unpredictability of these biases can ultimately lead to his downfall.

Anand leverages an ongoing cycle of societal oppression to shield his actions while carrying out his nefarious deeds in plain sight. Unlike many murderers who tactfully construct cover-ups to deceive the law, individuals like Anand possess a pre-existing veil. He effortlessly blends into society. As an upper-caste Hindu male operating within a community marked by sensitivity towards communal issues, he enjoys the camouflage of his privilege, much like how his victims are marginalized due to their lack of societal advantages. Furthermore, the fact that his wife seeks affection elsewhere serves as an additional advantage for Anand. This circumstance enables him to maintain a facade as a 'respectable man' who discreetly vanishes on weekends with his mobile library, evading any scrutiny or suspicion (Patra, 2023).

The OTT web series 'Dahaad' adeptly utilizes the serial-killer genre to delve into an India where the normalization of deeply ingrained biases—stemming from factors such as caste, class, religion, and gender—serves as the actual perpetrator. Protagonists Anjali Bhati and her accomplice Devi Lal strive to overcome apprehension, while antagonist Anand serves as a symbol of regressive societal attitudes A distinctive aspect of 'Dahaad' lies in its approach: it doesn't dismantle the patriarchal structure outright, but rather redefines the pathway toward accountability, spotlighting the responsibility of households and critiquing the practice of excusing rape perpetrators based on troubled childhoods. The series reclaims female agency by holding society accountable for the choices women are compelled to make (Poikayil, 2023). Narrative examples like Anjali's mother and Devi Singh's wife underscore the societal pressure to conform, which may, ironically, protect them from predators like Anand. Paradoxically, these women could have become Anand's victims without intervention. The character Parghi's hesitation to bring a female child into a perilous world further underscores this theme.

Further, the OTT web series 'Dahaad' masterfully intertwines post-structuralist ideas with the concept of the subaltern, offering a profound exploration of the interplay between power, identity, and representation in a highly stratified society. This integration is exemplified through the character of Anjali Bhati, whose journey epitomizes the struggles of a subaltern individual grappling with societal norms, illustrating post-structuralism's assertion that identities are fluid constructs shaped by cultural and social dynamics. Moreover, the series delves into power dynamics, notably embodied by the antagonist, Anand, who exploits his upper-caste privilege to commit crimes while maintaining a guise of normalcy. This portrayal underscores post-structuralism's perspective that power is not a static force but a shifting, dispersed entity within society, directly impacting the subaltern characters who occupy marginalized positions. Language and discourse are pivotal components, revealing how language constructs and reinforces societal norms, often oppressing subaltern characters. Moreover, the non-linear storytelling approach in 'Dahaad' challenges the idea of a singular narrative, aligning with post-structuralism's emphasis on multiple perspectives. These subaltern viewpoints invite viewers to critically engage with the intricate relationships between power, identity, and language in a socially complex and oppressive landscape, effectively resonating with the core principles of both post-structuralism and subaltern studies.

**Churail (Web Series):** 'Churails', a Pakistani web series, emerges as a profound exploration of the status of marginalized or subaltern, delivering a compelling narrative that transcends the boundaries of conventional storytelling. Set against the backdrop of Pakistani society, the series navigates the complexities of gender, class, and ethnicity, offering a nuanced portrayal of women on the fringes of societal norms. At the heart of 'Churails' is a daring and unapologetic representation of subaltern women who defy traditional roles, challenging stereotypes with resilience and agency. The series employs an intersectional lens, acknowledging the interconnected layers of identity that shape the experiences of marginalized (Qureshi, 2020). It recognizes that gender does not exist in isolation but intersects with factors such as class, ethnicity, and socio-economic status. This intersectionality is a crucial aspect of the series, as it allows for a more comprehensive exploration of the challenges faced by subalterns and dismantles the tendency to homogenize their experiences (Alix, 2020).

At the core of the narrative resides the establishment of a detective agency named 'Churails,' distinguished by the tagline 'Mard ko Dard Hoga' (Men will feel pain), an entity dedicated to redressing the transgressions of abusive and unfaithful husbands within the urban milieu. Governed by a diverse cohort of women native to Karachi, each characterized by distinct backgrounds and confrontations with individualistic challenges, this heterogeneous ensemble functions as a microcosm emblematic of the intricate diversity intrinsic to the subaltern (Rabel, 2020). From the affluent Sara, emancipating herself from the societal expectations tethered to the idealized role of a perfect housewife, to the tenacious and outspoken Jugnu, a provocateur of gender norms, the series meticulously sculpts characters that transcend reductionist representations. Integral to this ensemble are figures such as Batool, an unapologetically outspoken ex-convict, and Zubaida, an aspiring pugilist who assumes the role of the gang's ingénue (Kumar, 2020). Collectively, recruiting the likeminded women they metamorphose into commissioned vigilantes, extending their services to disillusioned housewives, with vibrantly coloured burqas assuming the symbolic mantle of superhero capes. Operating under the guise of a boutique store, the clandestine operation employs a confession-booth setup to obfuscate their identities, ingeniously incorporating a discreet aperture for the purpose of comforting apprehensive clients who arrive bearing remuneration and intricate details—a subtle commentary on the tendency of women to internalize the perceived transgressions of their spouses. Each character contributes to a collective voice that speaks against the marginalization of women based on their socio-economic standing or adherence to societal norms (Ramnath, 2020).

'Churails' is unapologetic in its critique of patriarchal structures and deeply ingrained misogyny within Pakistani society. The narrative boldly confronts these issues, providing a stark commentary on the restrictive roles assigned to women. Through the experiences of the protagonists, the series exposes the limitations imposed on women, shedding light on the systemic oppression they face. The characters' stories unfold as a critique of societal norms that perpetuate inequality, placing them in situations that mirror the struggles of many subaltern women. The 'Churails' collective encompasses a diverse spectrum of women, including middle-aged queer partners, technologically adept individuals proficient in hacking, sex workers, and transgender women. Additionally, the narrative introduces compelling male characters like Shams, Dilbar, and Boss, who steadfastly support the women from a secondary position, devoid of any manifestations of toxic masculinity. However, an intricate exploration would have been beneficial, considering the potential influence of romantic entanglements in motivating these male characters' allegiance to the female protagonists (Shrikrishna, 2020).

The strength of 'Churails' resides in its exploration of the intricate subtleties within Pakistani society, aspects that the broader community often hesitates to acknowledge. While there is a heightened awareness to present a cultural and Islamic facade, the series sheds light on the existence of alternate sexualities thriving in Pakistan (Azhar, 2013). The reluctance to openly acknowledge these realities reflects a pervasive hypocrisy, particularly among the male population, underscoring a societal dichotomy between public presentation and private acknowledgment.

The series effectively showcased the challenges faced by transgender individuals, using an incident involving Baby Doll, a transgender woman, as an illustrative example. Baby Doll faces discrimination when trying to enter an upscale hotel, despite presenting a valid citizen's identity card (CNIC). This incident highlights broader issues of transphobia, depicting trans individuals as inferior and excluded from privileged spaces. These spaces, reserved for specific social classes, reinforce stereotypes that marginalize trans women to public areas, portraying them as lacking respectability. However, Baby Doll's inclusion in the women-only group in 'Churails' asserts the equal status of trans women alongside cisgender women.

Babli and Pinky, former inmates, queer and romantic partners, are briefly depicted in an episode of the series where they use their newly acquired funds to attend a mujra. This portrayal raises concerns about perpetuating misconceptions regarding women's same-sex attraction, particularly for an audience with limited awareness of the LGBTQ+ community. The issue arises due to the association of Mujras with the objectification of the female body and the male gaze. Placing two queer women in a setting traditionally linked to male activity may not accurately represent queer attraction. However, the inclusion of desi butch representation is appreciated for diverging from predominantly Western-defined portrayals of butch identities. The depiction of the couple's moments is commendable for avoiding unnecessary sexualization, offering a genuine and sincere portrayal of their relationship (Taqdees, 2020).

Churails effectively depicts incidents often overlooked in society, using Batool's past, now a prominent member of the Churails gang, as a narrative lens. The series exposes the presence of sexual predators, rapists, and paedophiles within civil society, sometimes even within the same household. Batool, married off at a young age to an abusive, alcoholic sex maniac, faced a disturbing situation when she refused his unjust desires. In response, he turned to his five-year-old daughter, prompting Batool to take drastic action, a natural response from any mother. Ironically, instead of understanding and protecting women like Batool from pervasive injustices, Pakistani courts sentenced her to imprisonment for killing her abusive husband. This legal outcome highlights the lack of space for subaltern voices, denying them the opportunity to articulate their experiences, as articulated by Gayatri Spivak. Then there is Zubaida, another significant gang member and an aspiring boxer, who can be viewed as a subaltern figure as she represents marginalized individuals seeking autonomy within a society dominated by oppressive norms. Her quest for independence through boxing is met with resistance from her family, echoing the struggles of those marginalized within conventional societal structures. The cunning recall by her mother and subsequent exposure to societal norms reveal the internalization of oppressive standards. Zubaida's thwarted aspirations underscore how family dynamics, reflecting broader expectations, act as a significant impediment to the autonomy of subaltern individuals like her. (Amarkant, 2020).

The web series 'Churails' methodically explores the significant themes of racism and colourism, seldom broached by contemporary OTT productions. The central character, Jugnu, challenges societal norms by entering into matrimony with a Black man named Jackson. Jugnu's uncle, a retiring professor aspiring to author a book on 'Gender, Power, and Colonialism', paradoxically reveals deeply entrenched sexism and racism. Employing racial slurs, he unjustly attributes Jugnu's father's downfall to her interracial marriage with Jackson, claiming it tarnished the family's reputation. A distressing incident ensues where Jackson, being Black, accidentally damages a statue at a party, exacerbating tensions within Jugnu's privileged Pakistani family (Vakaao, 2020). 'Churails' not only addresses racial issues but also sheds light on women's condescending behaviour towards skin colour. In a scene, the salon owner instructs her assistant to use a shade to make the model appear fairer, emphasizing uniformity across the face and neck. What is particularly disturbing is the lack of objection from other women present. This underscores the pervasive racist and colour-discriminatory attitudes within Pakistani society, where dark skin is stigmatized, revealing that women themselves can be catalysts for such discriminatory practices.

Churails critically explores the decision to identify as 'gay' in Pakistani society and its ramifications on both the individual and their relationships within the societal and familial context. A married man engages in an undisclosed affair with a male lover, unbeknownst to his wife who is aware only of his infidelity. Seeking to expose him, the wife collaborates with 'Churails' and confronts them in a hotel room. The scene challenges stereotypes by depicting the husband and his male lover in an intimate embrace, countering objections that dismiss 'queer love' as mere 'lust'. However, the story takes a dark turn when the wife violently kills and consumes a portion of her gay husband, a disturbing act criticized for its violence and perceived homophobia. The killing of a gay husband by his wife, when analysed from a subaltern perspective, signifies the oppression faced by marginalized identities. The disdain directed at gay individuals, evident from both men and women, reflects societal norms deeply rooted in prejudice. Homosexuality, as depicted in the series, remains stigmatized, further marginalizing gay individuals. This act of murder highlights the severe repercussions marginalized groups may encounter when challenging or expressing their identities within oppressive societal frameworks. The portrayal underscores the pressing necessity for societal change and acceptance of diverse sexual orientations, urging a departure from discriminatory attitudes.

'Churails' masterfully interlaces the themes of sisterhood, solidarity, and resistance, presenting women not solely as a detective agency but as a cohesive force transcending socio-economic constraints. The narrative accentuates collective strength, actively challenging societal norms, and empowering subaltern women. The focal point is the prominence of economic inequality, conducting a thorough investigation into a society where agency and opportunities are influenced by one's financial status. The series deeply explores the challenges faced by financially disadvantaged women, compelling viewers to confront harsh realities and advocate for equity. Characters serve as conduits for societal reflection, effectively deconstructing the entrenched oppressed-oppressor binary within marginalized communities (Rizvi, 2020).

The series employs a distinctive approach that deliberately avoids simplistic tropes, fostering a profound understanding of subaltern experiences. 'Churails' specifically addresses the agency of subaltern women, portraying them as architects of their destinies and effectively countering victimhood narratives. By disrupting stereotypes and prompting a revaluation of societal norms, the series transcends conventional narratives. In conclusion, it provides a potent platform for subaltern voices, challenging perspectives, and catalysing social change. Through its intersectional lens, critique of patriarchal structures, and emphasis on solidarity, 'Churails' emerges as a groundbreaking force that empowers subaltern women, actively breaking barriers in digital storytelling for more inclusive and equitable narratives.

## CONCLUSION

The investigation into the portrayal of the 'subaltern' in the Indian web series 'Dahaad' and the Pakistani web series 'Churails' reveals a convergence of noteworthy similarities and defining distinctions. Both series share a commitment to a profound exploration of societal injustices, with a particular emphasis on elucidating the indomitable spirit inherent in marginalized individuals. Central to their thematic frameworks is the incorporation of intersectionality, acknowledging the convoluted interplay of various identity layers shaping the subaltern experience. Strongly confronting established societal norms, both 'Dahaad' and 'Churails' courageously address issues such as casteism, sexism, and discrimination, often relegated to the shadows of contemporary social discourse. The narratives feature empowered female protagonists as pivotal figures, transcending societal constraints and providing a compelling testament to the resilience and agency characterizing modern subaltern women. In their collective endeavour, these series actively contribute to breaking barriers, furnishing a platform that amplifies subaltern voices, fostering a deeper understanding of their experiences.

While both 'Dahaad' and 'Churails' share commonalities, the portrayal of the subaltern is shaped distinctly by the cultural contexts of India and Pakistan. In the case of 'Dahaad,' the focus gravitates towards the pervasive issues of caste and gender-based discrimination entrenched within Indian society. The narrative cautiously unpacks the biases and discriminatory practices endured by lower-caste women, with Sub-Inspector Anjali Bhati emerging as a symbol of unyielding resilience in the face of such systemic injustices. Her character serves as a focal point, challenging and navigating the persistent discriminatory practices that oppress lower-caste women and other subaltern groups. The antagonist, Anand, within the series embodies a chilling representation of societal complicity and shame, strategically preying on the vulnerabilities of marginalized women.

In contrast, the Pakistani web series 'Churails,' unfolds within a distinct cultural milieu, widening its thematic scope to encompass not only caste and gender-based issues but also focusing into the complexities of racism, colourism, and LGBTQ+ discrimination. This expansive approach offers a more comprehensive examination of subaltern experiences, extending the series' impact beyond immediate societal struggles. 'Churails' employs a distinctive narrative strategy, blending elements of a detective thriller with refined social commentary, resulting in a multi-layered storyline that transcends conventional explorations of subaltern narratives. As both 'Dahaad' and 'Churails' manifest a shared dedication to scrutinize the complexities inherent in societal norms and amplifying the voices of the subaltern, their distinctive cultural milieus and thematic emphases markedly contribute to their distinctive and compelling portrayals of the subaltern within their respective regional landscapes.

## REFERENCES

Agarwal, S., & Das, M. (2022). Indian Web Series - Revolutionising the Portrayal of Indian Women in Mainstream Media. *International Journal of Creative and Innovative Research in all. Studies, 4*(8), 95–100. http://www.ijciras.com/PublishedPaper/IJCIRAS1802.pdf

Alix. (2020, June 08). Does Gender Exist in Isolation? *Medium.* https://gender.medium.com/29-does-gender-exist-in-isolation-8688c972bf3

Amarkant, M. (2020, August 24). *Zee5 Churails Review: 10 Hard-Hitting Scenes that Portray the Myriad Shades of Womanhood | #Mainchurailhoon.* DiaryOfAnInsaneWriter. https://diaryofaninsanewriter.com/2020/08/24/churailspowerfulscenes/

Azhar, M. (2013, August 27). *Gay Pakistan: Where sex is available and relationships are difficult.* BBC News. https://www.bbc.com/news/23811826

Bharatvaraj, S. S. G., & Subramani, R. R. (2018). Manifestation of Subaltern Narratives in Cinema. *American International Journal of Research in Humanities, Arts and Social Sciences. 25*(1), 05-08. http://iasir.net/AIJRHASSpapers/AIJRHASS19-102.pdf

Bijolia, D. (2023, May 16). *Dahaad's Nuanced Portrayal of Casteism and Patriarchy Goes Where Few Web Series Dare.* Homegrown. https://homegrown.co.in/homegrown-creators/dahaads-nuanced-portrayal-of-casteism-and-patriarchy-goes-where-few-web-series-dare?fbclid=IwAR0Q5pnNt0vVbuY70uKNBt0SCnoPMCgelJkszISppQYBQzl9mfbz6RlP5D0

Bilal, M. (2023, May 25). *The true story of Dahaad and Cyanide Mohan, the serial killer who inspired Vijay Varma's character.* DailyO. https://www.dailyo.in/entertainment/the-true-story-of-dahaad-and-cyanide-mohan-the-serial-killer-who-inspired-vijay-varmas-character-39833

Bohra, R. (2021, 30 November). Impact of OTT on traditional mode of entertainment. *The Times of India.* https://timesofindia.indiatimes.com/readersblog/raghavi/impact-of-ott-on-traditional-mode-of-entertainment-39292/

Chakrabarty, D. (1998). Minority Histories, Subaltern Pasts. *Economic and Political Weekly, 33*(9), 473–479. https://www.jstor.org/stable/4406471

Chakraborty, P. (2022). Postcolonial Theory and the Challenge of Caste. *Interventions, 25*(2), 157–173. doi:10.1080/1369801X.2022.2029531

Chatterjee, P. (2023, May 05). Ranajit Guha, the Unconventional Historian. *The Wire*. https://thewire.in/history/ranajit-guha-the-unconventional-historian

Dhaktode, N. (2023). 'Can Subaltern Speak?': Discourse Without Considering Caste an Exploitative Institution. *Contemporary Voice of Dalit, 0*(0), 2455328X2211455. doi:10.1177/2455328X221145597

Dhar, D. (2021, November 09). *Redefining The 'Mainstream': Queer Representation and Content on OTT Platforms*. Feminism in India. https://feminisminindia.com/2021/11/09/redefining-the-mainstream-queer-representation-and-content-on-ott-platforms/

Eishita. (2022, July 14). *How OTT Platforms Expanded Representation In India & Helped Break Taboos*. ScoopWhoop. https://www.scoopwhoop.com/entertainment/how-ott-platforms-expanded-representation-in-india/

Fracchia, J. (2004). Subaltern Studies and Collective Memories in Piana degli Albanesi: Methodological Reflections on a Historiographical Encounter. *Asian Journal of Social Science, 32*(2), 247–273. https://www.jstor.org/stable/23654588. doi:10.1163/1568531041705112

Ghosh, S. (2023, June 28). Raima Sen: OTT contents now celebrate and amplify the voices of women from all walks of life. *The Times of India*. https://timesofindia.indiatimes.com/entertainment/bengali/movies/news/raima-sen-ott-contents-now-celebrate-and-amplify-the-voices-of-women-from-all-walks-of-life/articleshow/101341971.cms?from=mdr

Gopalkrishnan, S. (2019). Marginalised in the New Wave Tamil Film: Subaltern Aspirations in three films by Bala, Kumararaja and Mysskin. *Rupkatha Journal on Interdisciplinary Studies in Humanities,11*(2),1-9. https://rupkatha.com/V11/n3/v11n303.pdf

Gupta, C. D. (2008). *Seeing is believing: Selected writings on cinema*. Penguin India.

Hardiman, D. (2023, October 31). Ranjit Guha: A Tribute. *South Asian Studies, 0*(0), 1–6. doi:10.1080/02666030.2023.2270837

Hasan, S. (2005). India and Pakistan: Common Identity and Conflict. *Refugee Survey Quarterly, 24*(4), 74–80. https://www.jstor.org/stable/45054038. doi:10.1093/rsq/hdi085

Kamble, C., & Biswal, S. K. (2023, August 23). The Changing Role of Marginalised Women in Hindi Cinema: A Shift from Cinematic Negotiations to Empowerment. *Media Watch, 14*(3), 386–402. doi:10.1177/09760911231191266

Khurana, A. (2023, May 11). Dahaad Season 1 Review: Sonakshi Sinha and Vijay Varma both roar in this Cop Drama. *Entertainment Times*. https://timesofindia.indiatimes.com/web-series/reviews/hindi/dahaad/season-1/seasonreview/100101778.cms

Kroonenberg, S. (2023). Gramsci's Writing Body. On Embodiment and Subaltern Knowledge. *Interventions, 0*(0), 1–16. doi:10.1080/1369801X.2023.2191861

Kumar, A. (2023, May 31). 2 counts on which Amazon Prime's Dahaad won at showing caste—and 1 where it utterly failed. *The Print*. https://theprint.in/opinion/2-counts-on-which-amazon-primes-dahaad-won-at-showing-caste-and-1-where-it-utterly-failed/1603687/

Kumar, P. (2020, August 14). 'Churails' review: Justice League of women, by women, for women. *The Hindu*. https://www.thehindu.com/entertainment/reviews/churails-review-justice-league-of-women-by-women-for-women/article32351662.ece

Kumar, R. (2013). The making of the National Film Archive of India: Notes from the Archive of the Archive. *The Moving Image : the Journal of the Association of Moving Image Archivists*, *13*(1), 98–128. doi:10.5749/movingimage.13.1.0098

Lal, N. (2023, March 17). 'Hours of observing women in khaki led to realistic portrayal'. *Times of India*.

Lal, V. (2001). Subaltern Studies and Its Critics: Debates over Indian History [Review of A Subaltern Studies Reader 1986-1995; Nationalism, Terrorism, Communalism: Essays in Modern Indian History; Writing Social History; The Furies of Indian Communalism: Religion, Modernity and Secularization, by R. Guha, P. Heehs, S. Sarkar, & A. Vanaik]. *History and Theory*, *40*(1), 135–148. https://www.jstor.org/stable/2678074. doi:10.1111/0018-2656.00158

Liaqat, Q. (2023, March 20). The Notion of the 'Subaltern' and the Drone Victim Subjectivities in Pakistani Anglophone Fiction. *Interventions*, *0*(0), 1–19. doi:10.1080/1369801X.2023.2190906

Macwan, S. S. (2023, August 07). Ranajit Guha, Subaltern Studies, and the Indian English Novel. *Economic and Political Weekly*, *58*(31). https://www.epw.in/journal/2023/31/postscript/ranajit-guha-subaltern-studies-and-indian-english.html

Mallon, F. E. (1994). The Promise and Dilemma of Subaltern Studies: Perspectives from Latin American History. *The American Historical Review*, *99*(5), 1491–1515. doi:10.2307/2168386

Mehta, S. (2022, March 17). End of the stereotypical Bollywood 'heroine'? How OTT is empowering women-led narratives. *Khaleej Times*. https://www.khaleejtimes.com/arts-and-culture/end-of-the-stereotypical-bollywood-heroine-how-ott-is-empowering-women-led-narratives

Miller, R., Liu, K., & Ball, A. F. (2020). Critical Counter-Narrative as Transformative Methodology for Educational Equity. *Review of Research in Education*, *44*(1), 269–300. doi:10.3102/0091732X20908501

Mishra, M. S., & Deewan, G. (2019). Representation of Women in Post Colonial Hindi Movies. *Think India Journal*. *22*(41), 127-140. https://thinkindiaquarterly.org/index.php/think-india/article/view/19270

Morris, R. C. (Ed.). (2010). *Can the Subaltern Speak?: Reflections on the History of an Idea*. Columbia University Press., https://www.jstor.org/stable/10.7312/morr14384

Mubarki, M. A. (2018, December 06). Looking beyond Post-Colonial Modernity: Subaltern Masculinity and the Mumbai Cinema. *South Asia*, *41*(4), 723–743. doi:10.1080/00856401.2018.1527430

Mushtaq, W. (2020, November 02). *Churails: Why a feminist detective show was banned in Pakistan*. BBC News. https://www.bbc.com/news/world-asia-54629439

Noiriel, G. (1994). Foucault and History: The Lessons of a Disillusion. *The Journal of Modern History, 66*(3), 547–568. https://www.jstor.org/stable/2124485. doi:10.1086/244886

Pandey, G. (2008, October 10). Subaltern Citizens and their Histories. *Interventions, 10*(3), 271–284. doi:10.1080/13698010802444587

Patil, P. (2023, May 17). *Dahaad: A Crime Thriller Exploring the Hard-Hitting and Grim Realities of Modern Day India.* Feminism in India. https://feminisminindia.com/2023/05/17/dahaad-a-crime-thriller-exploring-the-hard-hitting-and-grim-realities-of-modern-day-india/

Patra, P. (2023, June 06). Anatomy of a villain: Why Vijay Varma's Anand in Dahaad is just a few steps away from greatness. *The Telegraph Online.* https://www.telegraphindia.com/entertainment/anatomy-of-a-villain-why-vijay-varmas-anand-in-dahaad-is-just-a-few-steps-away-from-greatness/cid/1942774

Piu, P. (2023, September 12). The journey of subalternity in Gayatri Spivak's work: Its sociological relevance. *The Sociological Review, 0*(0), 1–19. doi:10.1177/00380261231194495

Poikayil, G. (2023, May 29). Plight of Cyanide Mohan's victim's family proves true Dahaad's dissection of casteism, patriarchy. *Onmanorama.* https://www.onmanorama.com/news/kerala/2023/05/29/plight-cyanide-mohan-victim-family-proves-true-dahaad-dissection-casteism-patriarchy-amazon-prime.html

Prakash, G. (1994). Subaltern Studies as Postcolonial Criticism. *The American Historical Review, 99*(5), 1475–1490. doi:10.2307/2168385

Prakash, G. (1994). Subaltern Studies as Postcolonial Criticism. *The American Historical Review, 99*(5), 1475–1490. doi:10.2307/2168385

Prakash, G. (2012). Book Review: The Small Voice of History. *The Indian Economic and Social History Review, 49*(1), 147–148. doi:10.1177/001946461104900105

Qureshi, A. (2020, August 24). Review: *Why Churails Is A Revolutionary Show For Pakistan.* Feminism in India. https://feminisminindia.com/2020/08/24/churails-feminist-review-pakistan-show/

Rabel, G. (2020, November 21). 'Men Will Suffer': A Review of Churails. *The Gazelle.* https://www.thegazelle.org/issue/190/review-of-churails

Radhakrishnan, R. (1994). Postmodernism and the Rest of the World. *Organization, 1*(2), 305–340. doi:10.1177/135050849412004

Ramnath, N. (2020, August 11). 'Churails' review: Veils, deception and secrets in Zee5 web series set in Karachi. *Scroll.in.* https://scroll.in/reel/969985/churails-review-veils-deception-and-secrets-in-zee5-web-series-set-in-karachi

Ramnath, N. (2021, May 28). 'Maharani' review: This potboiler about Bihar politics never stops simmering.*Scroll.in.* https://scroll.in/reel/995992/maharani-review-this-potboiler-about-bihar-politics-never-stops-simmering

Rizvi, A. (2020, August 15). Watch Churails for the catharsis, not for the politics. *Images.* https://images.dawn.com/news/1185653

Roy, P. K. (2014). *India's era of digital activism*. Aljazeera. https://www.aljazeera.com/opinions/2014/1/6/indias-era-of-digital-activism

Santhanam, R. (2020, April 03). Do OTT platforms provide more space for sociopolitical content? *The Hindu*. https://www.thehindu.com/opinion/op-ed/do-ott-platforms-provide-more-space-for-sociopolitical-content/article31240686.ece

Sarkar, S., Shah, A. M., & Dutta, S. K. (2020). Bollywood in Transition: Chiragdeep's Reflection of the Film Industry. *Emerging Economies Cases Journal*, *2*(2), 79–86. doi:10.1177/2516604220968537

Sengupta, A. (2021, October 19). Subterranean parallels: Common threads between Satyajit Ray, Mrinal Sen and Ritwik Ghatak. *The Hindu*. https://frontline.thehindu.com/cover-story/subterranean-parallels-common-threads-between-satyajit-ray-mrinal-sen-and-ritwik-ghatak/article37018238.ece

Sethi, B. K., & Nayak, A. (2020). Speaking and Speaking Differently: Language as Resistance, Liberation and Celebration in Dalit Women's Autobiographical Narratives. *Contemporary Voice of Dalit*, *12*(2), 138–148. doi:10.1177/2455328X20924882

Sharma, S. (2021). *India's Zee is banking on a Pakistani show for its success*. Aljazeera. https://www.aljazeera.com/economy/2021/12/10/indias-zee-is-banking-on-a-pakistani-show-for-its-success

Shehzad, A. (2017). *India and Pakistan: A Land of Opportunity for Smartphones.* Mystory. https://yourstory.com/mystory/729539f5c4-india-and-pakistan-a

Shrikrishna, A. (2020). Asim Abbasi's 'Churails', a Pakistani web series on women divided by class privilege. *The New Indian Express*. https://www.newindianexpress.com/entertainment/hindi/2020/aug/27/asim-abbasis-churails-a-pakistani-web-series-on-women-divided-by-class-privilege--2188677.html

Spivak, G. C. (1982). The Politics of Interpretations. *Critical Inquiry*, *9*(1), 259–278. https://www.jstor.org/stable/1343284. doi:10.1086/448199

Spivak, G. C. (1985). Three Women's Texts and a Critique of Imperialism. *Critical Inquiry*, *12*(1), 243–261. https://www.jstor.org/stable/1343469. doi:10.1086/448328

Srinivasulu, K. (2017). The Caste Question in the Naxalite Movement. *Economic and Political Weekly*, *52*(21), 47–52. https://www.jstor.org/stable/26696125

Tamandehrou, A., & Khan, S. U. (2015). Assessing the importance of globalisation and new media technology in 21st century: An analytical overview. *International Journal of Multifaceted & Multilingual Studies*, *1*(4).

Taqdees, M. (2020, September 05). *Queer Representation in Churails: A Review*. Naya Daur. https://nayadaur.tv/05-Sep-2020/queer-representation-in-churails-a-review

The Subaltern in South Asian History and Society. Report of a Conference. (1983). *Economic and Political Weekly, 18*(9), 299–301. https://www.jstor.org/stable/4371898

Thomas, P. D. (2018). Refiguring the Subaltern. *Political Theory*, *46*(6), 861–884. https://www.jstor.org/stable/26617620. doi:10.1177/0090591718762720

Tiwary, I. (2023). Streaming and India's film-centred video culture: Linguistic and formal diversity. *International Journal of Cultural Studies*, *0*(0). Advance online publication. doi:10.1177/13678779231197696

Vakaao. (2020, November 01). *Churails Season 1 Episode 6 – Husn*. Vakaao. https://vakaao.com/news/churails-season-1-episode-6-husn

Vyavahare, R. (2022, September 17). Strong, real, imperfect: Is OTT redefining female characters? *TOI*. https://timesofindia.indiatimes.com/entertainment/hindi/bollywood/news/strong-real-imperfect-is-ott-redefining-female-characters/articleshow/94253705.cms

## KEY TERMS AND DEFINITIONS

**Colonialism:** Colonialism denotes the domination of one group of individuals by another, usually for the purposes of economic, political, and cultural control. This historical and social phenomenon has had far-reaching consequences, extending over centuries and across various continents.

**Discourse:** Discourse refers to a communication system involving language and symbols for conveying knowledge, ideas, and cultural practices. It encompasses spoken and written forms, shaping societal understanding. Discourse analysis in academia explores language structures, patterns, and power dynamics.

**Hegemony:** Hegemony, as defined by Antonio Gramsci, refers to the dominance or leadership of one social group over others, achieved not solely through force but by the shaping of consent through cultural and ideological means. Gramsci's concept emphasizes the role of cultural institutions and intellectual activities in maintaining social control.

**Historiography:** Historiography refers to the study of how history is written, and the methods, interpretations, and perspectives employed by historians. It involves the analysis of historical writings, the evolution of historical methodologies, and the examination of different schools of thought within the discipline of history. Historiography aims to understand the ways in which historical narratives are constructed, shaped, and influenced by the cultural, social, and political contexts in which historians operate. It provides insights into the changing nature of historical inquiry over time and highlights the diversity of approaches in interpreting and representing the past.

**Over-the-Top (OTT):** 'OTT' stands for 'Over-the-Top,' referring to content and services delivered directly over the internet, bypassing traditional cable or satellite providers. OTT platforms distribute streaming media, including movies, TV shows, and other video content, to viewers through internet connections. Popular OTT platforms include Netflix, Amazon Prime Video, Hulu, Disney+, and many others. The term 'Over-the-Top' suggests that these services go over the top of traditional distribution methods, allowing users to access content on various devices, such as smart TVs, computers, tablets, and smartphones, without the need for a traditional cable or satellite subscription.

**Postmodernism:** Postmodernism is a broad cultural, artistic, and intellectual movement that emerged in the mid-20th century as a response to and rejection of certain aspects of modernism. It is characterized by skepticism towards grand narratives, it questions authority, emphasizes individualism, and recognizes subjective knowledge, influencing diverse fields.

**Post-structuralism:** Post-structuralism, arising in the 1960s and 1970s, is a philosophical and humanities movement challenging structuralism. It questions established notions about language, texts, and social sciences, promoting diverse interpretations and perspectives.

**Societal Dichotomy:** It refers to a division or binary distinction within a society, often characterized by contrasting elements, categories, or groups. This division can manifest in various aspects of social life, such as gender roles, economic classes, cultural identities, or other polarized constructs. Societal dichotomies contribute to the formation of dualistic perspectives, reinforcing distinctions between different segments of the population. These divisions may lead to the establishment of hierarchies and unequal power dynamics, shaping social interactions, norms, and structures within a given community or society.

**Stereotypes:** Stereotypes are widely held and oversimplified beliefs or ideas about a particular group of people or things. These generalized perceptions often oversimplify the diversity within a group and can lead to unfair judgments or expectations based on preconceived notions rather than individual characteristics. Stereotypes can be based on various factors such as race, gender, nationality, or other social categories.

**Subaltern:** Subaltern refers to those who are socially, politically, and culturally marginalized, often belonging to lower socio-economic classes and facing various forms of discrimination. This includes but is not limited to issues related to caste, gender, race, and ethnicity. The subaltern often lacks agency within dominant power structures and their voices are frequently silenced or ignored.

# Chapter 2
# Rise of OTT Bengali Originals:
## Niche Content for the Diverse Audiences

**Lokesh Chakma**
*Visva-Bharati, India*

**Mausumi Bhattacharyya**
*Visva-Bharati, India*

## ABSTRACT

*This chapter delves into the dynamic landscape of over-the-top platforms, focusing on Bengali content, exploring their growth trajectory, unique initiatives, and content diversification. It highlights the emergence of ventures like Hoichoi, which has garnered widespread acclaim through its original and diverse subscription strategies. It underscores the role of quality content in attracting audiences and emphasizes the significance of stories. It discusses how regional content, including adaptations from literature, has succeeded, illustrating this with examples from Hoichoi's innovative projects. Furthermore, the chapter sheds light on the growing demand for women-centric narratives and how OTTs provide a platform for progressive portrayals of women's roles and issues. Despite the progress, the chapter acknowledges challenges, including the need for sustained originality and the expanding competition in the OTT sphere. In conclusion, it foresees a promising future for Bengali content on OTT platforms, driven by creative storytelling, audience preferences, and an evolving landscape.*

## INTRODUCTION

The emergence of Covid 19 pandemic raised a serious global threat as governments across the world executed lockdowns and followed strong measures like social distancing to curb the propagation of the virus leading to interruption in free and regular movements for millions of individuals. Confined within the four walls and physically being separated from families, well-wishers, and friends and the rise of unemployment, pay cuts alongside the relentless fear of catching and avoiding the disease were the emerging issues. Covid 19 brought numerous alterations at a fast leap and such formed rules were termed as the new normal (Bhattacharyya, 2020) The pandemic situation, and the imposition of lockdown specifically,

DOI: 10.4018/979-8-3693-3526-0.ch002

has changed the way we avail entertainment. The practice has gradually shifted from watching films in theatres to viewing them on OTT platforms. From films premiering exclusively on such platforms to accessing live telecasts of sporting events such as IPL have drawn a significant proportion of viewers toward OTT entertainment. The presence of locally produced web series in the list of the content offered by platforms has further accelerated acceptance of OTTs, especially among the technologically sound millennials and Gen Z. A High level of conversion to OTT from DTH or cable television subscriptions has been noticed among young viewers even before the pandemic (Subramanian, 2022)

Once the pandemic ebbed, the viewing tendency was further amplified with the addition of services offered by OTTs over cable television or DTH. The option of live streaming and viewing accessibility via smartphones has enabled audiences to view their preferred content in different languages and diverse content, viewing on-demand even when they are on the move. Customers are also shifting their preferences from paid television subscriptions to online-based streaming services. According to a report published by Boston Consulting Group content market of OTT platforms in India has reached a nuanced point. It is expected to touch the market proportion of $5 billion by 2023. In another research conducted by Phando Corp, the MAUs (Monthly Active Users) in the video segment of OTT in India have grown exponentially, at 300 percent, from 2019 and clocked 680 million in the year 2021. The consumption share reached 80 percent with the inclusion of Tier II and III cities and rural pockets (Subramanian, 2022).

In general, the OTT online streaming landscape can be separated into four sections: International OTT platforms trailing with local regional languages, local OTT players trying to retain a position over the regional OTT userbase, central regional OTTs mapping their niche in 3 or 4 languages and region-wise content aggregators. Apart from this, there are DTH services and telecom operators such as Jio Cinema, Vodafone Play, and Tata Sky Binge which offer regional OTT content through partnership (Agarwal, 2020). Hindi language content has a dominating presence across major Indian OTT platforms. However, there are large chunks of unmapped regional OTT platforms that are targeting local audiences, and their dominance is on the rise. Unlike the foremost OTT players, who target vast and varied audiences, the regional OTT operators have carved a niche for themselves, serving local viewers with local language content. (Rising Kashmir, 2022). Over the last few several OTT players like Disney+Hotstar, Netflix, Amazon Prime Video, ZEE5, Mubi, YouTube, Jio Cinema, and MX Player have surfaced as prime streamers of regional video content. There are also regional OTT players like Addatimes, Hoichoi, Manorama Max, Sun NXT, Prime Flix, and Ullu that are purely capitalizing on regional content. These platforms use the English language as a medium of default display on their interface but do not provide any English language content (Agarwal, 2020)

## Significance of Regional Content in OTT Platforms

The Indian content market is not homogenous and it is fragmented into thousands of tiny and varied markets and every OTT player is in search of that niche area to attract exclusive viewers. Hence majority of the players approve that the only way to tap these areas is through creating local content. Devendra Deshpande, CEO of Friday Filmworks, emphasizes the fact that original regional content possesses the boundless potential of success which is not restricted to Indian boundaries. Similar thoughts are also shared by Mr. Vijay Subramanium, Vice President of Content and Communication, Walt Disney Company India Limited, who points out that the addition of local content helps the OTT players to remain pertinent and the objective should focus on unique and pioneering content for the viewers. Co-founder of Arré, Mr. Ajay Chacko mentions that there is an enormous possibility in regional content and their platform is already banking a large chunk of staking over local content creation. He further resonates that the regional film and television markets are huge in India and he observes a similar tendency of regional content consumption in the digital format as well (Banerjee & Moniz, 2018)

There has been a boom in content production in diverse regional languages among OTT platforms in India. Ranging from vernacular content, regional language films, podcasts, and television shows, doubtlessly, there is a large segment of the Indian population who do not consume English language content and the market is enormous (Agarwal, 2020). According to FICCI's published report requirement for original content will expand in OTT platforms from 2,500 (2021) to over 4000 hours by the year 2024. The share of regional languages will increase by 54 percent, and there will be a further boom in regional OTT content due to dubbing and subtitling costs (FICCI, 2022). Psychologists consider the fact that the emotional thinking of an individual is decoded through his/her first language which generally forms the mother tongue for a clear expression. This is possibly the reason; why India possesses a diverse range of players catering to viewers with regional content over the years. With a population crossing 1.4 billion in India over 68.64 percent of the population residing in rural regions and semi-urban regions watching content majorly in native languages. The sector of Indian broadcasting, entertainment, and media specifically has been eyeing this segment of the population for creating the target audience userbase (Agarwal, 2020).

A report conducted by Research and Markets, titled OTT market in India 2022-27 mentions that the total financing of Indian OTT platforms has been US$ 672.59 million or INR 55.21 billion. Among the top spenders stand platforms like Disney+Hotstar, Netflix, and Amazon Prime Video, joined together accumulate a spending amount of US$ 384.35 million or INR 31.55 billion. The market approximation in 2022 presented that the top OTT platforms capitalized profoundly on the production of original content spending a collective cost of US$ 600 to 700 million without the inclusion of sports and film content. Furthermore, on April 2022, Amazon Prime Video declared the launching of more than 40 latest web series and films in different regional languages. In 2019 and 2020 Netflix spent an amount of US$ 400 million for producing original content and OTT platform SonyLIV has also been creating original content in regional languages in 2023. In another report published by Media Partners Asia, the predicted investment for original video content in 2022 was around US$1.3 billion and the report further predicts it to cross US$ 4.2 billion by the year 2027 (Cyrill, 2023)

To extract every possible profitable content from the heavily aggressive market of digital video, the three biggest competitors in the OTT sphere, namely, Disney+ Hotstar, Amazon Prime Video, and Netflix are spending large sums of capital to generate original content. Although the market might be highly driven every OTT player wants to stay afloat. According to the Vice-president of Amazon Prime Video, Asia Pacific, Gaurav Gandhi, the OTT viewing boom generally sums up due to a few amalgamated pointers like a special emphasis on high-content quality, cinematic worth, generation of regional and original content, the rise of a section of audience possessing disposable income and the growth of precepting subscribers (Mahalakshmi & Gopalan, 2021). As viewers are satisfied and spending money on content consumption, OTT players are also gladly spending finances to own content. Uday Sodhi, Digital Strategy Consultant at Kurate Digital Consulting mentions to generate a long-term investment, it is highly essential for an OTT platform to develop rich and unique content, and usually, a platform creates a digital repository within the first two to three years (Mahalakshmi & Gopalan, 2021)

Media experts opine that the streaming platforms that want to garner a wider reach across the country are required to emphasize on minimum of eight to nine regional languages. Content Head of Disney+Hotstar and HSM (Hindi-Speaking Market), Gaurav Banerjee mentions that as the taste of the consumer is changing, they try their best efforts for content accessibility in the desired language over all platforms which assists them in forming a profound emotional association with the viewers. Vice President, of Revenue and Study of Hoichoi, Mr. Soumya Mukherjee opines that the Hoichoi platform has increased its investment to double for producing content and bundling it more with the addition of originals and cinema premieres to satisfy the viewing demands of the Bengali-speaking audience (Jha, 2022).

Speaking on the rapid growth of original content in OTTs, the CEO and founder of Contiloe Pictures, Mr. Abhimanyu Singh mentions that such an initiative is the only mode for the platforms to flourish. He further mentions that every OTT platform will be required to have its collection of originals which will be the common norm and at the end of the day, if the content is strong, viewers will see it OTTs will receive more subscribers. Singh also mentions that the major dissimilarity between Indian television serials the Western ones is the way stories are being narrated. They generally have around 20 episodes while the Indian ones run for over 700 to 800 episodes. The emergence of the OTT space has transformed this rule and definition of the storytelling process. Before venturing into this space, Singh mentioned that he wanted to learn and comprehend this space before creating any content for the OTT medium. Singh also mentions that the streaming platforms have enabled them to create stability between cinema and television. He and his production team initially thought of not being able to deliver certain stories. Over television, the story was too short to execute, and hence the digital platform facilitated that space to them. Singh opines that in terms of the process of post-production, the digital space has taken up approaches that are comparable to films (Qureshi, 2020)

Sanjay Dwivedi, the Group Chief Financial Officer of Balaji Telefilms mentions the reasons behind the significant rise in original content production among OTTs. He adds that the considerable financing by telecom operators to intensify the penetration of the internet and technological advancements boosts the shift among consumers and approves the growth of digital set-up in India. He also mentions that such a transformation will influence the entertainment sector, especially the OTT platforms as it results in producing wider influence for reaching the target audience where the quantity and value of consumers are justly significant for maintainable business performance (PTI, 2021)

## Rise in Demand for Niche Content

There is a rising demand for niche content among OTT subscribers, and the reasons for this trend are several. First, OTT platforms empower content creators to narrate stories capable of evoking a wide range of reactions and sentiments. Situations are not required to be always cheerful and flawless. Over the years, platforms have developed the ability to showcase more such content in comparison to traditional platforms like cinema that possess static screen time or television which is limited to being the medium of family viewing. On the contrary, OTT has embraced the implementation of creative freedom to depict stories and roles in their natural form (Goyal, 2022). Second, the contents of OTTs can be viewed with ease, with a wide range of titles to choose from in different languages and formats. Such provisions have permitted the viewers to search for stories that have genuine and rich content dosed with shades of realism in contrast to films released in cinema halls which are still burdened by length and ruled by specific market forces (Goyal, 2022).

Following the steps of the OTT giants like Disney+Hotstar, Netflix or Amazon Prime, there has also been a growth of local language platforms in West Bengal like Hoichoi and Addatimes, which are now jostling for a share of the OTT market (Chakraborty, 2021) Hoichoi has already produced a hundred originals in Bengali. There are over 270 million Bengali-speaking audiences across the world, and presently the platform creates Bengali content for both Indian and Bangladeshi viewers. Besides, the original collection is also accessible in Hindi and circulated via several OTT partners for higher content reach among diverse audiences (The Telegraph, 2022). Hoichoi was launched in 2017 and Addatimes in 2016. Presently both platforms have reached millions of overseas and global Bengali speakers.

On average, the regular watch time among Hoichoi subscribers is 60 minutes, and on average count, each subscriber has been visiting the platform three times a day. The co-founder of Hoichoi, Mr. Vishnu Mohta, mentions that their focus is to divide content segment-wise equally into original shows and acquired cinemas. The streaming service also plans to release about 30 originals every year. The platform plans to create content via adaptations of traditional literature rope in renowned filmmakers on board and extend the limits on how stories are narrated (Jha, 2021). The focus here is to create a digital library of niche content to attract diverse audiences. Hoichoi is now available in Bangladesh and has plans to expand its reach to countries like Singapore, Malaysia, and Middle Eastern countries, where Bengali-speaking people work as laborers. To tap this market, Hoichoi plans to collaborate with different global telecom operators so that people living in foreign countries who might not have digital currency access can still avail of the services through packages received from telecom operators. Mohta opines that 40 percent of the direct subscription profits are received from international markets (Jha, 2021).

## Major Trends in the OTT Market

The report of Omdia remarks that the present congestion in the OTT market scenario of India signifies that the merging together of OTT platforms in this sector is foreseeable. The report also anticipates that as the market develops further, the activity of M&A (merger and acquisition) between the local OTT platforms will be observed in the approaching years. Furthermore, it mentions that the OTT market is becoming a delicate task and competitive space for both the national and international players to run their business smoothly. Since content remains the major discriminator among the OTTs, the investment spent on the content becomes the prime area of focus. Netflix and Amazon Prime Video started the trend of producing original content in Indian languages from 2017 onwards and since then, all the major OTTs have started raising their investments for original content creation (Farooqui, 2022)

Former Head of Marketing and Revenue of Applause Entertainment, Mr. Ashok A Cherian mentions that with the presence of all the large OTTs and telecom companies forming into platforms of video streaming, there is a great prospect for OTT platforms in present times. As internet usage rises, viewers can now consume large amounts of preferable content regularly. Production houses require unique, entertaining, homegrown, relatable, and superior content that can enthral the viewers every time. With the rise in digital viewing, there will be no set limit for the demand for original content. When original content is produced, production houses should confirm the adaptability and generate a wider commonality in the content to associate with the viewers spanning geographical boundaries (Banerjee & Moniz, 2018)

For OTT platforms, authentic and premium content particularly made in regional languages has been one of the major driving factors and discriminators as foremost players are trying to garner the attention of the audience. Market growth can be credited to quick developments among streaming platforms and the rising requirement for quality experiences in consuming content through diverse devices. RBSA Advisors conducted a report that mentions that the OTT market is predicted to peak at $12.5 billion by the year 2030 due to the rise in better digital connectivity, networking, and smartphone usage. The report also mentions that the next upsurge of OTT growth from Tier II, Tier

III, and Tier IV cities comprised of population speaking in local languages. The report further mentions that OTTs comprise of 7 to 9 percent market share in the $27 billion media and entertainment business sector of India. By the end of 2023, the report predicts an increase of 22 to 23 percent of growth in the market share (Tele.net, 2022)

Co-founder of CupShup, a Mumbai-based marketing agency, Sidharth Singh cites that a substantial amount of the Indian diaspora desires local content and OTT platforms can fill that void. Singh says that the regional OTT players can promote local talents as they are emphasizing more content. Moreover, the contents are distinctively located to link the language gap which the international OTT platforms have not been able to tap till now. Significantly, these OTT platforms are also putting creative insights into the content by cutting costs and creating shorter content as per the viewing demand of the consumers (Jha, 2022). The creation of common content and variation over the last few years by the OTT platforms has transformed the entertainment and contemporary genres, talents, languages, and formats that were once restricted to definite demographics (Kumar, 2023). Moreover, subscription pricing and content collection are the two characteristics that run the business of OTT platforms. In the last few years, companies have been testing both features to retain more subscribers on their platforms. with the rise of the pandemic that led to the massive growth in subscriptions, OTT platforms are assertively creating tactics to increase their content collection and reach to a wide number of audiences. Chief operating officer of Viacom 18 Digital Ventures, Mr. Gaurav Rakshit mentions that the consumers are choosing their content selection, which in itself is a massive transformation in the outlook of customers (Dey, 2021).

## The Potential Prospects for Major Production Companies

CEO of Dharma Productions, Mr. Apoorva Mehra opines that it will not be a mistake to acknowledge the fact that the OTT sector is speedily heading towards a distinctive advancement, bearing in mind, that the major portion of the content is being viewed via the digital medium in present times. These platforms provide numerous features attractive features like accessibility through smartphones or laptops which provides complete control in the hands of the viewer to stay connected or disconnect themselves from the platforms as related to timely watching on other platforms. Although every platform possesses its drawbacks and benefits, OTT platforms have certainly developed as the mediums that will remain constant in the digital era (Banerjee & Moniz, 2018)

Mr. Allu Aravind, promoter of the Aha OTT platform mentions that anything that is labeled as local, be it food or entertainment will have its significance. Individuals always remember the local flavors of cuisines, language, and other cultural standards. Co-founder of Hoichoi OTT, Mr. Vishnu Mohta articulates that viewers request local content that is depicted in their regional tone reflecting their stories and authenticities, leading to the creation of hyper-localisation (Ranipeta, 2023). Initially, the trend was dubbing a particular content into different regional languages. However, now the development is moving further as Mr. Ajit Thakur, CEO of the Aha OTT platform mentions that original content provides better investment returns and creation of such content also helps in connecting with the regional stories. Producing original content is also a cheaper alternative to acquiring feature films and labeling it as an IP (Intellectual Property) of the OTT player. Moreover, content creation helps in generating more professions and talents in the industry and delivers improved business metrics (Ranipeta, 2023)

## Path-Breaking Initiatives

In its operational 5th year in 2022, Hoichoi has declared twenty original shows alongside a Freemium segment for viewers who do not possess a premium Hoichoi subscription. The platform boasts a collection of over 100 original shows and more than 600 films in its digital library. Co-founder of Hoichoi, Mahendra Soni, mentions that the stunning response received from the OTT arena in the last four years from both Bengal and across India is overwhelming. This diverse audience engagement has been made possible through the creation of dynamic and relevant content. The platform also intends to curate remarkable content by collaborating with the finest industry talents (Sarkar, 2021). To reach viewers from remote areas globally, Hoichoi launched different strategies for partnerships and stress offline subscription payment methods in the fifth year of its operation. The platform also emphasizes innovative ideas like announcing gift cards in the digital format during Durga Puja curated to gift viewers with added benefits for streaming their content. Hoichoi has subscribers from around 150 countries, and the platform also aims to expand its presence in the Bengali-speaking diaspora across South Asia, including Singapore and Malaysia (Sarkar, 2021).

Moreover, Hoichoi has also launched its latest content segment titled Hoichoi Studios. Through this venture, the platform plans to generate edgy stories that were limited to online viewing. Such content will be made into the film format. The project aims to create three to five films for the theatre format annually from 2023 onwards apart from producing 25 to 30 originals annually and releasing the World premier films on the Hoichoi platform (Adgully, 2023). Hoichoi emphasizes more on involving the audience in the online ecosystem and does not focus on competition since it is not disparagingly significant in present times. Vishnu Mohta, co-founder of Hoichoi also mentions that the inclusion of more OTT players further helps in the growth of the OTT ecosystem raising cognizance and developing a practice among viewers (Sarkar G., 2023). OTT platforms catering to exclusive Bengali content including Addatimes, Platform8, and KliKK are also significantly contributing to the diversification and growth of Bengali Original content.

## Factors for Growth and Demand of Niche Content

According to Nimisha Pandey, Chief Content Officer of ZEE5, Hindi Originals, liking for stories that resonate with everyday life experiences of audiences in a profound way is going up. The authentic depiction is the exact keyword as viewers search for actual happenings that are accurately matched with the story. Previously there was not enough screen time available to showcase the shades of the darker side in films. However, now they can be justified as creative freedom enables content creators to explore uncomfortable themes in stories that can be presented in an engaging yet dependable way. Such contents are being viewed, and they are fulfilling performance pressure effectively. The viewers eagerly wait to see such films as they have the option to view other stories on diverse platforms. Stories that break away from the convention on how a hero should be presented and characteristics of imperfection are simpler to comprehend because such narratives have a realistic touch and enhance the entertainment factor as well (Goyal, 2022).

Manish Agarwal, Business Head, ZEE5, mentions that OTT facilitates the idea of curating personalized content and understanding. Personalization of language, for instance, enables Indian viewers to view content in their native language. Actor, businessperson, and producer Rana Daggubatti opine that in present times, the concept is to trace out new possibilities of storytelling, and one is not required to present every story on the silver screen, OTT platforms have opened this platform for diverse actors, creators and storytellers. Every region in India has specific subtlety content-wise, which the audiences would like to see in their preferred language. Contents that feature biographies, tales of folklore, and protagonists of a specific region have a common attraction across diverse platforms (Kashyap, 2019)

Although OTTs are paying attention to regional content through the creation of Originals to garner attention, another trend of creating family-themed shows is also on the rise. Chief business officer of ZEE5 India, Manish Kalra asserts that they want to emphasize family entertainment. He focuses on the fact that the viewer's demand for unique content brings a diverse range of openings for OTTs to try out unique, distinguished, and exclusive ideas, subjects, and presentations. OTT platform, AltBalaji has also strategies to bring out new content that targets the family section that will be shown on similar lines. Specific content will be comprehensive, made for families who socialize together at home and watch originals while specific content will be crafted for individual viewing or for viewers who prefer to watch content alone (Dey, 2021)

In present times, there are audiences for every story devoid of any format restrictions, linguistic barriers, or length and duration. The significant factor here is that it should be distinctive, possessing a unique storyline and gripping narrative. A common factor is that although the subjects and storylines are diverse, the collective component of darkness is similar in the OTT contents, which possibly signifies that reality is entrenched in the narratives. Most of the titles feature grey-shade characters. Harman Wadala, co-writer of Tabbar, streamed on Sony LIV, mentions that the emergence of the OTT platforms has enabled Indian viewers to consider themselves progressed enough to recognize the morally dark and melancholy realms infused with ethically opposed protagonists. Wadala believes there is no formula for creating unique content based on a dark theme. Writers are required to bring in fresher societal viewpoints and then meticulously bundle them up in a constricted and amusing tale (Goyal, 2022).

## Variation in Contents

The diversity of original and niche content can be traced from the analysis of some shows aired on the Bengali OTT platform, Hoichoi. Director Anjan Dutta debuts in the OTT space with 'Murder in the Hills', which narrates a gripping, dark, and muddled tale set in the hills of Darjeeling. The story revolves around the mysterious death of a popular Bengali film actor of the 90s. According to Anjan Dutt, the story has dark shades exploring the complexities of the characters. New and young faces have been featured in weaving violence, deeply ingrained mysteries, outrages, and deceits. With the twisting and turning of events, the hills become shady and overflown with gore and violence, narrating the past to hold the current events. Dutta further mentions that achieving this level through '*Murder in the Hills*' has been one of the most significant facets that the director has infrequently attained earlier (The Telegraph, 2022). '*Khyapa Sohor*' is set in the dark underworld of Kolkata, narrating the stories of gang wars. '*The Mahabharat Murders*' is directed by Soumik Halder, and the story is adapted from the novel *The Mahabharat Murders* written by Arnab Roy. It narrates the story of a serial murderer who believes that he is reincarnated as Duryodhana, a character from the ancient epic *Mahabharata* (Jha, 2021).

## Content Creation From Fiction Novels

The rise in OTT content viewing has led to the adaptation of digital content from more books and producers are looking over best-selling books to turn them into visual stories (The Tribune, 2021). 'REKKA' (*Rabindranath Ekhane Kawkhono Khete Aashenni*), streamed by Hoichoi, is an adaptation of a Bangladeshi novel of the same name by Mohammad Nazim Uddin. The series can be termed path-breaking for several reasons. Foremost it marks renowned film director Srijit Mukherjee's foray into the Hoichoi platform. The series casts celebrated actors like Anirban Bhattacharya, Rahul Bose, and Bangladeshi actress Azmeri Haque Badhon, who plays a pivotal role as the protagonist Muskan Zuberi. '*Indu*', another Hoichoi presentation starring actress Ishaa Saha in the lead, has attracted audiences from different sections. It is a family-oriented content with a dose of a thriller. It was directed by Sayantan Ghosal, and the story is by Sahana Dutta.

Hoichoi's original '*Mandaar*', a rework of William Shakespeare's Macbeth, has been widely acclaimed by audiences and elevated the standard of content creation in the arena of Indian OTT. Both the viewers and critics have acclaimed the work of 'Mandaar', which marks actor Anirban Bhattacharya's directorial debut in the OTT space (The Telegraph, 2022). '*Srikanto*' is s contemporary adaptation of Saratchandra Chattopadhyay's novel and features Rishav Basu and Sohini Sarkar in the lead roles. Kallol Lahiri's novel *Indubala Bhater Hotel* has also been reworked and made into a web series in Hoichoi. To make it more convincing, Tollywood actress Subhasree came out from her usual self and worked harder on the character of Indubala, for the web in Hoichoi. Indubala's character depicts a woman who hailed from Bangladesh initially. Her husband hails from Kolkata and she returns with him to the city before the partition. The storyline takes viewers on an emotional journey, evoking a sense of yearning as they search for their true place of origin. Food plays a significant part in the narrative which depicts the widowed Indubala's transition from a young girl in the pre-partition era who lived her youthful days in Bangladesh to her journey into being a woman facing difficulties in her married life and the execution of the pice hotel. Her serving of enchanted flavourful food in the hotel being operated in Kolkata takes the viewers and Indubala back to the roots in Khulna (Times of India, 2022)

## Streaming of Bangladeshi Content

The platform Hoichoi is also exploring and investing in originals from Bangladesh. For example, '*Boli*' highlights the tale of father and son. '*Bodh*' tells the story of a judge and his rising scruples after retirement. '*Sabrina*' depicts the subjugation of women, '*Karagar*', the tale is set in jail or '*Kaiser*', which narrates the story of a murderous detective hooked into the world of video games (Jha, 2021). *Buker Modhye Agun* directed by Taneem Rahman Angshu represents the shocking and strange demise of a superstar from the 90s era leaving behind several conspiracy theories. The suspense unravels after an officer starts to investigate the case. While thriller web series like *A Common Man*, *Mission Huntdown*, and *Odrishho* have also garnered a lot of viewership from both West Bengal and overseas. Sakib R. Khan, the country director of Hoichoi, Bangladesh, mentions that the concept of Hoichoi OTT is to provide the best amusement for the viewer. The platform has roped in the best actors and content creators to depict significant and appealing narratives for the audience who will be able to relate to and love the content. Furthermore, Vishnu Mohta, the co-founder of Hoichoi mentions that they collaborated and formed Hoichoi Bangladesh in the year 2019 when the idea of OTT was almost ambiguous in the country and have since been producing a few tremendously efficacious content (Niyogi, 2023)

## Emphasizing Content on Social Issues

OTTs have now developed into a platform where women-related narratives are depicted, creating a space for the generation of evocative discussions through the representation of confronting social subjects audaciously onscreen. Referring to subjects on body positivity, gender inequity, sexual violence, and mental health, content creators have dared to leap beyond the storytelling limitations producing stories that echo with the viewers on a deeper level. By focusing on these issues OTTs have not only brought entertainment but also empowered and refined the audience stirring to bring a transformation in reality (Ghosh, 2023). Sayantan Ghosal's directorial venture *Roktokorobi* aired on ZEE5 starring actors Raima Sen and Vikram Chatterjee depicts a tale of deep family secrets and highlights issues like infidelity, sexual abuse, and mental health (Niyogi, 2023)

## Focusing on Varied Women-Centric Content

Notwithstanding the path successfully traversed by OTT platforms like Hoichoi and Addatimes, a probable area of focus in the coming times could be shows exploring subjects related to gender inclusion and gender equality. Though '*Dupur Thakurpo*' a web series streamed in Hoichoi is a cult, its positioning, and presentation nudge one to view the relationship between in-laws, sisters, and brothers from a titillating perspective. Presented to be a comedy, it plays on the objectification of women, an inherent malice in society. Such a series can provide entertainment but not take forward the women's empowerment and growth agenda. Herein '*Astey Ladies*' seems to fit the bill more than the tale of *thakurpos*. This series, presented on Hoichoi, narrates the story of three upper-middle-class, highly mobile women running their beauty parlor named Salon De Paris. Hailing from diverse backgrounds, they are united by the love of the salon. With the property noticed by a fraud businessman, the three unite to fend off any ill attempts. Faced with dire financial needs, they start conducting petty robberies till detected by a smarter guy. Interestingly they were not handed over to the authorities nor punished for the act. Instead, they were offered the scope to participate in the conduct of a significant diamond theft. This storyline holds key indicators about the position of women, at least in the urban context. They are financially independent, networked, and mobile. They have a life of their own and independent views about whatever is relevant. They are not dominated by patriarchy and have a mind of their own. Importantly they are confident and can endure all challenges to reach their goals (Chattopadhyay, 2019)

With the rise in demand for varied storytelling, actress Raima Sen mentions that OTTs are expected to continue scouting and capitalizing on women-centric content. These narratives have not only garnered profits but have also emphasized the significance of distinct portrayals in the media scenario. Through the development of a space that brings up assorted stories and talented individuals, the industry is reforming the entertainment scenario and approving the influence of inclusion for future generations. OTTs have converted into a ray of hope for a generation of diverse women-oriented content. Depiction of multifaceted characters, riveting tales, and unwavering search for social problems, these contents are transforming the way female characters are depicted and distinguished in the entertainment sphere (Ghosh, 2023).

In 2018, Zee5 announced the launch of the platform's first original show in Bengali aimed at the native language-speaking premium audience. Titled '*Kaali*' portrayed a strong woman character played by Paoli Dam, who is compelled to struggle hard to save her son in the span of a night. In pursuit of the target, she does not cringe from getting involved in illegal activities to reach the drug mafias. She takes them on with grit in the fight to save her son. Immensely well accepted by viewers, this series positions women on an emotional and motivational level (Adgully, 2018). OTT platforms have provided content creators with a strong foundation to challenge outdated gender representation and move away from stereotypes. Such shows have garnered the attention of the viewers for their impenitent and multifarious depictions of women protagonists who challenge the expectations. By putting the female protagonists in central multifaceted roles, the stories explore their expeditions and shatter the conventional norms that limit the potential of female actors in the entertainment industry (Ghosh, 2023).

The growth of women-oriented content on OTTs has opened avenues for representing the understated voices that have been mostly disregarded by the conventional media. OTTs have provided a space for the depiction of stories that delve into the understanding of womenfolk with debilities, color, and belonging from the LGBTQIA+ community. OTTs have helped in prominently representing these deserving voices (Ghosh, 2023). Bengali web series such as *Rajneeti, Mohomaya, Paap, Tasher Ghawr, Hello*!, and *Nikhoj* among others portray women in strong and diverse roles breaking away from the stereotypical representation in films and television screens.

Such acts of bravery and courage depicted through central characters on OTT platforms can boost other women to stand up against the oppression they face in their own ways. More such stories on Bengali content offering OTT platforms are thus required and should focus on producers and content teams. It can have a healing effect on the rate of domestic violence and protest against societal harassment by encouraging women to protest and adapt ways to protect themselves from physical assault and harm. More women-oriented shows presenting women's personal, professional, social, and family identities can be explored, as can their emotional feelings, inherent dichotomies, likings and dislikes, and such, in future projects. With Hindi OTT platforms offering female characters who leave an impression on the viewer, it is time for replication of the model in the Bengali content space.

## Evolving Regulatory Scenario

With the rise in OTT viewing and consumption, the Indian government has undertaken several steps to form standards and guidelines for OTTs. The MEITY (Ministry of Electronics and Information Technology) has passed the IT (Information Technology) Rules, 2021 under the Intermediary Guidelines and Digital Media Ethics Code that oversees the digital media sector consisting of digital news in social media and OTT platforms. The guidelines have been enclosed under the IT Act, 2000, and will replace the 2011 rules for internet mediators. These rules emphasize putting an ethics code for the online news channels and OTT platforms that shall be directed by the MIB (Ministry of Information and Broadcasting). OTT platforms will now be required to follow a method of three-tier grievance redressal structure followed by a government supervision mechanism. Under these rules, the OTT players shall be mandated to classify their content according to five categories divided as per age criteria: U (Universal), U/A 7+, U/A 13+, U/A 16+, and A (Adult). Content rated as U/A 13+ and above requires the provision of a parental lock system in the OTT interface. Following the authorization of the new guidelines in May 2021, the MIB addressed the OTT players and digital media platforms requiring them to provide their overall particulars and observance status regarding the new rules within the time frame of 15 days. There have also been debates related to content censorship on OTT platforms. Presently, for OTT platforms, there is no programming code or any independent federation for administering digital content in India (Tele.net, 2022)

## The Drawbacks and Limitations

Despite having a massive potential market in the Bengal region and neighboring country Bangladesh, the share has largely stayed limited to these two platforms. Quality-wise and content-wise, both platforms lack in comparison to the country-wide famous OTT giants of the other significant OTT platforms down South India. Hence the most pertinent question that has been raised is, despite having a rich traditional heritage both in terms of literature and cinema, the OTT platforms have not been able to tap the digital trends because the seventh most spoken language globally is Bengali (Chakraborty, 2021)

Addatimes was the first OTT platform inaugurated in 2016. Rajiv Mehra, managing director and initiator of Addatimes, agrees with the point that the potential of the Bengali OTT market has not been grasped. Instead of creating random content, he emphasizes creating quality content to draw more subscribers. There was a subsequent boost in subscriptions after the web series '*Feluda Pherot*' was released by Srijit Mukherjee in 2016. On the other hand, on the other web platform Hoichoi, roughly twenty-five multiple-genre original shows are made annually, which is an enormous task. According to Soumya Mukherjee, Chief Operating Officer at Hoichoi, there is a huge potential, and the revenues of the platform are being doubled annually, and partnerships are formed strategically with telecoms, ISPs (Internet Service Providers), and others (Chakraborty, 2021)

Arindam Sil, the renowned actor, producer, and director, mentions that the film industry of West Bengal is still in the preliminary and average phase, especially in the field of constructing digital content. To garner more views easily, shoddy content was created, and that strategy did not work out. During his filming of '*Satyameva Jayate*', he experimented with the digital medium as an OTT release. Sil believes that there are specific contents that work exclusively in digital format as it provides staggering reach beyond cinema, and hence one needs to keep that idea at the time of creating OTT content (Chakraborty, 2021). BIGFlix was the first OTT platform inaugurated in 2008 by Reliance Entertainment. Even though it took around a decade for the OTT market to flourish. In the regional sector, major OTT players offer content in vernacular and local languages as add-ons. However, in such scenarios, too much content in Bengali has a minor contribution. Mehra opines that from the beginning, it was a challenging task to persuade Bengali audiences to watch content in their native language. Perhaps the inherent disbelief regarding such contents and hesitation on the quality brings such apprehensions. Mehra mentions that after '*Feluda Pherot*' was released in Addatimes, the subscription base extended gradually as such content appeals to everyone breaking all the social divisions. He also assumes that in the future, more literature-based content will be on-demand as people are looking for quality content (Chakraborty, 2021)

## The Silver Lining

According to Arijit Dutta, proprietor of Priya Cinema, maturity in the contents makes it subtle and distinctive for the OTT platforms in Bengal. He further mentions that audiences should comprehend the content variance among digital media, cinema, and television. Each platform has its specialty, and none can be replaced over the other. However, he also mentions that the Bengali content shown on most of the platforms would generally garner the attention of urban audiences. Many experts in the industry have highlighted that digital content does not possess the arresting silver screen offerings. Hence, the audiences do not receive the complete liberty of viewing as per their requirements. However, unlike the traditional platforms, in OTTs, there is no requirement to travel, buy tickets, or watch content at a specified time. Communication between the spectator and the content is nearly person-to-person, alongside an individual owning a smartphone is a probable target. Thus, theoretically, the platforms offer unmatched influence and variations in viewership and massive revenue. With time, the potential of the Bengali market can be realized (Chakraborty, 2021).

The Head of Sony Entertainment Television and Digital Business, Sony Pictures Networks India, Ashish Golwalkar, mentions that OTT provides content creators with a new prospect in storytelling. The audiences are generally accustomed to the 120-to-150-minute films or much longer formats seen on television. However, OTT platforms offer an opportunity to depict a story in 400 to 600 minutes every season. Hence the focus is laid down on listing, validity, roles, and their well-narrated graphs, shades, and their social ancestries. Golwalkar also mentions that most of the contemporary stories on OTT are popular not because they feature dark components but because we observe OTTs in their primary stage. Therefore, perhaps a significant chunk of the content showcased may have highlighted this area. However, once the OTT content area advances, the audiences will be exploring more assortments (Goyal, 2022). Therefore, before curating niche content for OTT platforms, it's important to grasp the aesthetics of storytelling beforehand and then present it to the audience in a compelling manner.

## The Way Forward

It can be acknowledged that content is the king, and it will remain so, especially in the domain of OTT, where innovative content will provide the OTT platforms with a driving edge and persistence in the coming days. Hence the OTT platforms should take up original content and provide the audiences with unique and stable shows to maintain the continuity and attract new subscribers. However, with the growth of competition in the OTT space and as the platforms continue to generate original content libraries, this uniqueness will become insufficient while differentiating one platform from the other. Thus, the platforms will have to match original content alongside niche viewers and personalization (Quiles, 2019). At present times, consumers want to view content that is available in their language and the content creators are reaching out to these audiences, leading to the growth of new content writers, content makers, and actors (Kumar, 2023). It is a bright time for content creators to generate content in the longer form as there is a scope for bringing the much-required development into the storytelling process. With the emergence of the Covid 19 pandemic, tremendous success in the growth of regional OTTs has been observed resulting in agreeing to take risky decisions among entertainment industry investors. The challenge that lies now is to continuously stay updated with the audience's fluctuating viewing patterns who crave to watch narratives that are deeply rooted and will only prefer content that is value for money and worth watching. The audience does not want repetitive or the same old content. Therefore, we can mention that the forthcoming scenario of entertainment in the country is not monumental, rather it can be termed as effortless, diverse, comprehensive, and all-inclusive similar to how good narratives should be (Kumar, 2023)

## REFERENCES

Adgully. (2018, November 13). *ZEE5 launches its first Bengali original web series 'Kaali'*. Adgully. https://www.adgully.com: https://www.adgully.com/zee5-launches-its-first-bengali-original-web-series-kaali-81623.html

Adgully. (2018, November 13). *ZEE5 launches its first Bengali original web series 'Kaali'.* Addgully. https://www.adgully.com: https://www.adgully.com/zee5-launches-its-first-bengali-original-web-series-kaali-81623.html

Adgully. (2023, May 17). *Hoichoi launches its new Content arm hoichoi studios.* adgully. https://www.adgully.com/hoichoi-launches-its-new-content-arm-hoichoi-studios-131865.html

Agarwal, M. (2020, March 27). *Mapping The Market: India's Regional OTT Video Streaming Landscape*. Inc.42. https://inc42.com: https://inc42.com/features/mapping-the-market-indias-regional-ott-video-streaming-landscape/

Banerjee, S., & Moniz, C. (2018, July 16). *The Rs. 5,000 Cr Opportunity for OTT Platforms in India.* Impact on Net. https://www.impactonnet.com: https://www.impactonnet.com/cover-story/the-rs-5000-cr-opportunity-for-ott-platforms-in-india-6074.html

Bhattacharyya, M. (2020, June). Growth of OTT in India: The Covid Factor. *E-Cine India*, 1-11. https://www.researchgate.net/publication/342520995_21A-Article-Mausumi-Bhattacharyya-Growth-of-OTT-in-India-The-Covid-Factor

Chakraborty, Y. (2021, January 13). *Is the OTT boom passing Bengal by?* Get Bengal. https://www.getbengal.com: https://www.getbengal.com/details/is-the-ott-boom-passing-bengal-by

Chattopadhyay, B. (2019, April 3). *Astey Ladies review: New Hoichoi series is a ton of fun in first half, but frustrating in second.* First Post. https://www.firstpost.com: https://www.firstpost.com/entertainment/astey-ladies-review-new-hoichoi-series-is-a-ton-of-fun-in-first-half-but-frustrating-in-second-6380621.html

Cyrill, M. (2023, April 11). Exploring India's Thriving OTT Video Segment: Trends, Growth, and Investor Interest. *India Briefing.* https://www.india-briefing.com: https://www.india-briefing.com/news/india-ott-video-segment-growth-trends-investments-27681.html/#:~:text=In%202022%2C%20market%20estimates%20showed,Hindi%2C%20Tamil%2C%20and%20Telegu

Dey, A. (2021, December 15). Family entertainment, regional content: OTT players craft strategy for 2022. *Fortune India.* https://www.fortuneindia.com: https://www.fortuneindia.com/enterprise/family-entertainment-regional-content-ott-players-craft-strategy-for-2022/106377

Farooqui, J. (2022, March 30). *OTT platforms in India spent Rs 3700 crore on original content in 2021.* Exchange4Media. https://www.exchange4media.com: https://www.exchange4media.com/digital-news/ott-platforms-in-india-spent-rs-3700-crore-on-original-content-in-2021-119384.html

FICCI. (2022). *Tuning into consumer: Indian M&E rebounds with a customer-centric approach. Ernst & Young Associates LLP*. FICCI. https://www.ey.com/en_in/media-entertainment/tuning-into-consumer-indian-m-and-e-rebounds-with-a-customer-centric-approach

Ghosh, S. (2023, June 28). Raima Sen: OTT contents now celebrate and amplify the voices of women from all walks of life. *Times of India.* https://timesofindia.indiatimes.com: https://timesofindia.indiatimes.com/entertainment/bengali/movies/news/raima-sen-ott-contents-now-celebrate-and-amplify-the-voices-of-women-from-all-walks-of-life/articleshow/101341971.cms?from=mdr

Goyal, S. (2022, January 15). Is The Darkness On OTT Forced Or Organic? *Outlook India.* https://www.outlookindia.com: https://www.outlookindia.com/website/story/entertainment-news-is-the-darkness-on-ott-forced-or-organic/409809

Jha, L. (2021, September 24). Bengali OTT service Hoichoi announces new line-up as watch-time doubles. *Live Mint.* https://www.livemint.com: https://www.livemint.com/industry/media/bengali-ott-service-hoichoi-announces-new-line-up-as-watch-time-doubles-11632484289520.html

Jha, L. (2022, August 2). Local OTT platforms battle rising costs of production. *Live Mint.* https://www.livemint.com: https://www.livemint.com/industry/media/local-ott-platforms-battle-rising-costs-of-production-11659381966885.html

Jha, L. (2022, March 28). Regional Languages to Account for 54% of OTT Content by 2024. *Live Mint.* https://www.livemint.com: https://www.livemint.com/industry/media/regional-languages-to-account-for-54-of-ott-content-by-2024-11648457191989.html

Kashmir, R. (2022, March 7). How the demand for local language content increased in the last 5 years. *Rising Kashmir.* http://risingkashmir.com: http://risingkashmir.com/how-the-demand-for-local-language-content-increased-in-the-last-5-years#:~:text=In%20recent%20years%2C%20we%20have,and%20receptive%20to%20new%20experiences

Kashyap, S. (2019, January 25). To win over Indian Audiences, OTT platforms need to understand the sensibilities of Bharat. *YourStory.* https://yourstory.com: https://yourstory.com/2019/01/content-language-india-ott-bharat

Kumar, S. A. (2023, April 4). The Multilingual Era Has Arrived And Here's Why It's A Blessing For Regional Stories. *India Times.* https://www.indiatimes.com: https://www.indiatimes.com/entertainment/binge/the-multilingual-era-has-arrived-and-heres-why-its-a-blessing-for-regional-stories-598030.html

Mahalakshmi, N., & Gopalan, K. (2021, January 1). *OTT players are splurging on content. What's their game?* Outlook Business. https://www.outlookbusiness.com: https://www.outlookbusiness.com/strategy-4/feature-17/ott-players-are-splurging-on-content-whats-their-game-6146

Niyogi, A. (2023, February 2). ZEE5 series Roktokorobi is a taut mystery thriller around family secrets and mental health. *Telegraph India.* https://www.telegraphindia.com: https://www.telegraphindia.com/entertainment/zee5-series-roktokorobi-is-a-taut-mystery-thriller-around-family-secrets-and-mental-hcalth/cid/1915890

PTI. (2021, August 15). Balaji Telefilms to invest in original content with affordable subscription for OTT platform. *New India Express.* https://www.newindianexpress.com: https://www.newindianexpress.com/business/2021/aug/15/balaji-telefilms-to-invest-in-original-content-with-affordable-subscription-for-ott-platform-2345046.html

Quiles, L. (2019, April 26). *Original Content as the Selling Point for OTT.* Intraway. https://www.intraway.com: https://www.intraway.com/blog/original-content-selling-point-ott/

Qureshi, N. (2020, March 26). *Original content is way forward for OTT platforms: Abhimanyu Singh of Contiloe Pictures.* Best Media Info. https://bestmediainfo.com: https://bestmediainfo.com/2020/03/original-content-is-way-forward-for-ott-platforms-abhimanyu-singh-of-contiloe-pictures

Ranipeta, S. (2023, March 3). *Move over Netflix, Prime Video, Hotstar! Regional OTT platforms are new favourites among Indian consumers.* CNBC TV. https://www.cnbctv18.com: https://www.cnbctv18.com/business/move-over-netflix-prime-video-hotstar-regional-ott-platforms-are-new-favourites-among-indian-consumers-16095301.htm

Sarkar, G. (2023, July 5). Bengali OTT players prioritise market growth over competition. *India Television.* https://www.indiantelevision.com: https://www.indiantelevision.com/iworld/over-the-top-services/bengali-ott-players-prioritise-market-growth-over-competition-180705

Sarkar, R. (2021, September 27). *Hoichoi announces 20 original shows.* Cinestaan. https://www.cinestaan.com: https://www.cinestaan.com/articles/2021/sep/27/31469/hoichoi-announces-20-original-shows

Subramanian, B. (2022, March 22). How India is 'Cutting the Cord' to drive an OTT revolution. *Business Insider.* https://www.businessinsider.in: https://www.businessinsider.in/advertising/media/article/how-india-is-cutting-the-cord-to-drive-an-ott-revolution/articleshow/90347924.cms

Tele.net. (2022, May). *OTT Action: Demand rises for high quality content and seamless streaming.* Tele.net. https://tele.net.in: https://tele.net.in/ott-action-demand-rises-for-high-quality-content-and-seamless-streaming/

The Telegraph. (2022, March 22). Hoichoi scores a century with its Originals. *The Telegraph.* https://www.telegraphindia.com/my-kolkata/lifestyle/hoichoi-scores-a-century-with-its-originals/cid/1857039

The Tribune. (2021, September 19). OTT Platforms, books, movies Thrive Together. *The Tribune.* https://www.tribuneindia.com/news/amritsar/ott-platforms-books-movies-series-thrive-together-313274

Times of India. (2022, December 7). Subhashree Ganguly's debut OTT project 'Indubala Bhater Hotel' to stream soon. *Times of India.* https://timesofindia.indiatimes.com: https://timesofindia.indiatimes.com/tv/news/bengali/subhashree-gangulys-debut-ott-project-indubala-bhater-hotel-to-stream-soon/articleshow/96049509.cms

## ADDITIONAL READING

Bala, S. Kaur, Manmeet., & Pandey, K. P. (2022). Cinema and OTT Wave. Kanishka Publishers.

Bhasin, P. (2022). Law Relating to Social Media Crimes, Intermediaries, Digital Media and OTT Platforms. OakBridge.

Pratap, Vipul., & Mittal, Rahul (2022). OTT Platforms & Digital Media. Ishaan Arts and Production.

Saxena, A. (2023). Cinema, OTT and social media. Kanishka Publishers.

## KEY TERMS AND DEFINITIONS

**HSM (Hindi-Speaking Market):** It refers to the regions in India where the Hindi language is the prime source of communication. The term is often used to identify the demography of television viewership and extract the viewership data.

**IP (Intellectual Property):** IP falls under the grouping of property which consists of impalpable conceptions developed by human intelligence. For example: patents, trademarks, business secrets, copyrights, etc. The definition and categorization of IPs can vary from country to country.

**ISPs (Internet Service Providers):** ISPs are the organizations that deliver internet access to individuals and companies enabling them to use the internet and charge a fee for their service. ISPs also provide assorted services like website domain registration, email and web hosting, etc.

**M&A (merger and acquisition):** It signifies the alliance of organizations or unions of their significant business possessions through monetary dealings among the corporations.

**MAUs (Monthly Active Users):** MAU is a prime indicator of performance, devised by organizations to tally the number of exclusive users who have visited any particular website over the last one month. Websites decode the MAUs by identifying the credentials of a website visitor's email address and username.

**OTT (Over the Top):** OTT refers to the abbreviation of Over The Top. It is a technology that helps to deliver content that can be streamed via the internet. Initially, viewers had to subscribe to cable operators to watch content on television. With the help of OTT technology users can directly avail content through OTT platforms and stream videos with the help of the internet

**Streaming services:** Streaming services provide online entertainment in the form of digital audio/video format. To avail of a streaming service, the user requires a digital device like a computer, smartphone, television, or tablet and an active internet connection.

# Chapter 3
# Study on the Growth of the OTT Platform During Lockdown and Its Future Scope

**Sakshi Raj**
https://orcid.org/0009-0006-1582-9008
*Christ University, India*

## ABSTRACT

*This research tries to review the growth of OTT platforms throughout the lockdown. It's vital to understand the extent of increase within the popularity of OTT platforms throughout lockdown to understand their future scope. It's evident that, since their launch, OTT platforms have solely discovered an upward curve in their usage, and because of the pandemic and lockdown there has been exponential increase in its popularity because of the modification in consumption patterns of individuals for diversion through numerous media platforms. This analysis conducted a survey and analyzed the opinions of individuals relating to OTT platforms, their consumption patterns, and its comparison with cinema to examine if OTT platforms were slowly taking on the foremost fashionable standard medium of diversion. It had been found that individuals used OTT over the other platforms outside of TV and YouTube to pass their time or for diversion.*

## INTRODUCTION

The research topic deals with the emergence and consumption within the duration of lockdown towards a selected sector i.e. OTT, over the top services. This sector belongs to an awfully huge trade, the show biz. Entertainment in layman's terms suggests that having fun and enjoying either with one's own company or even with their favourite ones. In today's world entertainment, is one among the foremost leading industries, which incorporates the sub-fields of theatre, film, fine art, dance, opera, music, literary publication, television, and radio. With the day to day development in technology, these industries are being improved at a quick pace. There are cases of cut- throat competition along with goodwill within the market. Showbiz is unpredictable because the main profits depend on the TRP of a selected TV

DOI: 10.4018/979-8-3693-3526-0.ch003

station, or the box-office collection of over 100-200cr. It is assumed to be a blockbuster and hit flick, whereas generally, other movies that couldn't reach this bar at the box office, are assumed to be average, in spite of the actual fact that those movies might have a good script or direction, etc. There are numerous ways that, (as mentioned above) as this industry is often perceived in varied forms. Currently, even the advertisers have understood this idea and are adamant in making new and attention-grabbing advertisements for constant. The customers look for creativity ability power} and out of the box concepts that please them and make them think and bear in mind about a complete product in an exceedingly explicit way. For instance, sometimes after we see 2 dominos dices, we tend to get reminded of the dish chain – Dominos'. Similarly, after we see 3 parallel vertical lines, we automatically identify them as Adidas.

With major players like Voot, AltBalaji, Zee5, and MX Player, India's streaming market is among the largest and fastest-growing in the world, according to the Global Web Index survey. The nation's video market, currently estimated to be worth over $700 million, is anticipated to increase to $2.4 billion by 2023, with the OTT sector seeing the fastest development.The big budget producers and directors will be forced to release movies directly onto OTT channels rather than at theaters as a result of the trend of people watching multiple content channels at once and the demand for newer and more exclusive content. For example, the Vidya Balan-starrer Shakuntala Devi biopic was premiere exclusively on Amazon Prime Video, along with the first time that Amitabh Bachchan had seen the global premiere of his film on a digital platform with his latest picture "GulaboSitabo," which also stars Ayushmann Khurrana, was unable to be seen in theatres when it was scheduled for release in April.Good content for over-the-top (OTT) services is highly sought after and during the lockdown.

It is truly said that this is the age of millennia, they neither hear their elders nor do they do what's suggested to them. The expansion of the population, youth, and marketing, everything else is interconnected. The OTT, which is evolving from niche to mass-based content and long-form is getting traction. (Girish menon, 2017) The number of OTT players increased from just 09 in 2012 to 32 in 2018. In 2017, the OTT industry in India achieved phenomenal growth of 160 per cent, as the top 16 OTT platforms saw their user bases grow from 63 million to 164 million between August 2016 and August 2017.(Rajat kathuria, 2019) .

India is the second-largest subscription TV market in the Asia Pacific region in terms of the number of subscriptions in TV households, which reached 154.3 million in 2016. This is expected to expand to 1.6% CAGR to reach 166.9 million by 2021. (Alpana kakkar, 2019)

Technology is ever-changing with clicks. From online banking strategies to e- wallets and what not, with one click everything is available at the doorstep.

When Netflix released the 15 episodes of a new season of *Arrested,* Development in the summer of 2013, reports showed that approximately 10% of viewers had seen the whole season within 24 hours and shared it among the millennial cohort, which marks Netflix at the top subscription list (matrix, 2014)

OTT has been enabled by technology advances such as smart phones, super-fast IP networks, open source platforms, innovative services, cutting edge functionalities and shift in consumer preferences towards their "freemium" based business models are seeing an ever increasing adoption rate. (Joshi Sujata, 2015) This freemium concept attracts consumers. They do not understand that nothing is free. Firstly, we have to pay for the internet packages and then only something can be surfed on the internet. So, ultimately, until and unless the consumer pays for the data, he/she won't be able to use any of the applications, which include internet connection, for example, Netflix, amazon, etc.

## OBJECTIVES

1. To find out and understand the increase in consumption of OTT platforms during lockdown.
2. To understand the future scope of the OTT platforms.
3. To study the shift in consumption from old entertainment to new.

## HYPOTHESIS

H1. During the pandemic, there has been a significant increase in consumption of OTT platforms.

H2 OTT platform will soon become a platform for releasing movies or maybe even surpass cinema. (which earlier release in cinema)

## REVIEW OF LITERATURE

According to (Purdy, 2018), the OTT market is a complex space dominated by three powerful players, - Netflix, Amazon Prime, Hulu - the Big 3. It is challenging for small players to find a place in such a competitive market where content rules, but consumers can be dragged if they focus on carefully chosen and thoughtfully organized content limiting to account sharing, getting the price right and reducing stir through improved customer relations. The survey result indicates that, although price is a dominant factor for purchasing OTT services, content drives most consumer choices. A mixture of good quality content tends to be what consumers value from the Big 3, with "Access to the vast library of content" proving to be the top reason users subscribe to these services. The promise of small players to provide the services, taking one of above mentioned factors, attracts the most big crowds for their subscription.

Article (Singh, 2020) OTT platforms have experienced an increase in consumption and subscribers can't, impacted by the COVID-19. The recent survey by InMobi states that 46% of viewers are watching more content online. According to the experts, OTT services could grow further in upcoming days as the television channels run out of content. "As these channels have been unable to shoot due to lock-down, they are running old content. In such a scenario people will gravitate towards OTT to watch fresh content," says Paritosh Joshi, media consultant and principal, Provocateur Advisor. To benefit from this increase in demand, platforms such as Amazon Prime video and Zee5, have made a few of their shows free to watch. Due to this, Zee5 has witnessed an 80% increase in subscriptions and over 50% times spend recently. Thrillers, urban drama and young romance are the top performing genres on the platform. The viewership for the movies has seen a significant increase across the platforms. Cinema streaming service MIBI has seen a 28% rise in viewership on its platform in March, compared to February.

Media play a fundamental role in the superstructure of every civilization. It is crucial to the spread of knowledge.(Patel, An Analysis of the Impact of Personal Communication Technology on the Public Policy Making Process In India, 2017).Technology was clearly the main factor influencing people's lives especially in the previous five years as, the world media had changed so quickly. The dramatic increase in the percentage of internet viewing was caused by the nationwide lockdown situation.(Patel, Social Media in the Indian Context: -New flavor of the Season, 2015). Online content viewing increased by

46%, according to a report by In Mobi, a smartphone marketing platform. The prime time consumption of OTT was also modified, moving from 10 pm to 12 am to 7 pm.(Financial Express, Apr 06, 2020).

According to (Kaushal, 2020) there has been a sharp increase of 80% in the subscription of OTT players during lock down. Due to the lockdown people have been forced to stay at home and with them being locked inside their home they have no other options left for entertainment than TV and digital. With the closed cinema halls and malls, people were forced to find other platforms for content. OTT players like ZEE5 and ALT Balaji saw growth in their subscriptions. Whereas Amazon Prime and Netflix did not reveal their figures. As we know that due to lockdown several big budget films are set to be launched on OTTs rather than cinema halls, this shows the tremendous growth of OTT as a platform. People are left with no option but to watch it on OTT, even for the hardcore cinema lovers. Due to the lockdown ZEE5 has seen an increase in users as well as consumption. Their original saw grew by 200%, movies by 236%. They saw an increase in their app downloads by 41% and an increase of 33%in their daily active users. There has been a rise of 45% in paid viewers. The head of program said that the two reasons for such positive increase are the lockdown and several initiatives taken by them for seamless entertainment. It is important to note that ZEE5 has premiered a film called Ghoomketu, which was originally meant to be released in theatres. Similar is the case with Alt Balaji, it had an increase of 17,000 new subscribers each day in lockdown which is 60% more compared to before lockdown. They have also been experiencing growth in all of their key markets and demographics. With the announcement made by Amazon Prime to release the much hyped films like Gulabo Sitaro and Shakuntala Devi biopic on OTT platform rather than in cinema due to the lockdown along with some other films it made clear that the scope of OTT platforms will have in future.

(Sharma, 2020)The ongoing COVID-19 crisis has forced the multiplexes to shut down. As a result, the production houses are now releasing onto OTT platforms directly. This has become a topic of concern amongst the multiplexes as the users might ship the traditional windowing pattern. INOX and PVR, both issued a public statement reacting to this issue. According to experts, this is a solution to a short term need of the film industry which is suffering from financial difficulties. It is a win-win situation for OTT players who need fresh content and the filmmakers who had their production ready but unable to release. However, this will not be a paradigm shift, in the long term, once things return to normal. The cinema generates more revenue for the production houses compared to the OTT platforms. Being this is a vital factor, the films will hit the cinemas earlier, once the market is free from the COVID-19 crisis. As a matter of fact, there may be a possibility of seeing an increase in multiplex services for a few days after returning things to normal.

(Singh D. P., 2019)The advent of the Internet is changing the trend of Indian Cinema very rapidly. Reduced prices of internet services laid the foundation for an increase in OTT services like Netflix, Hotstar, Zee5, Amazon Prime and Alt Balaji. Many ventures are producing video content solely for these OTT services. These new ventures are changing the way of Indian television and cinemas. The OTT service analyzes the video content watched by the users and provides them the content on similar genres. People like to enjoy the services for free. Hence, several OTT provide very limited content for free. Also, the subscription rates of these services are very low. The users enjoy the variety of international and national content. Action and comedy are the favourite genres of movies. According to the survey, users spend, on average, 2 hours a day watching OTT services. Web services are watched more often on OTT services. Genres like Young romance, action and urban-drama are in the top list. Most of the users watch the content at night. Entertainment is the biggest reason behind the use of OTT. The second biggest reason behind the use is the mobility of the gadget. Third is its availability of content on demand.

## THEORITICAL FRAMEWORK

Media Dependency Theory, developed by Sandra Ball-Rokeach and Melvin DeFleur in the late 1970s, is the best theory to use as it suggests that the media plays a crucial role in shaping individuals' perceptions and attitudes, and people become dependent on the media for various needs. The theory posits that media dependency increases when other sources of information and social interaction are limited. Applying this theory to the growth of Over-The-Top (OTT) platforms during lockdown and its future scope can help in understanding that with lockdown restricting outdoor activities, people became more dependent on OTT platforms for information and entertainment. OTT platforms offered a way for people to connect virtually, discuss shared content, and alleviate feelings of isolation during lockdowns.The increased dependency during lockdowns may have altered viewing habits permanently, leading to continued growth in OTT consumption.OTT platforms might continue to diversify their content to cater to a wider audience, thereby increasing dependency as they become a one-stop solution for various content needs. The growth of OTT platforms may extend globally, providing content that transcends geographical boundaries and cultural difference. Dependency on OTT platforms might raise concerns about over-reliance, potentially limiting exposure to diverse perspectives and information sources. Disparities in access to high-speed internet and digital devices may impede the growth of OTT platforms in certain regions, contributing to a digital divide. Media Dependency Theory provides a framework to understand how individuals became dependent on OTT platforms during lockdowns for information, entertainment, and social connection. The future scope involves changes in consumption patterns, challenges related to overreliance and digital divide, industry dynamics, and technological advances. The continued growth of OTT platforms will likely be influenced by these factors, shaping the way people consume and interact with media content in the future.

## RESEARCH DESIGN AND METHODOLOGY

The purpose of this study was to understand the growth of OTT platforms during the lockdown and its future scope. The quantitative method was selected for the same. The focus was on residents of Patna who view OTT for entertainment.

### Population and Sampling

The population is people of Patna who watch OTT for their entertainment. The age group selected is 15-45+. Since most respondents from this category were independent, indulging in anything they wish for is easier. Because of their entertainment consumption habits, they are aware of the research topic selected. The population has been filtered out through some screening questions.

### Tool Used for Data Collection

A structured questionnaire carrying 16 close-ended questions and was administered via Google form. The questionnaire had relevant queries to evaluate respondent insights.

## Data Interpretation and Analysis

The researcher has used a questionnaire as a tool for collection of data, The set of questions was prepared by the researcher after a detailed study on the topic and the set of questions aims to give us insights about the effect of television commercials on impulsive buying behaviour of consumers. The aim is to analyse the data gathered by 100 respondents across Patna corresponding to the objective of the research.

1. Demographics

The questionnaire was sent to the respondents via WhatsApp. It was filled by people in Patna of different age groups, gender and occupations. Those factors will be analysed in this section.

1.1 Age of respondents

Figure 1. shows that out of 100respondents. The age groups were classified into 3 segments. The maximum number of respondents was 52%, belonged to the age group 15-25, followed by 47% respondents in 26-35, 1% of respondents in the age group of 35-45.

*Figure 1. Age of respondents*

1.2 Gender of Respondents

Figure 2 shows the results of a survey in which people were asked about their gender. It can be seen that 52% of respondents are male, whereas 48% of users are female who have filled the survey.

*Figure 2. Gender of respondents*

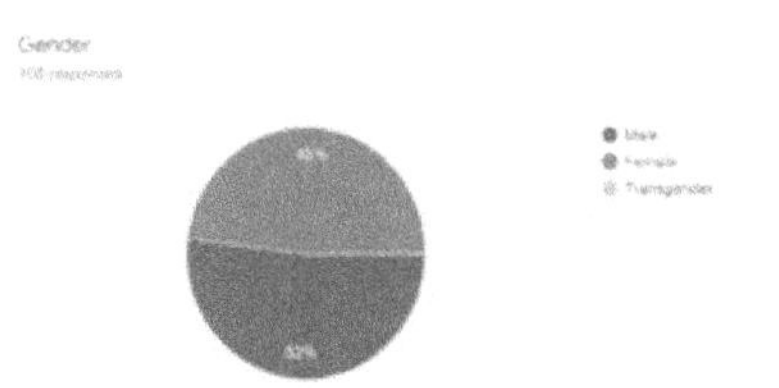


1.3 Profession of respondents

The above figure (Figure 3) show the professions of the people who were part of the survey. Almost 37% of respondents are perusing post graduation, 28% are working as employees .24% are under graduates. 10% are school students and the remaining 1% run their own business.

1.4 Screening question

*Figure 3. Profession of respondents*

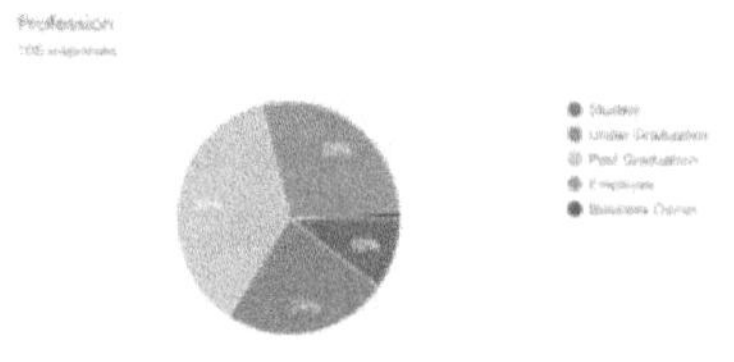

Figure 4 shows that 88% of respondents use OTT while 12 don't. This was a screening question to filter out the required population of who watch and who do not watch OTT.

*Figure 4. Screening question*

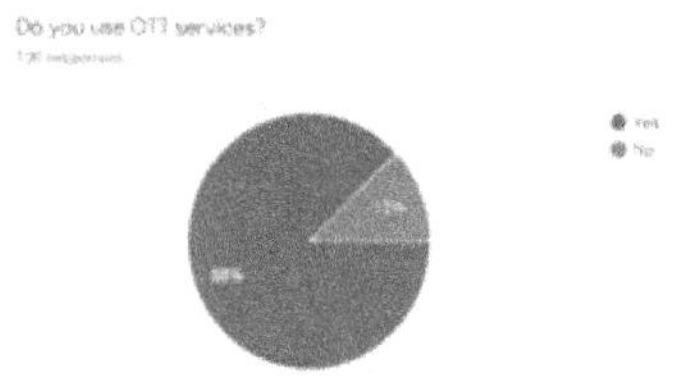

2. Consumption behaviour

This section of the questionnaire gathered data related to the OTT watching habits and patterns in lockdown of the respondents.

2.1 Percentage of population that watches OTT during lockdown

As seen in Figure 5, 61 respondents out of 100 watch OTT the most, 37 selected YouTube and the remaining 02 watches T.V most for the entertainment.

*Figure 5. Percentage of population that watches OTT during lockdown*

2.2 Awareness of the media

Figure 6 speaks to the respondents about their source of awareness for the OTT medium. The figure shows that 40% for example 40 respondents out of 100 came to know about OTT through their friends.

*Figure 6. Awareness of the media*

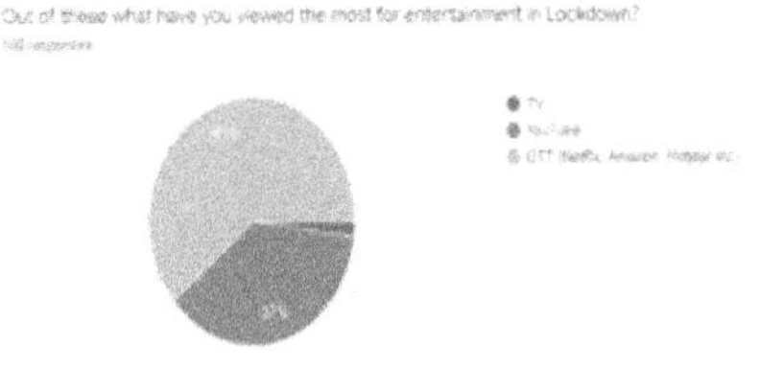


While 31% i.e., 31 respondents came to know through an advertisement. 26% responses connected to through others.

2.3 Hours spent on OTT

Figure 7 shows that to perceive how captivating OTT platforms are to users. The above inquiry was posed, the discovery states that 53.1% use it for less than 2 hours day by day. Over 32.7% (33) of them use for 2 to 4 hours. Just 14.3% of the clients expend it for more than 4 hours.

*Figure 7. Hours spent on OTT*

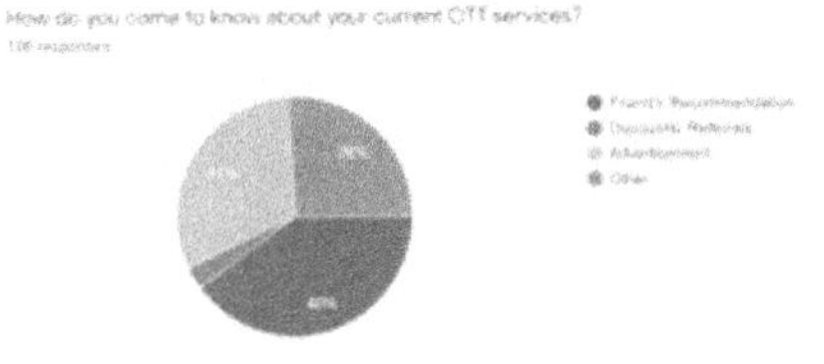


2.4 Impact of lockdown on OTT

Figure 8 shows the percentage of respondents of the impact on their OTT consumption. 63.3% of respondents have seen an increase whereas 32.7% of people are not affected by it and the remaining 4% of respondents have seen a decrease in their usage.

*Figure 8. Impact of lockdown on OTT*

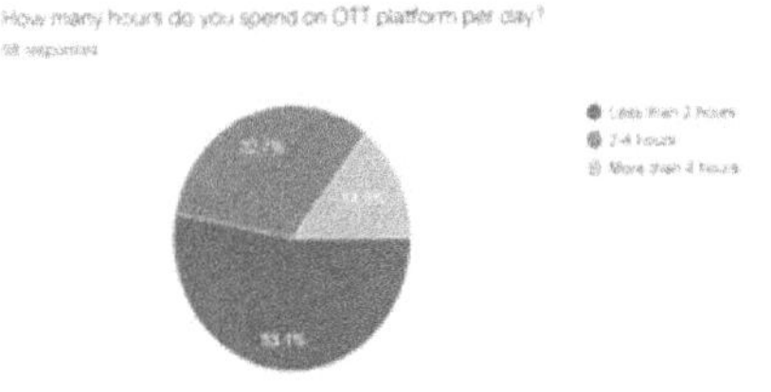

*Figure 9. Increased consumption*

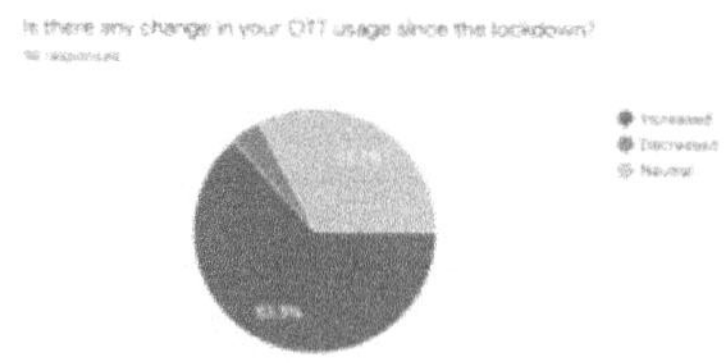


2.5 Increased Consumption

As seen in the figure, we see that 18 respondents said their consumption increased by 5 hours. 34 said by 1-2 hours, 12 said by 3 hours, 13 said it increased by 4 hours, 7 said by 8 hours and other 8 said that it increased by 9-10 hours.

*Figure 10. Mode of consumption*

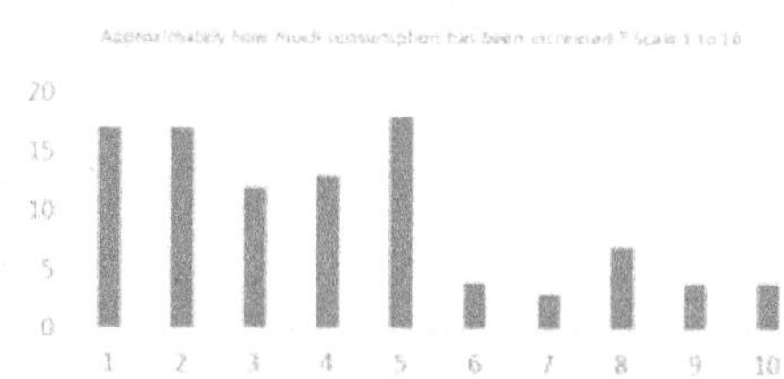


2.6 Mode of Consumption

From the above image, it is evident that a maximum of 88.7% of respondents consume OTT through their smart phones. 8.3% of respondents prefer TV and the remaining 3% consume through a laptop.

2.7 Started using a new OTT platform

Figure 11 shows who opted for a new OTT while being in the lockdown. With 46 of the population voted yes for it. 36% of the population voted for No and 18% of the population were not sure if they started using it or not.

*Figure 11. Started using a new OTT platform*

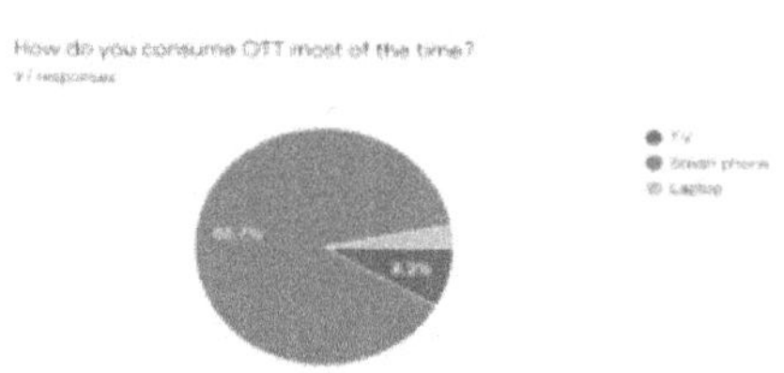

2.8 Consumer preference on OTT

Figure 12 shows the result of their preferences on OTT platforms. Over 41.8% of people prefer watching original/ exclusive content. While 26.5% refer to watching movies. 20.4% like to watch TV serials. 11.2% of respondents like other stuff on OTT.

*Figure 12. Consumer preference on OTT*

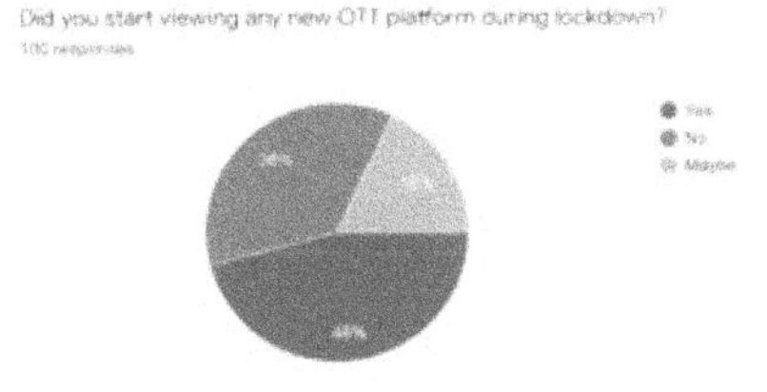


2.9 Waited for a movie to launch on the OTT platform

Figure 13 distribution of respondents based on respondents' preference of OTT. 50% of respondents responded said many times. While 37% of respondents responded said never. Then 13% of respondents said always.

*Figure 13. Waited for a movie to launch on the OTT platform*

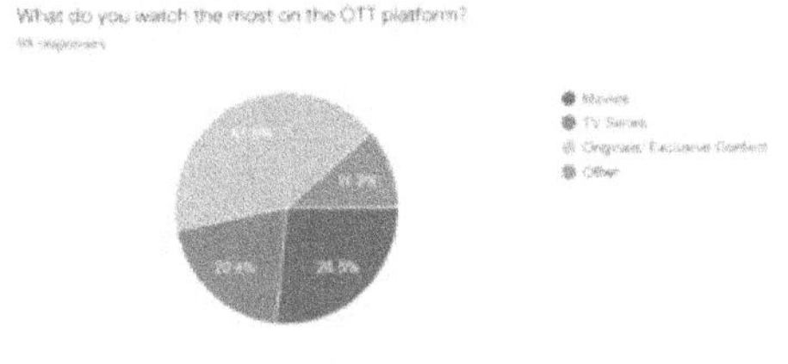


2.10 Movie release on OTT

Figure 14 the respondents were asked their views on movies released on OTT along with theatre. We see platforms a majority of 71% of respondents said yes, while the other 29% of respondents say no.

*Figure 14. Movie release on OTT*

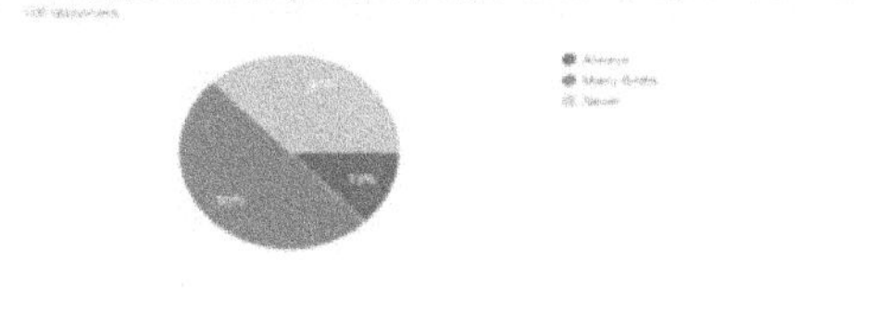

2.11 Cinema Vs OTT

Figure 15 states a futuristic question on their choice after the pandemic. 75% say they will choose OTT to watch movies instead of cinema, while 25% still prefer theatre.

*Figure 15. Cinema vs. OTT*

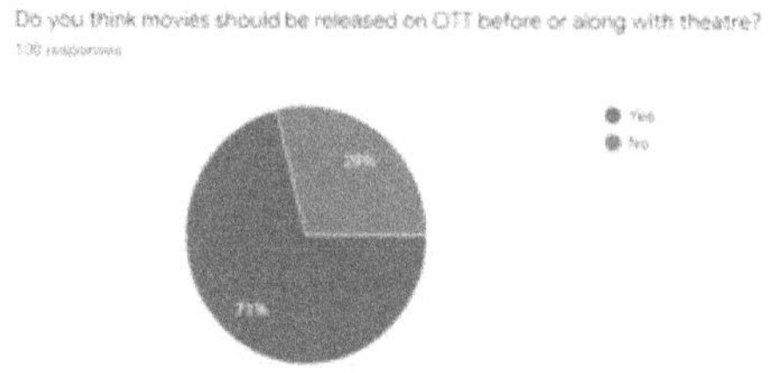


## Findings

Over the past few years, the number of over- the – top video streaming services has increased in India, which has led to the origin of different content consumption patterns. As content consumption has become difficult to control outside but recently, as of Covid 19, the majority of the consumption is occurring inside too, on a very high scale. The information gathered by the research shows that over 88% of respondents are aware of and use the OTT medium, From the following, we can derive that OTT is a wide-spread medium in India. Due to the prevailing situation of the lockdown people are constantly consuming different medium for their entertainment. The majority of people, 61%, resort to OTT for entertainment, followed by YouTube, 37% and T.V 2%. OTT platforms have become quite popular among the users as a considerable number of users 53.1% spends up to 2 hours per day. Also 32.7% of people spend up to 4 hours on it and 14.3% consumes it for more than 4 hours. Considering this, on an average, user spends approximately 3 hours on the medium per day. The pandemic and lockdown has led to a change in consumption patterns of the users. The researcher found out that 63.3% of people experienced an increase in their daily usage of OTT consumption. Users on an average experienced 3 to 4 hours of increase in growth of OTT and its future scope as consumption during the lockdown has increased. Along with this it also shows that OTT is easy to go platform for entertainment for majority of people during lockdown as they chose to spend more time using OTT rather than any other mediums. When it comes to the mode of consumption, people prefer smart phones (88.7%) over TV(8.2%) and laptop (3.1%). 46% of people started using a new OTT platform during the lockdown. This shows how lockdown has given a push to OTT for generating new customers. When asked about if the users ever wait for a movie to be released on OTT rather than watching it in a theatre, 50% said many times. The researcher discovered that 71% were positive and wanted the movies to be released on OTT platforms on the same date as in cinema halls/ theatres. Although 29% were still negative about it. Furthermore, when asked if a movie would be released in OTT and cinema, what would they prefer, only 25% said cinema and 75% chose OTT. This data implies that a majority of people (75%) would prefer watching a movie on OTT than to cinema hall (25%). This shows that people have started adopting OTT over theatre. Although OTT and theatre are two distinct platforms and provide distinct experiences to their customers and they cannot replace each other, the data shows that the majority of people want movies to be released on OTT and will also see movies on OTT as compared to the theatre.

### Limitations of the study

The samples collected by the researcher were limited to Patna.

## CONCLUSION

The data collected from the questionnaire survey was interpreted and analyzed above. Based on the analysis, the research on the topic 'Study on Growth of OTT Platform during Lockdown and its Future Scope' has been successfully completed. The hypothesis that during the pandemic there has been a significant increase in consumption of OTT platforms has been proven right. Since people have started consuming and adopting OTT more and more due to the lockdown. The researcher also proved the hypothesis that OTT would surpass cinema. We can see there is suddenly a rise in the usage of OTT over other media and people have positive responses to movies being released on OTTs and also prefer OTT over theatres in the future at the same time. This shows that OTT platforms can replace cinema halls but certainly create their own audience. We can say that in the future, this platform might give a tough competition and could become a threat for cinema halls or theatres.

## REFERENCES

Alpana Kakkar, K. N. (2019, may 15). Analysis of Product Placement in Web Series and Its Influence on Consumer Buying Behavior. *Journal of Business*. https://journalofbusiness.org/index.php/GJMBR/article/view/2755

Basu, R. (2020) *Television and OTT*. PWC. https://www.pwc.in/industries/entertainment-and-media/television-and-ott.html.

Financial Express. (2020). Now streaming, more than ever: Netflix, Amazon, others need to up their game. *Financial Express*. https://www.financialexpress.com/industry/technology/ now-streaming-more-than-ever-netflix-amazon-others-need-to-up-their-game/1918370/

Gaikwad, S. (2019). How Indian cinema evolved into being one of the most powerful modes of entertainment. *Businesstoday*. https://www.businesstoday.in/opinion/columns/how-indian-cinema-evolved-intobeing-one-of-the-most-powerful-modes-of-entertainment/story/307454.html

Kaushal, S. (2020). *OTT platforms in India See subscriber bases surge UP to 80% amid coronavirus lockdown*. Forbes. https://www.forbes.com/sites/swetakaushal/2020/05/26/good-news-amid-lockdownott-platforms-register-60-80-surge-in-subscription-base/#77d0fab7413a

Kohli, D. (2019). *Looking into the future – Ott trends*. YourStory. https://yourstory.com/2019/02/looking-into-future-top-ott-trends-2019

Roshan. (2017). *5 benefits of Video on Demand over traditional TV*. muvi. https://www.muvi.com/blogs/5-benefits-video-demandtraditional-tv.html

Sharma, K. (2020) Will movies releasing directly on OTT platforms become a norm post-Covid? Experts think otherwise. *Business insider*. https://www.businessinsider.in/advertising/media/article/will-movies-releasingdirectly-on-ott-platforms-become-a-norm-post-covid-experts-thinkotherwise/articleshow/75808931.cms

Singh, D. (2020) How is coronavirus impacting the streaming platforms with an increasing appetite of viewers. *Financial express*. https://www.financialexpress.com/brandwagon/how-is-coronavirus-impacting-thestreaming-platforms-with-an-increasing-appetite-of-viewers/1919916/

Patel, K. (2017). An Analysis of Impact of Personal Communication Technology on Public Policy Making Process In India. in*ternational Journal of Emerging Trends in Information & Knowledge Management, 1*(1). Retrieved 06 12, 2020.

Patel, K. (2015). Social Media in the Indian Context: -New flavor of the Season. *Global Journal Of Multidisciplinary Studies, 4*(6), 2680-270.

# Chapter 4
# The Rise of Digital Celebrities:
## Analyzing the Impact and Sustainability of Stardom on OTT Platforms

**Archan Mitra**
https://orcid.org/0000-0002-1419-3558
*Presidency University, India*

**Sayani Das**
https://orcid.org/0000-0003-0076-7466
*Institute of Mass Communication, Film, and Television Studies, India*

## ABSTRACT

*This article studies celebrity status in digital media, focusing on OTT platforms and social media. New fame paths have changed famous dynamics. This has changed content development, audience engagement, personal branding, and professional-personal boundaries. This study compares digital celebrities to film and television stars, and finds a complex world where renown is achieved by competence, deliberate online presence, and active audience connection. This study examines digital celebrities' professional trajectories, their capacity to migrate into traditional media, and OTT platform fame's future. The results show that digital fame requires constant ingenuity, flexibility, and digital pattern understanding. With more material and specialist personalities, over-the-top (OTT) platforms have a promising stardom potential. This study analyses modern celebrity culture and its potential effects on digital stardom.*

## INTRODUCTION

With the development of the internet and the spread of digital media channels, the entertainment sector has undergone tremendous changes. The advent of Over-the-Top (OTT) platforms, which have upended conventional methods of content distribution and consumption, is one significant phenomenon (Lobato, 2018). The emergence of these platforms has given rise to a new class of celebrities known as digital celebrities (Hearn & Schoenhoff, 2016). The emergence of digital media has brought about a significant transformation in the realm of stardom and celebrity. This study examines the emerging phenomenon

DOI: 10.4018/979-8-3693-3526-0.ch004

of digital celebrities, analysing their impact and the evolving nature of fame within the realm of Over-the-Top (OTT) platforms and social media. This study explores different facets of digital stardom, encompassing the emergence of over-the-top (OTT) platforms and their influence on the formation and cultivation of stardom. Additionally, it examines the content strategies employed on these platforms and the consequent proliferation of niche stars as a result of increased content diversity.

Moreover, the authors explore the pivotal significance of social media in fostering audience engagement and cultivating online branding. This discourse pertains to the impact of social media on the creation and promotion of content, as well as the transformation of celebrity culture due to evolving patterns of consumption. An examination of digital celebrities and traditional film and television stars allows for a comparative analysis that sheds light on the distinct challenges and opportunities associated with digital stardom. Furthermore, this study explores the strategies employed by digital celebrities to cope with the challenges associated with their fame, including the establishment of personal and professional boundaries, as well as their efforts to sustain their careers in the face of ever-evolving digital trends. Additionally, we examine their ability to achieve success in both traditional media and over-the-top (OTT) platforms, while also considering the potential trajectory of their stardom in the future.

In order to better understand the increasing prominence of digital celebrities in the changing media and entertainment industry, it is crucial to explore the historical intricacies of celebrity culture. A retroactive analysis provides a clear distinction between the conventional path to fame, which was mostly achieved through film and television, and the present trend that focuses on digital platforms. This comparison not only emphasizes the influence of technology progress on celebrity, but also accentuates the change in audience involvement and methods of consuming content. Academic publications, such the ones authored by Marwick (2013) and Turner (2014), offer useful perspectives on the development of celebrity culture, highlighting the importance of changes in media throughout history.

Expanding upon these fundamental principles, this study presents a conceptual framework that outlines the distinct attributes of digital celebrities. This model seeks to encompass the complex and diverse aspects of achieving fame in the internet realm. It takes into account several factors including the dynamics of creating content, interacting with audiences, implementing online branding strategies, and the capacity to quickly adapt to ever-changing digital trends. A framework like this helps to thoroughly analyze the influence of digital superstars and also predicts the long-term success of their fame in a constantly changing digital environment.

This research aims to go beyond academic discussions and make a positive contribution to society. The study aims to analyze the phenomenon of digital celebrities and their impact on current culture. It intends to improve media literacy and provide a deeper understanding of the wider consequences of this emerging form of fame. The knowledge obtained from this research could be crucial in influencing future media strategies, promoting a more knowledgeable and discerning consumer of digital information. The main objective of this chapter is to clarify the social consequences of the digital celebrity phenomenon, with the goal of making a beneficial contribution to the comprehension and advancement of media and communication in the digital era.

## Objectives of the Study

The objective of this study is to examine the evolution of stardom and celebrity within the realm of digital media, with a specific focus on over-the-top (OTT) platforms. The aim is to discern the unique characteristics associated with this emerging type of celebrity.

The objective of this study is to examine the tactics utilised by digital celebrities in various aspects of their work, including content creation, branding, audience engagement, and management of personal-professional boundaries. Furthermore, this research aims to compare these strategies with those employed by traditional film and television stars.

The objective of this study is to examine the obstacles and prospects linked to the durability of digital celebrities' professional trajectories, their achievements in transitioning to conventional media, and the forthcoming patterns of fame on over-the-top (OTT) platforms.

## UNDERSTANDING THE OTT ECOSYSTEM

### The Emergence of OTT Platforms

The emergence of Over-the-Top (OTT) media service platforms has had a significant impact on the television and broadcasting industry. The observed expansion of these platforms in recent years can be attributed to advancements in Internet technology and shifts in viewing behaviours (Hussein & El-sayed, 2021). In contrast to conventional methods of broadcasting, over-the-top (OTT) platforms utilise the Internet as a means of delivering content to users, thereby introducing a novel paradigm of media consumption (Cunningham & Silver, 2019). The emergence of over-the-top (OTT) platforms can be attributed to a notable driving force, which is the change in consumer behaviour. There is currently a growing consumer demand for increased autonomy in determining the content, timing, and manner of their viewing experiences. Over-the-top (OTT) platforms cater to this demand by offering viewers the capability to stream content at their preferred time and place (Lotz, 2017).

The proliferation of over-the-top (OTT) platforms can largely be attributed to advancements in technology and enhancements in Internet connectivity. The proliferation of smartphones and the widespread availability of high-speed Internet have facilitated consumer access to over-the-top (OTT) platforms. The expanded reach of these platforms has facilitated an increase in their growth, as evidenced by Ramasubramanian's research in 2021. The adoption of innovative business models is identified as another significant factor contributing to the rise of OTT platforms. The majority of over-the-top (OTT) platforms function under a subscription-based framework, which has demonstrated considerable efficacy. The economic viability of these models enables platforms to allocate resources towards the development of premium, unique content, thereby augmenting their attractiveness to consumers (Wayne, 2018). The over-the-top (OTT) industry is currently in its early developmental phase, yet it exhibits considerable potential for expansion. The growth and transformation of OTT platforms are anticipated to persist as technology advances and consumer behaviour evolves, thereby influencing the manner in which individuals engage with media..

### OTT Platform Content Strategies

The rapid expansion of Over-the-Top (OTT) platforms has resulted in a transformation of content strategies in order to accommodate a diverse audience (Ramasubramanian, 2021). Over-the-top (OTT) platforms, including Netflix, Amazon Prime, and Hulu, have made substantial investments in the creation of exclusive, premium content with the aim of attracting and retaining their audience (Wayne, 2018). One of the primary content strategies employed by over-the-top (OTT) platforms is the creation and distribution of original content. The presence of original content functions as a distinctive marketing advantage, which significantly contributes to the increase in subscription figures for the platforms. For example, the television series House of Cards and Stranger Things produced by Netflix have played a significant role in drawing in audiences and fostering a sense of allegiance to the brand (Tryon, 2015).

Over-the-top (OTT) platforms also leverage the worldwide distribution network facilitated by the Internet, producing content that caters to a diverse range of international viewership. According to McDonald and Smith-Rowsey (2016), the company allocates resources towards the creation of content in multiple languages and cultures with the aim of appealing to a global audience. The utilisation of data-driven content strategy is a distinguishing characteristic of over-the-top (OTT) platforms. The authors, Hosseinichimeh and Rahmani (2020), assert that the utilisation of viewer data allows for the comprehension of consumption patterns and viewer preferences. This understanding subsequently influences content production and acquisition decisions. This methodology facilitates the development of highly customised content, thereby enhancing viewer engagement and improving rates of viewer retention.

OTT platforms have also embraced a multi-genre approach as a significant strategy to accommodate diverse viewer preferences. According to Lotz (2017), the platforms offer a diverse array of content, encompassing various genres such as dramas, documentaries, and children's shows. This extensive selection caters to the diverse needs and preferences of different demographic segments. Furthermore, over-the-top (OTT) platforms have acknowledged the significance of regular content updates in order to maintain viewer engagement. Regular updates of content not only serve to sustain the interest of viewers, but also present avenues for cross-promotion and recommendation, ultimately leading to heightened engagement among viewers (Van Esler, 2019). In summary, the effectiveness of OTT platforms is significantly dependent on their ability to develop inventive content strategies, which are guided by technological advancements, viewer data, and evolving viewer preferences.

## The Role of OTT Platforms in Creating and Fostering Stardom

The advent of Over-the-Top (OTT) platforms has brought about a significant transformation in the television and film industry, offering a novel medium for up-and-coming artists and a supplementary channel for established celebrities (Cunningham & Silver, 2019). OTT platforms have emerged as significant players in the contemporary media landscape by placing a strong emphasis on the creation of original content, thereby assuming a pivotal role in the cultivation and promotion of stardom. First and foremost, over-the-top (OTT) platforms possess the capability to cultivate emerging talent and establish them as prominent figures in the entertainment industry. Given the substantial investments made by these platforms in the creation of original content, there exists a significant potential for emerging artists to effectively exhibit their artistic abilities. The aforementioned phenomenon has resulted in the emergence of novel individuals within the entertainment sector (Tryon, 2015). For instance, over-the-top (OTT) platforms such as Netflix and Hulu have served as a platform for emerging artists who have subsequently achieved widespread recognition. Notable examples include Millie Bobby Brown, who gained prominence through her role in the Netflix series Stranger Things, and Elisabeth Moss, who rose to fame for her portrayal in the Hulu series The Handmaid's Tale.

In addition, over-the-top (OTT) platforms play a role in cultivating and promoting established celebrity status. Esteemed actors have employed these platforms as a means to delve into diverse roles, frequently diverging significantly from their conventional public persona. These opportunities not only contribute to the preservation of their popularity but also offer a wider platform for them to demonstrate their versatility (McDonald & Smith-Rowsey, 2016). Over-the-top (OTT) platforms have emerged as a significant avenue for actors to expand their reach to a global scale, thereby facilitating their access to international audiences (Wayne, 2018). The exposure of actors on over-the-top (OTT) platforms to a wide array of cultures and fan bases through global distribution plays a significant role in their attainment of international stardom.

In addition, over-the-top (OTT) platforms employ data analytics techniques to customise content based on viewer preferences, thereby indirectly influencing the rise of stardom. According to Hosseinichimeh and Rahmani (2020), platforms can enhance an actor's popularity by strategically selecting cast members who align with the audience's preferences in terms of character types and storylines. Therefore, over-the-top (OTT) platforms have emerged as influential entities within the realm of celebrity cultivation, offering avenues for burgeoning talent to ascend and established stars to advance their professional trajectories.

In this part, the researcher explores the rise and tactics of OTT platforms, which are essential for understanding the context in which digital celebrities function. This section offers crucial insights into the distinctive attributes and obstacles encountered by digital celebrities through an examination of the inventive distribution methods, content tactics, and global reach of OTT platforms (Hussein & Elsayed, 2021; Wayne, 2018). Comprehending this is crucial for examining the strategies employed by digital celebrities in creating content, engaging with their audience, and adjusting to the OTT landscape. This will help achieve the study's objective of investigating the transformation of fame in the digital era.

## FACTORS DRIVING STARDOM ON OTT PLATFORMS

### Content Diversity and the Emergence of Niche Stars

According to Holt and Perren (2009), the proliferation of over-the-top (OTT) platforms has made it easier to develop and disseminate a wide variety of material that appeals to a broad swath of the population as well as the interests of that people. People have found that this new type of entertainment offers a more unique experience because the content that is produced for OTT platforms is distinct from the content that is seen in TV and movies. The inclusion of diverse content is a fundamental aspect of Over-the-Top (OTT) platforms, as it allows for the provision of services that appeal to a wide range of viewers (Ramasubramanian, 2021). The diverse array of content available on these platforms offers opportunities for the representation of specialised narratives and, as a result, the emergence of individuals who have gained prominence within specific interest areas.

The emergence of over-the-top (OTT) platforms has provided an avenue for a wide range of diverse narratives that were historically underrepresented in mainstream media. This phenomenon can be attributed to the company's dedication to ensuring a wide range of content options in order to cater to the diverse preferences of their worldwide audience (McDonald & Smith-Rowsey, 2016). The presence of a diverse range of narratives facilitates the opportunity for actors with specialised appeal to gain prominence, thus contributing to the development of niche celebrities. The emergence of specialised celebrities can be ascribed to the capacity of over-the-top (OTT) platforms to deviate from conventional narrative conventions. Unlike traditional media, these platforms have the advantage of not being restricted to a fixed programming schedule, allowing them the freedom to explore a wide range of genres and narratives (Wayne, 2018). This phenomenon has facilitated the emergence of specialised performers who demonstrate exceptional abilities in particular genres or character archetypes, providing them with an opportunity to exhibit their talents.

Furthermore, the utilisation of data-driven content strategies by over-the-top (OTT) platforms contributes to the expansion of niche stars. By comprehending the preferences and consumption patterns of viewers, these platforms have the ability to recognise potential niche areas of interest and tailor their offerings accordingly (Hosseinichimeh & Rahmani, 2020). Consequently, actors who demonstrate exceptional skills in these specialised domains have the potential to attain substantial levels of popularity. Over-the-top (OTT) platforms have additionally facilitated the proliferation of specialised individuals who have gained prominence, owing to the platforms' dedication to fostering inclusivity and representation. Lotz (2017) argues that the provision of a platform for underrepresented stories and communities creates opportunities for actors who embody these communities, thereby facilitating the rise of individuals who may have otherwise struggled to find a place within the dominant media landscape. The presence of varied content on over-the-top (OTT) platforms has facilitated the rise of specialised individuals who have gained prominence within specific interest areas, thereby enhancing the overall breadth and diversity of the media environment.

## Case Study: Mirzapur and its Impact on the Indian OTT Landscape

### Overview of Mirzapur

"Mirzapur" is an Indian crime thriller web series that can be streamed on Amazon Prime Video. Debuted in 2018, it swiftly gained prominence as one of the most widely discussed shows in the Indian OTT arena. The series is renowned for its realistic depiction of criminal activities, absence of order, and intense competition, all taking place in the setting of Mirzapur, a city located in Uttar Pradesh.

### Impact on Main Actors' Careers

The principal actors of "Mirzapur," particularly Pankaj Tripathi and Ali Fazal, had a substantial boost in their careers as a result of the show's success. Pankaj Tripathi, portraying the character of Kaleen Bhaiya, a ruthless and authoritative mafia leader, garnered significant acclaim and recognition. His performance garnered him significant praise from critics and elevated his profile, resulting in more prominent opportunities in both Indian cinema and other OTT platform endeavors.

Ali Fazal, who played the role of Guddu Pandit, a young individual aspiring to succeed, also witnessed a rise in his profession. The series enhanced his prominence and resulted in a wider range of substantial acting prospects.

### Audience Reception and Engagement

The television series "Mirzapur" was well-received by audiences, particularly among the younger age groups. The show's authentic and unadulterated storytelling really connected with viewers who were looking for entertainment that went beyond the traditional tales found in Bollywood.

Accurate public access to specific numerical data regarding viewership statistics may be limited. However, one can assess the show's popularity by examining its level of involvement on social media platforms and the level of anticipation for its upcoming seasons. The series garnered widespread appeal among diverse audiences, both domestically in India and on an international scale, showcasing the extensive worldwide influence of OTT platforms.

## Role of Amazon Prime Video

Amazon Prime Video played a pivotal role in this transition. Amazon showcased its dedication to creating top-notch, indigenous content that appeals to the varied and changing preferences of Indian audiences by investing in "Mirzapur." Amazon's decision was a deliberate move to enter the growing Indian OTT market, which is defined by a youthful, tech-savvy populace seeking content that reflects their experiences and goals.

The triumph of "Mirzapur" serves as a prime example of how OTT platforms such as Amazon Prime Video are revolutionizing the way content is consumed in India. The platform provides a medium for storytelling that questions conventional narratives and appeals to a demographic looking for content that is more relevant, diverse, and provocative.

## Conclusion

The example of "Mirzapur" serves as evidence of the ever-changing nature of the Indian over-the-top (OTT) streaming industry. This highlights the impact of digital platforms on both the consumption of content and the careers of actors, as well as the evolution of storytelling. The popularity of these shows demonstrates a notable change in the Indian entertainment sector, with OTT platforms emerging as influential entities in shaping current mainstream culture.

## The Use of Social Media and Audience Engagement

The utilisation of social media has emerged as an indispensable instrument for effectively captivating and involving target audiences across various domains, encompassing but not limited to the realms of entertainment, politics, marketing, and other fields that entail public discourse. Social media platforms such as Facebook, Twitter, Instagram, and LinkedIn have facilitated enhanced connectivity between brands, influencers, and organisations and their respective audiences, thereby establishing a novel level of interaction (Perrin, 2015). One key benefit of social media is its ability to facilitate interactive communication between users. In contrast to conventional media, which typically operates through a unidirectional dissemination of information, social media platforms enable audiences to actively participate by expressing their responses, sharing personal viewpoints, and directly interacting with content creators (Kaplan & Haenlein, 2010). The interactive nature of this platform facilitates a more robust connection between providers of content and their respective audiences, thereby augmenting the level of engagement.

The utilisation of social media platforms also facilitates the customization of content. According to Hargittai and Litt (2011), platforms have the ability to customise the content displayed to users based on their data, resulting in improved relevance and increased user engagement. In addition, it is worth noting that social media platforms provide a platform for the creation of user-generated content, which enables audiences to actively engage in the process of content creation. This dynamic fosters a sense of community and belonging among users (Voorveld & van der Goot, 2013). Moreover, social media possesses the capacity to facilitate the rapid dissemination of content, thereby greatly amplifying its scope. The potential for social media content to become viral has the capacity to significantly enhance audience engagement and foster broader discourse surrounding the content (Berger & Milkman, 2012). It can be argued that social media has significantly transformed the manner in which audience engagement is conducted, primarily through the facilitation of interactive communication, customization of content, and facilitation of widespread dissemination of information. Therefore, it has emerged as a crucial instrument for content providers aiming to establish and sustain robust relationships with their target audiences.

### Branding and Online Presence

In the contemporary era of digitalization, the concepts of branding and online presence have become closely interconnected, exerting substantial influence on one another. The significance of establishing a strong online presence cannot be overstated as it plays a pivotal role in the achievement of a brand's success, enhanced visibility, and increased customer engagement (Labrecque, vor dem Esche, Mathwick, Novak, & Hofacker, 2013). The process of branding primarily entails the development of a distinct and exclusive identity that serves to distinguish a particular business from its competitors. Consistency in reflecting this distinct identity is imperative across all online platforms, encompassing the company's website as well as its various social media profiles. The maintenance of consistency in brand image serves to strengthen the perception of the brand and enhance its ease of recognition (Aaker, 2010).

An optimal online presence extends beyond the mere existence of a website or social media platform. Consistent engagement with the audience, achieved through the provision of captivating content, timely responses to comments or inquiries, and frequent updates regarding the brand's offerings, is essential. According to Kietzmann, Hermkens, McCarthy, and Silvestre (2011), active engagement plays a crucial role in enhancing brand visibility, establishing credibility, and cultivating a more profound relationship with customers.

Search Engine Optimisation (SEO) is an essential element of a brand's digital presence. According to Jansen and Spink (2006), enhancing the website's content by strategically incorporating pertinent keywords can lead to an enhanced position on search engine results pages. Consequently, this can augment the brand's visibility and subsequently attract a larger number of visitors. Social media platforms offer a distinct opportunity for brands to engage in direct interaction with their customers. According to Hanna, Rohm, and Crittenden (2011), brands have the ability to cultivate a devoted online community by generating and disseminating captivating content, promptly addressing customer inquiries, and actively engaging in discussions.

Furthermore, the impact of online reviews and ratings on a brand's reputation is substantial. It is recommended that brands adopt a proactive approach in managing their online reputation. This can be achieved by actively encouraging satisfied customers to provide positive reviews and promptly addressing any negative feedback (Gensler, Völckner, Liu-Thompkins, & Wiertz, 2013). The establishment and maintenance of a proficient online presence is imperative for the successful development of a brand in the contemporary era of digital technology. The enhancement of brand visibility and the cultivation of customer relationships are both integral factors in fostering enduring brand success.

## THE IMPACT OF STARDOM ON CONTENT PRODUCTION AND CONSUMPTION

### Influencing Content Creation and Promotion

The development and dissemination of content have emerged as crucial factors for businesses, influencers, and organisations aiming to establish a robust digital footprint. The content creation and promotion process is influenced by a range of factors, such as audience comprehension, search engine optimisation, and the utilisation of social media platforms. The initial stage in content creation involves comprehending the target audience. The process entails the examination of audience demographics, preferences, behaviours, and needs in order to create content that effectively connects with them (Bleier & Eisenbeiss, 2015). The process may encompass audience segmentation and persona development in order to ensure that the content is customised to suit the distinct interests and requirements of various audience segments.

The practise of Search Engine Optimisation (SEO) has a substantial impact on the creation and promotion of content. The enhancement of content visibility on search engine results pages is achieved through the optimisation of relevant keywords, which subsequently leads to an increased likelihood of reaching a broader audience (Jansen & Spink, 2006). The practise of content promotion entails the distribution of content across various digital platforms. Social media platforms have become prominent instruments for the promotion of content. According to Kaplan and Haenlein (2010), social media platforms facilitate targeted advertising, enable two-way communication with audiences, and provide a convenient means of content sharing, thereby expanding its scope.

User-generated content (UGC) also plays a significant role in the promotion of content. According to Smith, Fischer, and Yongjian (2012), promoting the creation and dissemination of content related to a brand or product can have a substantial impact on enhancing visibility and engagement. In the realm of content promotion, influencer marketing emerges as a pivotal facet. Engaging in partnerships with influential individuals who possess a substantial number of followers can facilitate the expansion of one's reach, potentially leading to a more attentive and involved audience (Freberg, Graham, McGaughey, & Freberg, 2011). Therefore, content creation and promotion in the contemporary digital environment are influenced by a range of factors, such as audience comprehension, search engine optimisation (SEO), utilisation of social media platforms, user-generated content (UGC), and the practise of influencer marketing.

## Growth of the Indian OTT Video Market

Growth of the Indian OTT Market: By 2025, the OTT video market in India, estimated to be worth INR 93.67 billion, is expected to have grown significantly from its 2020 valuation of INR 22.62 billion. The substantial increase in content consumption patterns signifies a notable change and emphasizes the impact of digital celebrities in bringing consumers to these platforms.

Here is a graphical representation of the growth of the Indian OTT video market:

- The graph shows the market values for the years 2018, 2020, 2023, and the projected value for 2025.
- In 2018, the market was valued at $0.8 billion.
- By 2020, it increased significantly to INR 22.62 billion.
- The projection for 2023 is $5.03 billion.
- By 2025, the market is expected to reach INR 93.67 billion.

OTT Audience in India: Showing a 13.5% year-over-year rise, the OTT audience in India as of 2023 reached a new high of 481.1 million. This growth, however at a slower pace in comparison to the 20% growth of the previous year, indicates the increasing influence of OTT platforms and the significance of celebrity status in attracting viewers to digital content.

*Figure 1. Growth of the Indian OTT video market (2018-2025)*
*(The Global Statistics)*

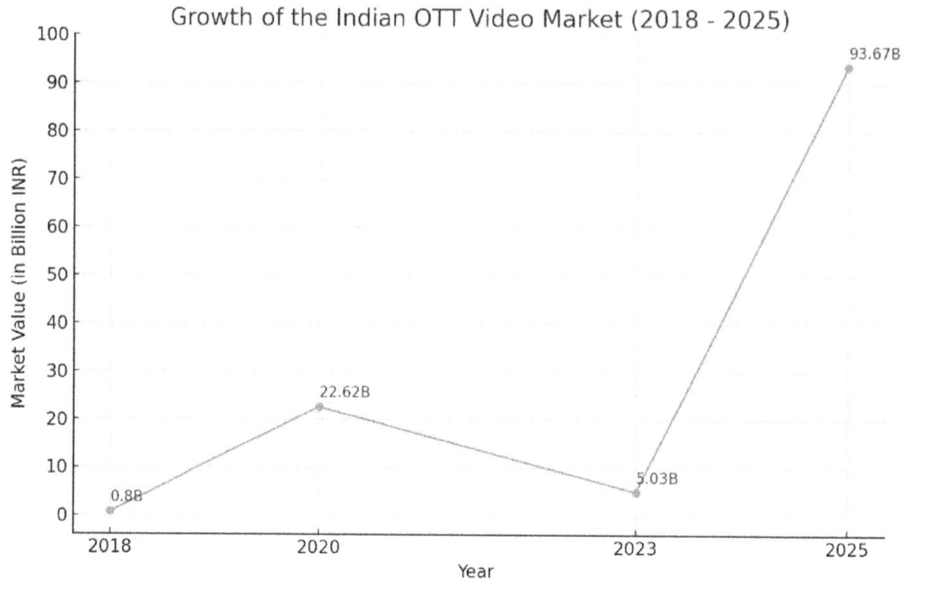


*Figure 2. OTT Audience reach in India*
*(https://dazeinfo.com/2023/11/08/indias-ott-audience-reached-481-million-in-2023-only-7-6-are-directly-paying-to-streamin g-platforms/)*

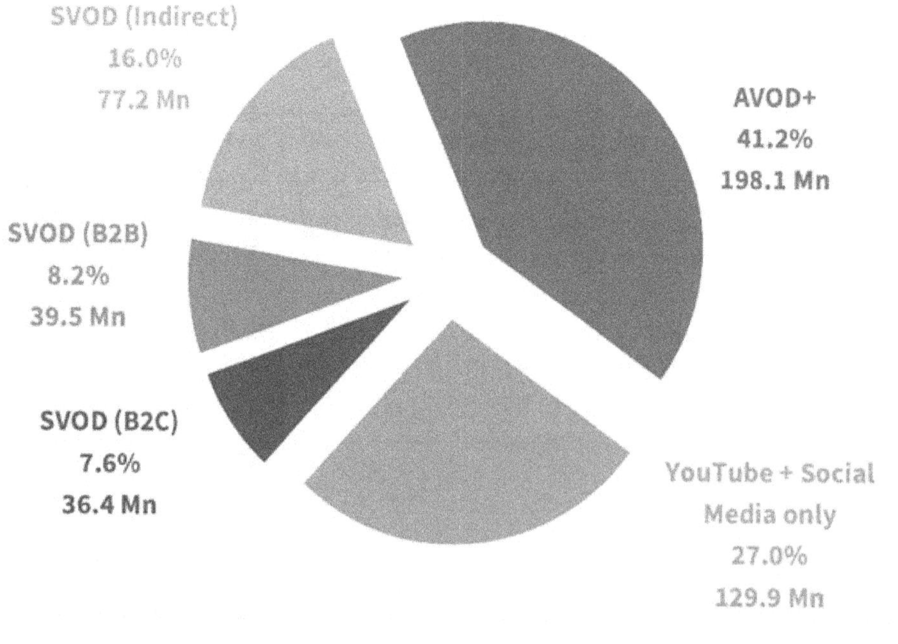

- Total audience: 481.1 million (48.11 crore)
- Year-over-year growth: 13.5%
- OTT penetration rate: 34%
- Segmentation of OTT audience by consumption type:
- SVOD (B2C): 36.4 million (directly paid for OTT subscriptions)
- SVOD (B2B): 39.5 million (subscriptions via telecom bundles)
- SVOD (Indirect): 77.2 million (access through family/friends)
- AVOD+: 198.1 million (exclusively watch free streaming platforms)
- YouTube + Social only: 129.9 million (Source: Dazeinfo)

Market Share of Major OTT Players: As of 2023, Disney+ Hotstar (41%), Eros Now (24%), Amazon Prime Video (9%), Netflix (7%), and ZEE5 (5%), were the top four OTT platforms in India. The rise in popularity of these platforms can be partially ascribed to the digital celebrities who are linked to their unique content, influencing the preferences and engagement of the audience.

Segments of Subscription Video-on-Demand (SVOD): In India, there are 36.4 million individuals who directly subscribe to at least one OTT platform, and an additional 39.5 million who have subscribed through telecom bundles. This categorization offers valuable insights into the impact of celebrity status on consumer preferences for high-quality content on these channels.

## Changing Consumption Patterns

The impact of technological advancements and the emergence of the digital era have significantly shaped consumption behaviours in diverse industries. The aforementioned alterations have been particularly noteworthy within the domains of media, retail, and food consumption, among various others. The emergence of Over-the-Top (OTT) platforms has significantly transformed consumption patterns within the media industry. According to Wayne (2018), online streaming platforms have largely supplanted traditional media consumption methods, such as television and radio, by providing access to on-demand content. Various platforms, such as Netflix, Amazon Prime, and Hulu, afford users the opportunity to engage with media content in accordance with their individual preferences and schedules, resulting in the emergence of the binge-watching phenomenon (Pittman & Sheehan, 2015).

The rise of e-commerce in the retail industry has led to notable shifts in consumption behaviours. The advent of online shopping has provided consumers with the convenience of making purchases from their own residences, resulting in a notable transition away from traditional physical retail establishments. In addition, the extensive utilisation of mobile devices has facilitated the emergence of mobile commerce, which grants consumers the ability to engage in shopping activities while being mobile (Huang & Benyoucef, 2013). The consumption patterns of food have undergone notable transformations, primarily attributed to the rising prevalence of online food delivery services. The emergence of online food delivery platforms such as Uber Eats and DoorDash has facilitated consumer access to a diverse range of culinary choices from various restaurants, enabling the convenient delivery of these meals directly to their residences (Morganosky & Cude, 2000). The alteration of consumption patterns propelled by digital advancements has significantly reshaped the terrain of diverse industries. With the ongoing evolution of technology, it is anticipated that there will be corresponding changes in consumption patterns, thereby requiring businesses and service providers to continuously adapt.

### Comparing Digital Celebrities to Traditional Film and Television Stars

The emergence of digital media has given rise to a distinct category of celebrities, which sets them apart from conventional film and television personalities. Digital celebrities, also known as influencers, have emerged as prominent figures in contemporary society, attaining widespread recognition through their utilisation of social media platforms and the creation of online content. This phenomenon signifies a notable transformation in the dynamics of fame generation and consumption (Marwick, 2015). Conventional film and television celebrities commonly achieve recognition and popularity by means of established mass media platforms. The management of their careers is frequently entrusted to industry professionals, such as agents, publicists, and production companies. The construction and preservation of their public image is meticulously managed by a group of experts, while their engagements with supporters are typically tightly regulated and constrained (Turner, 2004).

In contrast, digital celebrities frequently attain recognition through the creation and dissemination of content on various social media platforms. The trajectory of their ascent to prominence typically follows an organic pattern, commencing with a modest group of supporters that gradually expands through the dissemination of information by means of interpersonal communication and online platforms. According to Abidin (2016), celebrities tend to project a public image that is characterised by a sense of personal authenticity. This is often achieved through the direct management of their own social media accounts and active engagement with their fan base. Digital celebrities frequently maintain a heightened level of engagement and interactivity with their fan base. They engage in regular communication with their fan base through various means, such as commenting, liking, sharing, and even direct messaging. According to Khamis, Ang, and Welling (2017), the utilisation of two-way communication facilitates a heightened sense of intimacy and connection, a phenomenon that is not commonly observed among conventional celebrities.

The monetization strategies employed by digital celebrities exhibit notable variations. In contrast to traditional celebrities who primarily derive their income from their involvement in film or television productions, endorsements, and public appearances, digital celebrities frequently capitalise on their fame by means of sponsored content, advertisements on their platforms, crowdfunding, and merchandise sales (Djafarova & Rushworth, 2017). In short, digital celebrities and traditional film and television stars embody distinct routes to achieving fame, each possessing distinctive attributes, prospects, and obstacles. The ongoing evolution of the digital landscape raises intriguing questions regarding the potential transformations of these various manifestations of celebrity.

## NAVIGATING THE CHALLENGES OF DIGITAL FAME

### Maintaining Relevance in a Competitive Landscape

In the dynamic and rapidly evolving digital environment, the task of sustaining relevance poses a considerable challenge for brands, influencers, and businesses. There are various approaches that can be employed to sustain relevance in a competitive setting, such as continuous innovation, customer engagement, adaptability to changes, and personal branding (Eisenhardt & Martin, 2000). The importance of continuous innovation cannot be overstated in a highly competitive environment. In order to distinguish themselves from competitors and adapt to evolving customer demands, brands must consistently strive to innovate their products, services, and customer experiences (Teece, 2007).

One additional crucial strategy involves actively interacting with customers. Businesses have the opportunity to engage with their customers, collect feedback, and enhance their products and customer experiences by utilising various digital platforms such as social media and email marketing (Kietzmann et al., 2011). The ability to effectively adjust to various types of changes, such as those related to technology, economy, or society, is of utmost importance in order to sustain relevance. In light of the swift rate of technological progress, it is imperative for businesses to remain up-to-date with current trends and incorporate them into their strategic approaches. This may entail the adoption of novel technologies, the utilisation of data analytics, or the exploration of nascent markets (Bharadwaj, El Sawy, Pavlou, & Venkatraman, 2013).

The establishment of a personal brand holds significant importance within the realm of influencers and digital celebrities. It is imperative for individuals to develop a distinct and genuine online identity that effectively connects with their target audience. According to Labrecque, Markos, and Milne (2011), the maintenance of relevance can be achieved through consistent content creation, active engagement with followers, and adherence to the brand's identity. In short, the ability to remain pertinent in a highly competitive environment necessitates the integration of ongoing innovation, active customer involvement, adaptability, and the cultivation of one's personal brand. Although the implementation of these strategies may pose challenges, they have the potential to result in long-term competitive advantage and achievement.

## Dealing With the Pressures of Fame

Celebrities must create a support network of close friends, family, and dependable professionals to help them cope with the rigours of fame. They must also create separation between their private and public lives to safeguard their privacy and mental health. It's crucial to prioritise self-care activities, get expert advice from therapists and counsellors, and keep your perspective and honesty. In spite of intense public scrutiny and high expectations, celebrities can navigate the demands of fame, manage stress, and maintain long and fulfilling careers by using these techniques (Luthar et al., 2013; Kerr et al., 2017; Ferguson et al., 2020; Parker et al., 2018).

## Managing Personal and Professional Boundaries

In the contemporary digital age, the ability to readily disseminate one's personal affairs to a global audience has become effortlessly accessible. Consequently, the establishment and maintenance of distinct personal and professional boundaries have emerged as indispensable practises for both celebrities and influencers (Marwick & Boyd, 2011). Celebrities, influencers, and individuals occupying prominent positions in the public sphere frequently encounter the challenge of effectively managing the delicate boundary between their public image and personal affairs. The complexity of this situation is exacerbated by the growing desire of audiences for authenticity, a phenomenon that frequently manifests in the sharing of personal aspects of one's life (Abidin, 2016).

In order to uphold these boundaries, it is imperative to establish precise delineations between the facets of an individual's life that are open to public scrutiny and those that ought to be kept confidential. This may entail restricting the dissemination of information pertaining to familial connections or personal affiliations, as well as maintaining confidentiality regarding specific pastimes or areas of interest (Dyer,

2004). An additional approach involves the establishment of separate environments to facilitate personal and professional engagements. As an illustration, certain influencers may opt to uphold distinct social media profiles to delineate their professional and personal spheres. The aforementioned differentiation can serve the purpose of establishing followers' anticipations regarding the nature of the content that will be presented on each respective account (Senft, 2013).

Furthermore, it is crucial to take into account the potential ramifications of disclosing personal information. The utilisation of personal anecdotes and firsthand encounters can establish a profound connection between influencers and their audience. However, excessive disclosure of personal information may result in overexposure and leave individuals vulnerable to potential exploitation by external parties (Khamis, Ang, & Welling, 2017).

Lastly, the establishment of boundaries also encompasses the delineation of work hours and the preservation of mental well-being. Numerous prominent figures in the entertainment industry and social media sphere experience a pervasive sense of obligation to maintain a constant public presence, thereby increasing their susceptibility to experiencing burnout. Establishing defined work hours, allocating periods of rest, and prioritising self-care are essential components for sustaining a well-rounded equilibrium between professional and personal life (Kerr, Delle Fave, & Vella-Brodrick, 2017). In summary, the task of effectively managing personal and professional boundaries presents a formidable yet indispensable aspect of navigating a public persona. Celebrities and influencers are faced with the challenge of effectively managing their privacy and mental well-being, all the while upholding a genuine connection with their audience, within the intricate realm of their profession.

## THE SUSTAINABILITY OF STARDOM IN THE DIGITAL ERA

### Longevity of Digital Celebrities' Careers

The durability of digital celebrities' professional trajectories is contingent upon a multitude of factors. In contrast to conventional celebrities who typically rely on a dedicated team of professionals to oversee their professional endeavours, digital celebrities typically assume multiple roles as managers, content creators, and public relations specialists (Khamis, Ang, & Welling, 2017). The success and durability of individuals or entities in this context are contingent upon their capacity to remain pertinent, adjust to evolving patterns, uphold a distinctive perspective, and consistently involve their target audience (Marwick, 2015). This frequently necessitates a significant level of commitment and consistent generation of fresh material. Nonetheless, digital celebrities who adeptly navigate these challenges can experience enduring careers that span over several years, or even multiple decades.

### Success in Traditional Media Crossover

Certain digital celebrities have demonstrated proficiency in transitioning to traditional media platforms, such as film or television, while others have encountered challenges in navigating this crossover (Deuze, Witschge, & Brunow, 2016). The proficiencies necessary for achieving success in the realm of digital media, such as the capacity to actively participate in interactive and reciprocal communication with supporters, may not always be applicable or easily adaptable to conventional media platforms. Moreover, there exists a notable disparity in the norms and expectations between traditional media industries and the digital domain. Nevertheless, individuals who possess the ability to effectively navigate these aforementioned challenges can reap the advantages of heightened exposure and enhanced opportunities within the realm of traditional media.

### The Future of Stardom on OTT Platforms

The emergence of a novel form of celebrity is being witnessed alongside the increasing popularity of over-the-top (OTT) platforms. Over-the-top (OTT) platforms present content creators with the potential to access a worldwide viewership, surpassing the geographical constraints typically associated with conventional broadcast media (Cunningham & Craig, 2019). Consequently, there has been a notable increase in the variety of content, facilitating the emergence of individuals who possess specialised appeal and may not conform to the conventional archetype of a celebrity. The future prospects of stardom on over-the-top (OTT) platforms appear auspicious, owing to ongoing technological advancements and evolving audience preferences. This presents a wide array of opportunities for content creators to attain fame and acknowledgment.

## CONCLUSION

The advent of the digital revolution has undoubtedly transformed the paradigm of fame, giving rise to a novel cohort of public figures who exploit over-the-top (OTT) platforms and social media to establish worldwide connections with their audiences. The investigation of this phenomenon unveiled discernible characteristics that differentiate digital celebrities from their conventional counterparts. Digital celebrities have become influential individuals who excel in creating content, representing brands, and exerting influence, effectively utilising digital platforms to engage with audiences, promote products or services, and cultivate their personal brand. The examination of digital celebrities' strategies revealed that audience engagement, authenticity, and adaptability are crucial factors contributing to their achievements. These strategies, although different from those employed by conventional celebrities, often coexist in an intricate dynamic of interaction and impact with traditional media. Additionally, the management of personal and professional boundaries has emerged as a pivotal aspect of the careers of digital celebrities, thereby highlighting the distinctive challenges associated with digital stardom.

The examination of the durability of digital celebrities' professional trajectories has unveiled a milieu characterised by swift transformations and an ongoing imperative for novelty. Notwithstanding these challenges, digital celebrities who have achieved success have demonstrated exceptional adaptability, thereby transcending into conventional media platforms. When considering the future trajectory of stardom on over-the-top (OTT) platforms, there is a promising outlook characterised by a growing range of content and the emergence of specialised stars. In summary, the phenomenon of digital stardom represents a notable transformation in the methods by which fame is attained and sustained. The strategies, challenges, and opportunities associated with digital celebrity will continue to evolve alongside the ongoing development of digital media. Comprehending these dynamics is imperative for individuals seeking to navigate this emerging terrain of renown and impact. This investigation establishes a basis for future research in the dynamic realm of digital fame.

## Potential Areas for Future Research

Extension into Comparative Studies: Upcoming studies might compare digital celebrities in various cultural or geographic contexts to see how regional variations affect the emergence and longevity of digital celebrity on OTT platforms.

Longitudinal Analysis: It would be possible to trace the career paths of digital celebrities over time and gain insight into the elements that influence their long-term success or downfall by conducting a longitudinal study.

Impact of Changing Technologies: Future research on the effects of new media formats and platforms on digital celebrity culture may be particularly intriguing as technology keeps developing.

## Research Constraints

Data Scope: The study's use of both depth and breadth may be constrained, especially if it makes significant use of secondary sources or excludes empirical, first-hand information from OTT platforms or digital celebrities.

Changing Industry Dynamics Quickly: Because the digital and over-the-top (OTT) landscapes move swiftly, study findings may become old quickly, requiring ongoing revisions.

Generalizability: It's possible that the conclusions and findings are unique to the OTT platforms and digital celebrities under investigation; they might not apply to other digital media platforms or cultural contexts.

## REFERENCES

Aaker, D. A. (2010). *Building strong brands*. Simon and Schuster.

Abidin, C. (2016). Visibility labour: Engaging with influencers' fashion brands and #OOTD advertorial campaigns on Instagram. *Media International Australia, Incorporating Culture & Policy*, *161*(1), 86–100. doi:10.1177/1329878X16665177

Berger, J., & Milkman, K. L. (2012). What makes online content viral? *JMR, Journal of Marketing Research*, *49*(2), 192–205. doi:10.1509/jmr.10.0353

Bharadwaj, A., El Sawy, O. A., Pavlou, P. A., & Venkatraman, N. V. (2013). Digital business strategy: Toward a next generation of insights. *Management Information Systems Quarterly*, *37*(2), 471–482. doi:10.25300/MISQ/2013/37:2.3

Bleier, A., & Eisenbeiss, M. (2015). Personalized online advertising effectiveness: The interplay of what, when, and where. *Marketing Science*, *34*(5), 669–688. doi:10.1287/mksc.2015.0930

Cashmore, E. (2006). *Celebrity culture*. Routledge. doi:10.4324/9780203962916

Cunningham, S., & Craig, D. (2019). *Social media entertainment: The new intersection of Hollywood and Silicon Valley*. NYU Press.

Deuze, M., Witschge, T., & Brunow, D. (2016). Prefiguring the Post-Demise of Print: Digital Life After Newspapers. *Journal of Media Innovations*, *3*(1), 109–127.

Djafarova, E., & Rushworth, C. (2017). Exploring the credibility of online celebrities' Instagram profiles in influencing the purchase decisions of young female users. *Computers in Human Behavior*, *68*, 1–7. doi:10.1016/j.chb.2016.11.009

Driessens, O. (2013). Celebrity capital: Redefining celebrity using field theory. *Theory and Society*, *42*(5), 543–560. doi:10.1007/s11186-013-9202-3

Dyer, R. (2004). *Heavenly bodies: Film stars and society*. Routledge.

Eisenhardt, K. M., & Martin, J. A. (2000). Dynamic capabilities: What are they? *Strategic Management Journal*, *21*(10-11), 1105–1121. doi:10.1002/1097-0266(200010/11)21:10/11<1105::AID-SMJ133>3.0.CO;2-E

Freberg, K., Graham, K., McGaughey, K., & Freberg, L. A. (2011). Who are the social media influencers? A study of public perceptions of personality. *Public Relations Review*, *37*(1), 90–92. doi:10.1016/j.pubrev.2010.11.001

Gensler, S., Völckner, F., Liu-Thompkins, Y., & Wiertz, C. (2013). Managing brands in the social media environment. *Journal of Interactive Marketing*, *27*(4), 242–256. doi:10.1016/j.intmar.2013.09.004

Giles, D. C. (2000). *Illusions of immortality: A psychology of fame and celebrity*. Macmillan. doi:10.1007/978-1-137-09650-0

Hanna, R., Rohm, A., & Crittenden, V. L. (2011). We're all connected: The power of the social media ecosystem. *Business Horizons*, *54*(3), 265–273. doi:10.1016/j.bushor.2011.01.007

Hargittai, E., & Litt, E. (2011). The tweet smell of celebrity success: Explaining variation in Twitter adoption among a diverse group of young adults. *New Media & Society*, *13*(5), 824–842. doi:10.1177/1461444811405805

Hosseinichimeh, N., & Rahmani, A. (2020). A computational approach to managing and delivering video contents on OTT platforms. *Computers & Industrial Engineering*, *140*, 106262.

Huang, Z., & Benyoucef, M. (2013). From e-commerce to social commerce: A close look at design features. *Electronic Commerce Research and Applications*, *12*(4), 246–259. doi:10.1016/j.elerap.2012.12.003

Hussein, M., & Elsayed, H. (2021). Over the Top (OTT) Platforms: A New Era in the Television Industry. *The Journal of Business and Retail Management Research*, *16*(1), 1–12.

Jansen, B. J., & Spink, A. (2006). How are we searching the World Wide Web? A comparison of nine search engine transaction logs. *Information Processing & Management*, *42*(1), 248–263. doi:10.1016/j.ipm.2004.10.007

Kaplan, A. M., & Haenlein, M. (2010). Users of the world, unite! The challenges and opportunities of Social Media. *Business Horizons*, *53*(1), 59–68. doi:10.1016/j.bushor.2009.09.003

Kerr, G., Delle Fave, A., & Vella-Brodrick, D. (2017). The art and science of a sustainable performing career in music. In *The science of wellbeing* (pp. 201–227). Springer.

Khamis, S., Ang, L., & Welling, R. (2017). Self-branding, 'micro-celebrity' and the rise of Social Media Influencers. *Celebrity Studies*, *8*(2), 191–208. doi:10.1080/19392397.2016.1218292

Kietzmann, J. H., Hermkens, K., McCarthy, I. P., & Silvestre, B. S. (2011). Social media? Get serious! Understanding the functional building blocks of social media. *Business Horizons*, *54*(3), 241–251. doi:10.1016/j.bushor.2011.01.005

Labrecque, L. I., Markos, E., & Milne, G. R. (2011). Online personal branding: Processes, challenges, and implications. *Journal of Interactive Marketing*, *25*(1), 37–50. doi:10.1016/j.intmar.2010.09.002

Labrecque, L. I., vor dem Esche, J., Mathwick, C., Novak, T. P., & Hofacker, C. F. (2013). Consumer power: Evolution in the digital age. *Journal of Interactive Marketing*, *27*(4), 257–269. doi:10.1016/j.intmar.2013.09.002

Lotz, A. D. (2017). *Portals: A treatise on Internet-distributed television*. Michigan Publishing, University of Michigan Library.

Marwick, A. E. (2015). Instafame: Luxury selfies in the attention economy. *Public Culture, 27*(1 75), 137-160.

Marwick, A. E., & boyd. (2011). To see and be seen: Celebrity practice on Twitter. *Convergence (London)*, *17*(2), 139–158. doi:10.1177/1354856510394539

McDonald, P., & Smith-Rowsey, D. (2016). *The Netflix effect: Technology and entertainment in the 21st century*. Bloomsbury Publishing USA.

Morganosky, M. A., & Cude, B. J. (2000). Consumer response to online grocery shopping. *International Journal of Retail & Distribution Management*, *28*(1), 17–26. doi:10.1108/09590550010306737

Perrin, A. (2015). *Social media usage: 2005-2015*. Pew Research Center.

Pittman, M., & Sheehan, K. (2015). Sprinting a media marathon: Uses and gratifications of binge-watching television through Netflix. *First Monday*, *20*(10). Advance online publication. doi:10.5210/fm.v20i10.6138

Ramasubramanian, V. (2021). Disruption in the Indian OTT industry. *International Journal of Information Management*, *56*, 102–106.

Senft, T. M. (2013). *Microcelebrity and the branded self. A Companion to New Media Dynamics*, 346-354.

Smith, A. N., Fischer, E., & Yongjian, C. (2012). How does brand-related user-generated content differ across YouTube, Facebook, and Twitter? *Journal of Interactive Marketing*, *26*(2), 102–113. doi:10.1016/j.intmar.2012.01.002

Teece, D. J. (2007). Explicating dynamic capabilities: The nature and microfoundations of (sustainable) enterprise performance. *Strategic Management Journal*, *28*(13), 1319–1350. doi:10.1002/smj.640

Tryon, C. (2015). TV got better: Netflix's original programming strategies and the onset of "binge" viewing. *Media Industries Journal*, *2*(2), 104–117.

Turner, G. (2004). Understanding celebrity. *Sage (Atlanta, Ga.)*.

Van Esler, M. (2019). Market strategies of video on demand platforms. *International Journal on Media Management*, *21*(1), 41–66.

Voorveld, H. A., & van der Goot, M. (2013). Age differences in media multitasking: A diary study. *Journal of Broadcasting & Electronic Media*, *57*(3), 392–408. doi:10.1080/08838151.2013.816709

Wayne, M. L. (2018). Netflix, Amazon, and branded television content in subscription video on-demand portals. *Media Culture & Society*, *40*(5), 725–741. doi:10.1177/0163443717736118

# Section 2
# User Experience

# Chapter 5
# Riding the Dragon:
## The Emergence and Impact of Over-the-Top Media in China

**Mohamad Zreik**
https://orcid.org/0000-0002-6812-6529
*Sun Yat-sen University, China*

**Runping Zhu**
https://orcid.org/0000-0003-3302-7376
*Lanzhou University, China*

## ABSTRACT

*This chapter examines the growth and influence of over-the-top (OTT) media within China's unique economic and legal context. It examines the business models, revenue sources, and revolutionary shift in content consumption and creation patterns brought about by major platforms like iQiyi, Tencent Video, and Youku. The regulatory practices in China are highlighted, along with the potential and difficulties that they provide to the OTT scene. Questions concerning media pluralism, content quality, and algorithmic recommendation systems are highlighted as the authors look at the broader societal ramifications, including the effect on cultural trends, celebrity culture, and the film and television industry.*

## INTRODUCTION

Media delivered through the internet rather than through conventional means of distribution like cable or satellite TV is known as "over-the-top" (OTT) media (Hutchins, Li, & Rowe, 2019). These services provide users with a wide selection of movies, TV shows, and live broadcasts that can be accessed on-demand, allowing for a more individualized and adaptable viewing experience (Abreu et al., 2017).

The importance of OTT media in China is huge. With its rapidly digitizing population and advanced technological infrastructure, the country has taken the digital world by storm in the past decade. There will reportedly be more than 989 million internet users in China by 2024, a sizable proportion of whom would be avid consumers of digital material (Junzhi, 2021). As a result of this shift, OTT platforms have

DOI: 10.4018/979-8-3693-3526-0.ch005

amassed hundreds of millions of users and solidified their position inside China's digital entertainment ecosystem (Craig & Cunningham, 2019).

OTT media has revolutionized Chinese media consumption by granting viewers greater control over what, when, and how they consume media. There is a wide range of viewing habits that can be accommodated by OTT services, from watching a movie on a smartphone during a commute to binge watching a series at home (Tsai, 2023). The rise of OTT media coincides with a general trend toward more nuanced and sophisticated programming. Some original Chinese productions on iQiyi, Tencent Video, and Youku have received critical praise on a global scale (Lin, 2022).

The development of OTT media in China also has far-reaching effects on the media system as a whole in the country. These mediums have altered the mechanics of content creation, advertising, and even cultural transmission, redefining what it means to be entertained (Evens & Donders, 2018). Therefore, knowing the history and significance of OTT media is crucial to comprehending the present and future of China's digital culture and economy.

The primary goal of this chapter is to provide readers an in-depth look at how and why OTT media has taken off in China. The study's overarching goal is to learn how OTT media, as shown by giants like iQiyi, Tencent Video, and Youku, has altered the content industry as a whole. The chapter aims to illuminate the distinctive business strategies and revenue streams that these platforms have developed to succeed in China's economic and regulatory context. In addition, this chapter intends to investigate how the rise of OTT media has affected cultural movements, celebrity culture, and the Chinese cinema and television business.

The chapter is organized to comprehensively address these many concerns of OTT media in China. It starts with a brief background on OTT media in the country, placing its rapid expansion in the context of broader trends in the Chinese internet industry. This lays the groundwork for further investigation into the business structures and income streams that have contributed to the success of OTT platforms.

In the pursuit of a comprehensive understanding of the rise of the OTT media in China and its consequent societal impacts, the methodology employed is critical to elucidating the findings presented in this chapter. This research adopted a qualitative content analysis and a review of secondary sources, including industry reports, academic journals, and regulatory publications. Initially, the study conducted an extensive literature review to establish a theoretical framework and identify gaps within the existing body of knowledge. Subsequently, a content analysis was performed on a curated selection of OTT platforms to discern prevalent themes and narratives. This multi-layered approach ensured a robust analysis of the OTT media landscape, enabling the synthesis of a broad spectrum of data into coherent findings that reveal the depth and nuances of China's digital media evolution.

The chapter then continues on to an examination of the effects of OTT media on the creation and consumption of content. This section examines the ways in which audiences interact with OTT content, including the rise of binge-watching, the ability to customize content on these platforms, and the production of successful new shows and movies.

The chapter's focus is on the potential and problems offered by China's regulatory framework and how they affect the OTT landscape. The chapter concludes with a discussion of the broader social ramifications of OTT media's meteoric ascent in China. It does not only examine the ways in which technology has changed society and culture, but also poses vital concerns regarding the role of media pluralism, content quality, and algorithmic recommendation systems. The chapter's format is all-encompassing since its goal is to provide a thorough analysis of the rise and influence of OTT media in China.

# HISTORICAL DEVELOPMENT OF OTT MEDIA IN CHINA

## The Emergence of OTT Platforms: iQiyi, Tencent Video, and Youku

Over-the-top media's meteoric rise in China can be traced back to the decade of the 2000s, which was a watershed moment in the history of the country's internet (Rowe, 2005). Chinese internet users were given a new option for watching and sharing videos in 2006 with the introduction of Youku, one of the first video-sharing websites. Following this, in 2010, Tencent, a major player in China's digital ecosystem, released Tencent Video. It took use of its large preexisting user base to become a dominant player in the OTT market (Wang & Yu, 2021).

Baidu, China's most popular search engine, released iQiyi in the same year. iQiyi set itself apart by prioritizing licensed content and professionally produced videos, satisfying the growing appetite for high-quality internet video (Keane, Yecies, & Flew, 2018). Later, the platform led the way in the creation of original content, a key factor in its meteoric rise to prominence.

Broader tendencies in China's internet industry have supported the expansion of these platforms. China's huge population, growing internet access, and spike in smartphone usage have all created ideal conditions for the development of OTT media in the country. The shift from analog to digital television has also contributed to the expansion of these services.

The rise of digital, on-demand platforms like iQiyi, Tencent Video, and Youku in China signals a change in the country's media environment. They have altered the habits of Chinese viewers significantly by introducing new methods of information delivery and consumption. Because of their impact, established media organizations have revised their tactics to thrive in the Internet era (Keane et al., 2020). Content licensing and competition have been stumbling blocks in the growth of these platforms. However, they have evolved and grown to show the promise of OTT media in China and the energy of the country's digital sector.

The distinction between OTT platform viewers and traditional television audiences is pivotal for content creators and advertisers alike. OTT services offer a more granular understanding of their viewer base thanks to advanced analytics and data-collection capabilities. Unlike traditional television, which relies on broader demographic data and ratings, OTT platforms can track user engagement at a highly detailed level—ranging from viewing habits to content preferences. This enables a more precise audience profiling, revealing a viewer base that tends to be younger, tech-savvy, and more open to niche content compared to the general television audience. Advertisers are drawn to OTT platforms due to the potential for targeted advertising and the ability to reach specific demographics with higher accuracy. Furthermore, the engagement metrics provided by OTT platforms allow for an in-depth analysis of viewing patterns, informing content strategy and advertisement placement in a manner traditional TV cannot match. Hence, understanding the skew towards particular audience profiles on OTT platforms is critical not only for content personalization but also for advertisers seeking to maximize the impact of their marketing efforts.

## Contextualizing Their Growth in the Chinese Internet Sector

The expansion of OTT platforms in China mirrors the development of the country's internet industry as a whole. The rise of the internet and mobile-first consumer culture in China, as well as improvements to the country's telecommunications infrastructure, have created ideal conditions for the development of OTT media in the country.

By 2022, China had more than 989 million people using the internet. This massive audience has been an important factor in the expansion of the OTT market (Ramon & Rojas-Torrijos, 2023). The proliferation of internet access among the Chinese population has led to a rise in the viewing of online videos. As a result, companies like iQiyi, Tencent Video, and Youku jumped on the bandwagon to meet the surging demand for video streaming services.

Smartphones and the development of 4G and 5G networks in China have also played important roles (Zreik, 2023a). Smartphones made it possible to access digital media at any time and place, and increased internet speeds made streaming video and audio a seamless experience. This shift to mobile-first digital consumption drove a boom in interest in OTT services, which provided viewers with a degree of customization and adaptability that was unavailable through more conventional broadcast channels (Menon, 2022).

State-led initiatives have also directed the growth of China's internet industry. The expansion of the digital economy in China can be attributed in large part to the government's push for digitalization, which has included substantial investments in internet infrastructure and pro-internet regulations (Su & Flew, 2021). These rules have opened up new doors for OTT services but have also presented new obstacles, most notably in the form of content restriction.

The presence of integrated tech titans like Tencent and Alibaba, owners of Tencent Video and Youku, respectively, has also contributed to the expansion of OTT content. These internet behemoths have been able to give their OTT platforms a boost by cross-promoting them and creating synergies with their other businesses, which range from e-commerce to social networking (Fung & Chik, 2021).

Complementing the historical perspective on OTT media's rise, it is crucial to recognize the shortcomings of traditional media that have precipitated this shift. Traditional broadcast mediums were constrained by fixed schedules and limited geographic reach, offering little in terms of accessibility and convenience. The content variety was often narrow, catering to mass appeal rather than diverse tastes, and interaction with viewers was unidirectional and limited. This rigidity stands in stark contrast to the adaptable and user-focused nature of OTT platforms, which tailor the viewing experience to individual preferences and allow for a richer engagement. Additionally, traditional media required substantial investment in infrastructure, a barrier that OTT platforms circumvent with their internet-based delivery models. This flexibility and efficiency have not only fueled OTT media's growth but also underscored the deficiencies of traditional systems, marking a definitive turn in the media landscape.

## BUSINESS MODELS AND REVENUE STREAMS

### Subscription Services

As China's OTT media ecosystem developed, platforms looked at new ways to monetize their offerings to ensure their long-term viability. For example, iQiyi, Tencent Video, and Youku have all seen subscription services become an integral part of their business models (Wang & Lobato, 2019).

Users of subscription services pay a periodic charge (often by the month or the year) to have unrestricted access to the service's media archive. This scheme coincides with the 'on-demand' character of

OTT platforms, allowing viewers to watch shows whenever and on whatever device they like. The fact that viewers rarely have to endure annoying commercial breaks is another perk.

Due to the widespread expectation of free internet content in China, the uptake of subscription services was originally delayed (Che & Ip, 2017). Viewers only started to understand the benefit in subscribing when OTT companies began investing extensively in original and exclusive content. To entice paying customers, streaming services often offer exclusive content, release movies and shows ahead of or at the same time as traditional media outlets, and provide better viewing quality (e.g., in high definition or without ads).

And to accommodate a wide range of viewer tastes and budgets, Chinese OTT services frequently offer tiered subscription tiers. There might be different tiers of memberships, with the most expensive ones granting early access to new releases, special content, and perhaps even offline viewing (Ott & Mack, 2020).

The subscription model has been essential to the continued success of these services. Subscription fees offer a platform with a reliable source of money that may be used to fund expensive initiatives like content acquisition, original production, and platform upkeep and improvement.

## Advertising

Advertising is another important revenue stream that has greatly aided the business models of OTT platforms in China. Like traditional television, OTT services make money by appealing to advertisers with large user bases and diverse content libraries. Brands and marketers in China can benefit greatly from the OTT platforms' reach and targeting capabilities due to the country's increasingly digital and mobile populace.

There is a wide range of formats used for advertising on OTT services. Commercials are typically presented in two formats: either before the content begins playing, or in the middle of it. Banner advertisements and sponsored content are also commonplace on platforms. Users who haven't paid for a subscription may see more ads, which may make them want to switch to a plan that doesn't display ads (Abreu et al., 2017).

OTT platforms are able to give highly personalized advertising because of their extensive data analytics and user profiling. Advertisers may target specific audiences based on demographics, interests, and contexts because to the quantity of data collected by these platforms about users' viewing patterns, preferences, and behavior (Arigo et al., 2018). Advertisers love this level of precision in targeting, and OTT platforms typically benefit from increased ad revenues as a result.

In addition, OTT platforms in China have tested with cutting-edge ad formats. For example, the usage of interactive commercials has grown in popularity since it enables viewers to buy the items shown in the ad without leaving the film. Some of the most important OTT platforms, like Youku, are subsidiaries of larger tech conglomerates that offer a wide range of digital services; by integrating their advertising, these companies may reap the benefits of synergies across their various offerings (Su, 2019).

There must be a happy medium between advertising and user experience, though. Ads that are too frequent or too disruptive to the viewing experience can drive people away. In order to increase profits without compromising viewers' satisfaction, OTT services are always testing new forms of advertising.

## Virtual Gifts

The distinctive digital culture of China is reflected in the fact that virtual gifts have become a popular and growing revenue source for OTT platforms there. This practice originates in China's live-streaming sector, where fans demonstrate their appreciation by purchasing virtual goods and sending them to their favorite streamers.

Many OTT services now use variations on this concept to attract users and monetise their content. Virtual gifts are a way for fans of shows or actors on iQiyi, Tencent Video, and Youku to demonstrate their appreciation. The giver receives kudos and bragging rights from the recipient's fanbase thanks to a visual representation of the gift in the platform's user interface (Zhang, 2022).

It's not uncommon for viewers of a hit reality show or concert to offer one other virtual flowers, applause, and even more out-of-the-ordinary gifts like virtual vehicles and houses. The platforms then turn these digital presents into actual money. In addition, some platforms distribute a piece of the revenue to the content creators as an incentive to increase their interaction with the audience.

China's robust fan culture and widespread adoption of digital payment methods make the country an ideal testing ground for the virtual presents business (Zreik, 2023b). Microtransactions are becoming increasingly common due to the convenience of digital payments and the fact that fans are prepared to spend money to express their support for their favorite shows or stars.

Furthermore, OTT platforms might use virtual gifts as an interactive method of attracting and retaining users (Nigam, 2022). Gift-giving and -receiving provide a social and gamified element, which increases audience engagement and loyalty.

Regulation and user spending behavior are two areas where the virtual gifts concept could use improvement. To prevent unfair practices and wasteful spending, platforms should have proper safeguards and procedures in place.

## SHIFT IN CONTENT PRODUCTION AND CONSUMPTION

### New Patterns of Audience Engagement: Binge-Watching and Personalization

There have been substantial shifts in both the creation and consumption of content in China with the introduction of OTT media, with new forms of audience involvement emerging as a major trend. Binge-watching and customization are two trends indicative of the growing popularity of digital media over more traditional forms.

With the proliferation of OTT media channels, "binge-watching," or the viewing of numerous episodes of a TV series in a single sitting, has become increasingly common (Rahman & Arif, 2021). Before the digital age, TV shows only aired at specified times, making it difficult to fit viewing into busy schedules. However, OTT platforms like iQiyi, Tencent Video, and Youku have given viewers more control over when and how much content they consume thanks to their extensive libraries of on-demand offerings (Chung, 2022). Thanks to this flexibility, viewers can now "binge" on a complete season or series in one sitting.

The penetration of OTT platforms in China is a critical indicator of their influence and reach within the country's vast population. Recent data suggest that OTT services have achieved significant penetration rates, with industry reports indicating that more than half of China's internet users regularly access OTT content. For instance, a report by the China Internet Network Information Center (CNNIC) in

2021 stated that approximately 60% of internet users in China used an OTT service, signaling a robust audience base. Moreover, with China's internet population exceeding 900 million, this translates to a massive segment of the population engaged with OTT media, substantially outpacing traditional broadcast viewership in certain demographics. The implications of this are profound, not only for content creators and distributors but also for advertisers and policymakers. Understanding the extent of OTT's reach illuminates the transformative impact it has on consumption patterns, cultural dissemination, and the digital economy at large.

The production and dissemination of material could be affected by this tendency. For example, several Chinese OTT services now release an entire season of a show all at once, rather than spreading it out over the course of several weeks (Hutchins, Li, & Rowe, 2019). Binge watching's rise in popularity has also influenced how material is made, with producers putting more effort into building tension and introducing new storylines at the end of each episode.

Personalization has emerged as another distinguishing characteristic of OTT viewing alongside binge-watching. OTT platforms use sophisticated algorithms and user data to select and offer material that will appeal to each viewer. In addition to tailoring the information you're shown, you can also adjust the interface you use, the way your media plays, and even the ads you see.

By making it easier and more relevant to find material that appeals to a specific person, personalization increases audience involvement and happiness. Data on viewer preferences can affect decisions about what types of content to develop or license, therefore it also has significance for content production.

## Creation of Original Series and Films

OTT media services like iQiyi, Tencent Video, and Youku have expanded from content distribution to content creation as they have grown in popularity. This change not only increases the variety of their content but also gives them more discretion over their collection and makes it easier to meet the specific needs of their users.

The fierce competition amongst OTT platforms has helped fuel the production of original content. Offering unique, high-quality material has become an important differentiation as media outlets fight for audience attention and allegiance. Having their own original shows and movies is a great way for streaming services to stand out from the competition and attract new viewers.

The original material produced by many Chinese OTT platforms has been quite successful. Both "The Story of Yanxi Palace" on iQiyi and "The King's Avatar" on Tencent Video have become huge hits in China and around the world (Xiao, 2019). These events not only increase the platforms' visibility, but also strengthen their brand identities and encourage greater participation from users.

By producing their own content, platforms may better manage the standard, timing, and monetization of their offerings. They have the freedom to try out different forms, stories, and genres, and use analytics to inform their work. A platform might, for instance, elect to put more money into historical dramas or reality shows, depending on user tastes.

Furthermore, these platforms can promote Chinese culture and narratives internationally through the creation of original series and films. Because of their widespread availability, OTT services are a promising medium for disseminating Chinese media and entertainment abroad (Keane et al., 2020).

However, it is difficult and expensive to create new material from scratch. It requires a lot of money for pre-production, production, post-production, and promotion. Also, it's vulnerable to the fickleness of Chinese viewers and the country's strict content rules.

Despite these obstacles, China's OTT platforms should prioritize the production of original series and films. Original content production is anticipated to be a central emphasis of their growth and differentiation strategies as they continue to navigate the competitive and rapidly evolving OTT landscape.

### Promotion and Dissemination of Chinese Culture

OTT media's meteoric ascent in China has opened up a new channel for the domestic and international diffusion of Chinese culture. These platforms are becoming important mediums for presenting Chinese culture, history, and contemporary life as they develop and disseminate material.

OTT services in the United States appeal to a wide variety of viewers. They create works in many different styles, from period pieces to contemporary urban tales to fantasy epics to reality television (Benshoff, 2015). Aspects of Chinese culture and society are frequently reflected in this material. Popular Chinese historical dramas, for example, often feature stories of emperors, warriors, intellectuals, and commoners and explore themes connected to Confucian principles, social hierarchies, and political intrigue as they take viewers through many eras of China's rich past (Ott & Mack, 2020).

In contrast, urban dramas and reality shows of the present day portray modern Chinese life, displaying the country's growing urbanization, the hopes and fears of its expanding middle class, and broader social themes including gender roles in the family and the workplace.

Chinese over-the-top platforms are crucial in exporting Chinese culture to the rest of the world. As these services grow in popularity around the world, more and more people are exposed to Chinese media for the first time. The global success of Chinese productions like "The Untamed," "Nirvana in Fire," and "The Story of Yanxi Palace" has resulted in a greater interest in and admiration of Chinese art and literature (Zhao, 2019). In addition to facilitating cross-cultural understanding, these platforms work with foreign production companies to provide material that fuses Chinese and international traditions.

## REGULATORY ENVIRONMENT IN CHINA

### Government's Approach to Internet Governance and its Effect on the OTT Landscape

When compared to other countries, the Chinese government's strategy for internet governance is comprehensive and novel. It is based on the notion of cyber sovereignty, which holds that nations have the same authority over the internet within their boundaries as they do over their actual land. To put this idea into action, the government can and likely will restrict or block access to material it deems offensive or damaging on the internet (Shen, 2016).

Cyber sovereignty for China's OTT media ecosystem means a regulatory climate that affects how companies like iQiyi, Tencent Video, and Youku do business. In China, for instance, if an OTT platform wants to stream movies, TV shows, or music videos, it must first get a number of different licenses and approvals. In addition, each platform must ensure that non-local material makes up no more than a predetermined share of its entire library (Keane et al., 2020). The form of OTT content in China has been influenced by these strict rules, with indigenous material dominating the market.

Since the government can slow down or completely block access to OTT services as part of its internet governance, good relations with the government are crucial to the smooth operation of these platforms.

The regulatory climate may be harsh, but there is good news for OTT services. Because of the emphasis placed on domestic content, there has been an uptick in original, culturally relevant shows produced by local talent (Hutchins, Li, & Rowe, 2019). As a result, this has helped these sites gain traction with Chinese audiences. China's OTT ecosystem is unique and full of challenges and opportunities, but entry barriers for global companies have been created by the country's tight rules.

### Opportunities and Challenges: Content Censorship, Data Privacy, and Market Competition

While censorship in China is generally regarded as a barrier by international media businesses, it really presents opportunity for domestic OTT services and Chinese content producers. By limiting access to streaming of foreign content, the Chinese government is, in effect, encouraging the development of homegrown content (Zhao & Keane, 2013). This censorship has sparked a proliferation of Chinese films, TV shows, documentaries, and other types of media on sites like iQiyi, Tencent Video, and Youku. As a result, these platforms have expanded their market share by attracting a dedicated following of users interested in regional programming.

However, censorship is a problem for these websites as well. Strict government regulation of media has the potential to stifle artistic expression, decreasing the availability of new perspectives and approaches in media. Moreover, the fact that even harmless information may be reported for suppression adds an air of uncertainty for platforms and content creators alike (Qiang, 2011).

The protection of personal information is also an important part of the law in China. In recent years, the Chinese government has enacted stricter regulations meant to protect citizens' personal information. Users benefit from this, but OTT services that leverage user data for algorithmic recommendations, tailored advertising, and original content development may face difficulties as a result (Zhang, 2019). It can be costly for platforms to implement the data protection processes and policies necessary to comply with these rules.

Despite its benefits to shoppers, China's competitive market can be difficult for OTT services to navigate. Domestic platforms face competition not only from one another but also from the state-owned broadcasters that continue to dominate China's media landscape. Another competitive threat for OTT services has been the proliferation of short-video platforms like Douyin (also known as TikTok outside of China) (Lin, 2022). While these challenges exist, they are balanced out by the fact that competition encourages these platforms to constantly develop and adapt to changing customer demands, resulting in a vibrant and ever-evolving OTT scene in China.

## BROADER SOCIETAL IMPLICATIONS

### Impact on Cultural Trends and Celebrity Culture

The explosion of OTT media in China has far-reaching consequences for Chinese society, especially in terms of the influence it has on cultural norms and celebrity culture. OTT platforms are altering the nature of fame and celebrity in the digital era as they produce and disseminate a broad variety of material, influencing public debate, tastes, and trends.

Due to their function as content producers and curators, OTT services have an outsized impact on social movements. Certain ideas, aesthetics, or narratives may become more prevalent in popular culture as a result of the shows and films they choose to make or include. The success of a reality show, for instance, may have an impact on the genres of music and dance that are widely performed.

The rate of adoption and growth of OTT platforms in China is not only indicative of a technological shift but also reflects changing consumer behaviors and preferences. In terms of factual support, data from the China Internet Network Information Center (CNNIC) highlights a year-on-year increase in OTT service users, with a notable surge in subscriptions post-2019. A 2022 market analysis by iResearch indicated that the OTT market saw a compounded annual growth rate (CAGR) of 12% over the past five years, outstripping the growth of traditional television viewership. Additionally, the revenue from China's OTT market is projected to reach a significant figure, with estimations from the same report suggesting a climb to over $50 billion by 2025, thereby underscoring the economic weight of this sector. These metrics underscore the burgeoning dominance of OTT media, confirming its central role in the current media ecosystem and projecting its potential for future expansion.

In evaluating the qualitative aspects of programming on OTT platforms, it becomes essential to identify what content resonates with the audience and drives engagement. A thematic analysis of popular series and films on platforms such as iQiyi, Tencent Video, and Youku reveals a propensity for narratives that blend traditional Chinese motifs with contemporary issues, reflecting a society in transition. Moreover, the success of genres such as period dramas, romantic comedies, and reality TV has been significant. These genres not only dominate the viewership charts but also set the trends for discussions on social media platforms. For instance, the acclaim received by shows like 'The Longest Day in Chang'an' underscores the audience's appetite for high production values and historical authenticity, whilst series like 'Ode to Joy' highlight the shift towards more relatable and urban-centric storylines (Marzuki et al., 2023). This content evaluation is critical for understanding the factors behind OTT's compelling appeal and the nuanced preferences of its viewer base.

OTT platforms have an impact on popular culture beyond just the content they provide through their algorithms and recommendations. These platforms can amplify trends or introduce viewers to new genres and formats by curating content based on user preferences and viewing patterns.

Another major repercussion for society is how OTT has changed the celebrity culture. In the old school of the entertainment business, being famous meant making it on the big screen or on the evening news. The proliferation of OTT services, however, has leveled the playing field and opened the door to a broader spectrum of people (Ulin, 2013).

By starring in original series or films, or by appearing in reality shows or variety programs, many actors and actresses have become household names thanks to OTT services. Some of these superstars have built huge followings, which has a ripple effect on consumer culture.

OTT platforms also facilitate the rise of non-traditional celebrities like influencers, vloggers, and live-streamers who attract sizable online followings and participate actively in the communities where their content is consumed. This change is indicative of a larger trend in the media industry, where celebrity and power are becoming less centralized and more distributed.

## Influence on the Chinese Film and Television Industry

China's film and television business has been significantly impacted by the rise of over-the-top media platforms in the country. The production, financing, and promotion of movies and television shows have all been affected by the rise of these sites.

Cinemas and broadcast television were once the primary means of disseminating films and television programs in China. However, iQiyi, Tencent Video, and Youku's ascent to prominence in the market have altered the distribution environment. These sites provide a new channel for TV and movie distribution, opening up access to a global audience that isn't bound by time zones or broadcast schedules (Wang & Yu, 2021). This has widened the content's potential audience and made it more accessible.

More unique and diversified films and television shows are being made as a result of the demand for premium, exclusive content by OTT services. Many services have made sizable bets on their own original content, either by forming partnerships with seasoned producers or by encouraging the growth of in-house creative talent. As a result, the production industry has become more competitive, leading to increased quality and new forms of expression.

OTT services have also changed the way film and TV productions are financed. These sites have amassed huge user bases and substantial capital, making them major backers of film and TV productions. Crowd-funding tools have been implemented on a few of these sites, letting people pool their resources to support a common cause (Craig & Cunningham, 2019).

The promotion of movies and television shows has also been greatly affected by OTT services. By collecting information on their users, they can promote material to the people most likely to be interested in it, based on their watching habits. These sites also serve as a promotional hub where creators, actors, and spectators may communicate in real time.

The rise of OTT media has provided opportunities and threats to China's film and television industry. It has caused a change in revenue from things like ticket sales and advertising to things like subscription fees and in-app purchases, disrupting the status quo in the process. Concerns regarding market competition and content diversity are also raised by the dominance of large OTT platforms (Vogel, 2020).

## Media Pluralism, Content Quality, and the Role of Algorithmic Recommendation Systems

OTT media's meteoric expansion in China raises serious questions about the future of media plurality, content quality, and the function of algorithmic recommendation systems. The term "media pluralism" describes the wide variety of outlets, perspectives, and subject matter available to the general audience (Vermeulen, 2022). Even though OTT media has greatly increased the quantity and variety of information available to viewers, worries regarding media pluralism continue to be raised. iQiyi, Tencent Video, and Youku are just a few of the key platforms that control the majority of the market's content production, promotion, and distribution. This centralization of control has the potential to reduce the variety of opinions and viewpoints expressed in the media.

The quality of the content being distributed is also an important factor. Because of the intense competition in the OTT media sector and the sheer volume of content being created, some worry that platforms will put quantity ahead of quality. In addition, catering to the largest possible audience might sometimes mean sacrificing originality in favor of a broader appeal (Craig & Cunningham, 2019).

There are major repercussions associated with the algorithmic recommendation systems that are integral to OTT platforms. These algorithms analyze user data to make personalized recommendations, making it easier for viewers to find what they're looking for in the broad content landscape. They can improve viewers' interest and contentment, but they also pose serious questions. Artificial intelligence (AI) has the ability to isolate users in "echo chambers" or "filter bubbles," where they only see content that confirms their existing beliefs and values.

In addition, viewers may not completely grasp how their viewing recommendations are derived because of the opaque nature of many algorithmic systems. Data privacy and manipulation worries may be heightened by this lack of disclosure.

The study set out with clear objectives to explore the ascent of OTT media in China and its broader societal implications. The findings indicate that OTT platforms have significantly influenced cultural trends, suggesting a shift in celebrity culture and a diversification of content, which aligns with the initial objectives to determine the extent of digital media's impact on social norms. Furthermore, the rise of OTT services has reconfigured the traditional film and television industry, in line with our objectives to assess industrial shifts. The findings illuminate how these platforms have become prominent players in content distribution and production, echoing the study's aim to trace the evolution of media distribution channels.

In situating these findings within the annals of literature, it is evident that there is a convergence with scholars such as Wheeler (2014) who observed the democratization of celebrity through digital platforms, and Patel, Khadia and Awasya (2020) who discussed the transformation of content consumption dynamics due to OTT media proliferation. Conversely, our findings diverge from the viewpoints of Montero and Finger (2021), who posited limited impact of digital platforms on traditional media industries. Instead, we found that OTT platforms have not only coexisted but also actively disrupted and redefined the norms of media production and distribution. This engagement with both corroborating and conflicting literature demonstrates the study's comprehensive examination of the OTT media landscape and enriches the discourse with new empirical insights from the Chinese context.

## CONCLUSION

Over-the-top media's meteoric ascent in China represents a sea change in the international media industry. This chapter has examined how the Chinese OTT media sector is a synthesis of distinct historical trends, novel business models, shifting patterns of audience participation, and tight regulatory environments.

China's expanding prominence in worldwide digital content production and distribution is reflected in the history of OTT media in the country, which has shifted from platforms largely supplying licensed material to those investing extensively in original productions. Rapid urbanization, increased digital connectivity, and the rise of a tech-savvy middle class in China are all factors that have contributed to the growth of these platforms.

Subscription services, advertising, and virtual presents are only some of the components of business models that have been adapted to meet the needs of the Chinese market. These successful revenue streams demonstrate the marketability of OTT platforms and emphasize their expansionary potential.

There has been a dramatic change in the ways that content is made and consumed. Binge-watching, customizing material, and creating hit original shows and movies have all had a significant impact on how people consume media. In addition to serving as a means of cultural export, the development and dissemination of Chinese narratives has provided a distinctive picture of Chinese culture around the world.

The development of OTT media in China, however, is not without obstacles. OTT platforms face a challenging regulatory framework, notably with regards to content filtering, data privacy, and market competitiveness. The rise of OTT media has far-reaching social consequences, including effects on popular culture, the celebrity industry, and the cinema and television production industry at large. Extra complexity is added by worries about media diversity, content quality, and the function of algorithmic recommendation systems.

The growth of OTT media in China is expected to continue in the coming years. Because of China's massive population, rapidly expanding digital infrastructure, and the country's novel combination of commercial innovation and regulatory adaptability, the OTT landscape in China is ripe for future expansion and evolution.

For this growth, however, to be long-lasting and beneficial for all, the obstacles must be overcome. The goal of regulatory rules should be to promote innovation while yet preserving control. In their quest for profit, OTT services must avoid jeopardizing content quality or variety. Algorithmic recommendation systems' influence on the viewing experience should be disclosed and accounted for.

The global media landscape will be affected by the rise of OTT media in China. Digital platforms have the ability to change content generation, distribution, and consumption, as seen by the success of Chinese OTT platforms both at home and abroad. From China's example, other nations and regions can learn something useful, which they can then apply to their own situations.

## REFERENCES

Abreu, J., Nogueira, J., Becker, V., & Cardoso, B. (2017). Survey of Catch-up TV and other time-shift services: A comprehensive analysis and taxonomy of linear and nonlinear television. *Telecommunication Systems*, *64*(1), 57–74. doi:10.1007/s11235-016-0157-3

Arigo, D., Pagoto, S., Carter-Harris, L., Lillie, S. E., & Nebeker, C. (2018). Using social media for health research: Methodological and ethical considerations for recruitment and intervention delivery. *Digital Health*, *4*, 2055207618771757. doi:10.1177/2055207618771757 PMID:29942634

Benshoff, H. (2015). *Film and television analysis: An introduction to methods, theories, and approaches*. Routledge. doi:10.4324/9780203129968

Che, X., & Ip, B. (2017). *Social networks in China*. Chandos Publishing.

Chung, J. M. (2022). Introduction to Metaverse and Video Streaming Technology and Services. In *Emerging Metaverse XR and Video Multimedia Technologies: Modern Streaming and Multimedia Systems and Applications* (pp. 1–60). Apress.

Craig, D., & Cunningham, S. (2019). *Social media entertainment: The new intersection of Hollywood and Silicon Valley*. NYU Press.

Evens, T., & Donders, K. (2018). *Platform power and policy in transforming television markets*. Palgrave Macmillan. doi:10.1007/978-3-319-74246-5

Fung, A. Y., & Chik, G. H. (2021). When East Asian media industries are faced with digitalization: Transformation and survival strategies. In *The Routledge companion to media industries* (pp. 312–323). Routledge. doi:10.4324/9780429275340-31

Hutchins, B., Li, B., & Rowe, D. (2019). Over-the-top sport: Live streaming services, changing coverage rights markets and the growth of media sport portals. *Media Culture & Society*, *41*(7), 975–994. doi:10.1177/0163443719857623

Junzhi, D. (2021). The development of e-commerce in China during the COVID-19 pandemic on the example of the textile industry. *BRICS Journal of Economics*, *2*(3), 54–69. doi:10.38050/2712-7508-2021-3-3

Keane, M., Yecies, B., & Flew, T. (Eds.). (2018). *Willing collaborators: Foreign partners in Chinese media*. Rowman & Littlefield.

Keane, M., Yu, H., Zhao, E. J., & Leong, S. (2020). *Chinas Digital Presence in the Asia-Pacific: Culture, Technology and Platforms*. Anthem Press. doi:10.2307/j.ctv20pxxzt

Lin, L. (2022). *Convergent Chinese television industries: An ethnography of Chinese production cultures*. Springer Nature. doi:10.1007/978-3-030-91756-2

Marzuki, S. Z., Zreik, M., Buyong, S. Z., Goyipnazarov, S., Mirzaliev, S., & Bakhramovna, I. A. (2023). An Importance Satisfaction Analysis of Hosting Events in Hangzhou, China: A Pilot Study. *Advances in Social Sciences Research Journal, 10*(6.2), 107–117. doi:10.14738/assrj.106.2.14996

Menon, D. (2022). Purchase and continuation intentions of over-the-top (OTT) video streaming platform subscriptions: A uses and gratification theory perspective. *Telematics and Informatics Reports*, *5*, 100006. doi:10.1016/j.teler.2022.100006

Montero, J., & Finger, M. (2021). *The rise of the new network industries: Regulating digital platforms*. Routledge. doi:10.4324/9781003141327

Nigam, A. (2022). Online gaming and OTT consumption: An exploratory study of generation Z. *Journal of Promotion Management*, *28*(4), 420–442. doi:10.1080/10496491.2021.2008576

Ott, B. L., & Mack, R. L. (2020). *Critical media studies: An introduction*. John Wiley & Sons.

Patel, M. K., Khadia, R., & Awasya, G. (2020). A study: OTT viewership in "lockdown" and viewer's dynamic watching experience. *International Journal on Transformations of Media. Journalism & Mass Communication*, *5*(2), 10–22.

Qiang, X. (2011). Liberation technology: The battle for the Chinese internet. *Journal of Democracy*, *22*(2), 47–61. doi:10.1353/jod.2011.0020

Rahman, K. T., & Arif, M. Z. U. (2021). Impacts of Binge-Watching on Netflix during the COVID-19 pandemic. *South Asian Journal of Marketing*, *2*(1), 97–112. doi:10.1108/SAJM-05-2021-0070

Ramon, X., & Rojas-Torrijos, J. L. (2023). European Public Service Media, Disability Sports and Cultural Citizenship in the Digital Age: An Analysis of Agenda Diversity in the Tokyo 2020 Paralympic Games. *Journalism and Media*, *4*(1), 289–303. doi:10.3390/journalmedia4010020

Rowe, P. (2005). *East Asia modern: shaping the contemporary city*. Reaktion books.

Shen, H. (2016). China and global internet governance: Toward an alternative analytical framework. *Chinese Journal of Communication*, *9*(3), 304–324. doi:10.1080/17544750.2016.1206028

Su, C. (2019). *Changing dynamics of digital entertainment media in China* [Doctoral dissertation, Queensland University of Technology].

Su, C., & Flew, T. (2021). The rise of Baidu, Alibaba and Tencent (BAT) and their role in China's Belt and Road Initiative (BRI). *Global Media and Communication*, *17*(1), 67–86. doi:10.1177/1742766520982324

Tsai, L. L. (2023). A deeper understanding of switching intention and the perceptions of non-subscribers. *Information Technology & People*, *36*(2), 785–807. doi:10.1108/ITP-04-2021-0255

Ulin, J. (2013). *The business of media distribution: Monetizing film, TV and video content in an online world*. CRC Press. doi:10.4324/9780240824543

Vermeulen, J. (2022). To nudge or not to nudge: News recommendation as a tool to achieve online media pluralism. *Digital Journalism (Abingdon, England)*, *10*(10), 1671–1690. doi:10.1080/21670811.2022.2026796

Vogel, H. L. (2020). *Entertainment industry economics: A guide for financial analysis*. Cambridge University Press. doi:10.1017/9781108675499

Wang, J., & Yu, X. (2021). Three currents of platformization in China. *The South Atlantic Quarterly*, *120*(4), 777–794. doi:10.1215/00382876-9443336

Wang, W. Y., & Lobato, R. (2019). Chinese video streaming services in the context of global platform studies. *Chinese Journal of Communication*, *12*(3), 356–371. doi:10.1080/17544750.2019.1584119

Wheeler, M. (2014). The mediatization of celebrity politics through the social media. *International journal of digital television, 5*(3), 221-235.

Xiao, H. F. (2019). *Youth economy, crisis, and reinvention in twenty-first-century China: Morning sun in the tiny times*. Routledge. doi:10.4324/9780429326905

Zhang, C. (2019). China's new regulatory regime tailored for the sharing economy: The case of Uber under Chinese local government regulation in comparison to the EU, US, and the UK. *Computer Law & Security Report*, *35*(4), 462–475. doi:10.1016/j.clsr.2019.03.004

Zhang, X. (2022). An Engagement-Based Model: Chinese Online Video Streaming Services for Chinese Viewers. In *Streaming and Screen Culture in Asia-Pacific* (pp. 163–180). Springer International Publishing. doi:10.1007/978-3-031-09374-6_9

Zhao, E. J. (2019). *Digital China's informal circuits: Platforms, labour and governance*. Routledge. doi:10.4324/9781315173795

Zhao, E. J., & Keane, M. (2013). Between formal and informal: The shakeout in China's online video industry. *Media Culture & Society*, *35*(6), 724–741. doi:10.1177/0163443713491301

Zreik, M. (2023a). From Boom to Bust: A Study of China's Economy in the Wake of COVID-19 Outbreak in H1 2020. *BRICS Journal of Economics*, *4*(1), 147–171. doi:10.3897/brics-econ.4.e101050

Zreik, M. (2023b). Stirring Up Soft Power: The Role of Chinese Cuisine in China's Cultural Diplomacy. In K. Kankaew (Ed.), *Global Perspectives on Soft Power Management in Business* (pp. 292–306). IGI Global. doi:10.4018/979-8-3693-0250-7.ch015

## KEY TERMS AND DEFINITIONS

**Algorithmic Recommendation Systems:** Tools used by OTT platforms that analyze user data and behaviors to provide personalized content suggestions, aiming to enhance user experience and increase content consumption.

**Binge-Watching:** The practice of watching multiple episodes of a television series or other content in rapid succession, a behavior facilitated by OTT platforms that offer on-demand access to complete series.

**Content Curation:** The process of gathering, organizing, and presenting digital content based on thematic and quality criteria. For OTT platforms, this involves selecting movies, TV shows, and other content types that align with their audience's preferences.

**Content Monetization:** The methods and strategies used by content creators and distributors to generate revenue from their content. This includes subscription fees, advertising, and pay-per-view or transactional video on demand.

**Digital Disruption:** The changes that occur when new digital technologies and business models affect the value proposition of existing goods and services. In the context of OTT media, this refers to how digital streaming services have disrupted traditional television and movie distribution models.

**Echo Chamber:** A situation in which information, ideas, or beliefs are amplified or reinforced by transmission and repetition inside an "enclosed" system, where different or competing views are censored, not presented, or underrepresented.

**Media Pluralism:** The degree to which there is diversity in media content and viewpoints available to the public. It relates to the existence of a multiplicity of independent and autonomous media outlets.

**Over-the-Top (OTT) Media Services:** Digital media service offerings directly to viewers via the Internet, bypassing traditional platforms such as broadcast, cable, or satellite television providers.

**Streaming:** The continuous transmission of audio or video files from a server to a client, allowing users to consume content without downloading it entirely first.

**Virtual Gifts:** Digital items that can be purchased and given to content creators on streaming and social media platforms as a form of monetary support or appreciation.

# Chapter 6
# The Role of South African Streamed Soap Operas in Contemporary Cultural:
## Imperialism and Implications on the Project of Nation Building in Zimbabwe

**Jennings Joy Chibike**
*Lupane State University, Zimbabwe*

## ABSTRACT

*This chapter is bent at, through audience perspectives, examining the role of over-the-top media; precisely, streamed South African soap operas in the creation of hybrid cultures and the impact of these cultures on the nation building exercise in Zimbabwe. This qualitative chapter utilises the uses and gratifications theory and the reception theory as theoretical toolkits. Its empirical data is drawn from in-depth interviews with sampled Zimbabwean journalists who stream South African soap operas. The chapter fleshes out that South African soap operas have cultivated new and hybrid cultures in Zimbabwe—among them acceptance of the LGBT community, acceptance of ATR, the use of different languages during communication processes and the appreciation of traditional dressing among others. Empirical data from the participants further exhibits that these hybrid cultures have subsequently become an ideological engineering hub for nation building motives as they carry discourses that can be used for unifying the Zimbabwean people.*

## INTRODUCTION

Imperialism and nation building are two contemporary salient dilemmas that most African countries by virtue of having colonial residue are pondering with. Colonialism weakened the socio-economic and political fabric in African states, in the process, creating fissures permitting the easy seepage of knowledge, practices and beliefs which make the inhabitants of the content susceptible to imperialism.

DOI: 10.4018/979-8-3693-3526-0.ch006

In the same vein, colonialism engineered tribalism, ethnic and religious conflicts which led to the dissection of nations. Contemporary scholarship also points not only to the West spearheading cultural imperialism but also other African countries being imperialist to others. Against this milieu, this study moves away from European imperialism and grapples with African imperialism which has not attracted expansive scholarship. The baffling scholarly questions in need of academic interrogation being, what hybrid cultures are introduced by African countries to others and how do they aid nation building. Thus, situated within the precincts of media studies, this chapter, through views and perspectives of audiences, is poised at examining the role that streamed South African soap operas play in mutating already existing cultures and introducing new cultures, and their effect on the nation building process in Zimbabwe. The specific questions being answered, through views of journalists who are also audiences of streamed South African soap operas, in this chapter, being;

1) How do streamed South African soap operas birth hybrid cultures in Zimbabwe?
2) What is the influence of hybrid cultures birthed by soap operas to the nation building project in Zimbabwe?

This chapter is lauded for endeavouring to offer a fresh perspective to the examination of the intersection of media, cultural imperialism and nation building in Zimbabwe through adopting an audience trajectory in examining the consumption of over the top media and its implication in the nation-state project, something prior scholars have totally ignored. Seeing the scarcity of scholarship in this regard, there is need for a serious examination of over the top media and its influence on important discourses of imperialism and nation building in a country such as Zimbabwe, whose government unequivocally states that it has started fanning the flames of nation building which were at the verge of being extinguished by colonial residue.

## BACKGROUND OF THE STUDY

The geographic area called Zimbabwe was birthed after a protracted liberation war pitting Zimbabweans and white British colonial masters. The British, fearing the collective stamina and potential of a united native man in Zimbabwe, devised mechanisms to weaken and divide them. Among other things they cultivated, irrigated and natured ethnicity and tribalism which grew into being a major divisive forces. The locals began to realize their differences specifically based on geographic location, tradition and language. This gave rise to the Shona people, the major ethnic group which resides in northern Zimbabwe and speaking Shona and, the Ndebele people of Southern Zimbabwe who speak a language called isiNdebele. The natives besieged by a barrage of ethnic and tribal problems struggled to fight colonialism and their efforts at nation building were thwarted. The white minority settlers introduced Christianity in the process dismantled a unified belief in African Traditional Religion (ATR) which was deemed as satanic, evil, cultic and barbaric. This further debilitated social cohesion in Zimbabwe. Interestingly, the settlers did not only stop there but further introduced education. This according them was a panacea for Zimbabwean men's uncivilised conduct and thought. Consequently, the educated began shunning the uneducated in the process further sowing the seeds of division. During the liberation war, Zimbabwean armies, the Zimbabwe People's Revolutionary Army (ZIPRA) and the Zimbabwe National Liberation Army (ZANLA) were divided on tribal grounds with ZIPRA being a brain child of Zimbabwe African

People's Union (ZAPU) a Ndebele political outfit. ZANLA was under the tillage of Zimbabwe African National Union (ZANU) a Shona dominated political party. Ethnic tension and tribalism reached a turning point when 20 000 Ndebele people were butchered by a Shona army trained in North Korea in what is commonly referred to as the Gukurahundi Genocide. The tension emanating from this genocide culminated to the unity accord of 1987 which led to Shona and Ndebele politicians uniting under a party called ZANU PF and consequently marking an end to the genocide and tribal feud. In an independent Zimbabwe, the project of nationalism was kick started in a bid to eliminate tribalism and creating a unified nation. The government among many measures used the media as a tool of advancing nationalism objectives. Through state controlled media the government ushered in programs that were pro-nationalism and enhanced the identity of Zimbabwe as a peace loving and unified country while concealing the continued ethnic marginalisation of the Ndebele from the socio-political and economic discourse of the country, government corruption, electoral malpractices, economic dilapidation, and human rights violations among other things. All private, independent and foreign owned media which mirrored a real reflection on the Zimbabwean situation were banned from operating and had their licenses terminated. Thus, in Zimbabwe, foreign media and private media is deemed as not good enough to enter the communicative space as it carries imperialist western ideologies which are said to be false and anti-nation building. What is however interesting is that due to the advent of digital technologies foreign media and its products are now easily accessed in Zimbabwe by local audiences. In light of this, the chapter grapples with the impact of these foreign programs on Zimbabwean culture and nation building project.

## LITERATURE REVIEW: UNDERSTANDING SOAP OPERAS AND THEIR RITUALIZED USE

Widely believed to have mushroomed in the 1930s, soap operas have attracted wide spread examination by scholars. Some studies have examined how they are used to educate people (Khalid and Ahmed, 2014). Chivandikwa (2010) examined how they are instrumental in cultural imperialism among female audiences. This study takes a detour and focuses on the role of streamed South African soap operas, through audience perspectives, in cultivating hybrid cultures through cultural imperialism and how this subsequently impacts the nation building venture in Zimbabwe.

According to Hayward (1997) the creation of soap operas is anchored of women's lives and what they might possibly enjoy. In the same vein, Mankekar (2002) is of the belief that soap operas mainly portray issues that largely affect women thus bringing to rest the debate on reasons why females are attached to soap operas than males. The other major issue is that though males sometimes watch soap operas (Geraghty, 2005) soap operas are primarily meant to cement gender discrepancies as they are prone at cementing femininity (McDonald, 1995). Therefore, this chapter is poised at examining how television programs pregnant with women issues alter the cultures of Zimbabwean audience regardless of gender and how this affects nation building. The importance of this lies in that the chapter offers novel literature on the intersection of cultural products said to be targeting women, hybrid cultures and nation building in Zimbabwe.

Soap operas play an important role in the society for example they are instrumental in the development, peace building and modernisation agendas (Khalid and Ahmed, 2014). According to Geraghty (2005) and Buckingham (1987) soap operas carry narratives that speak to societal problems and also offer cues to these problems. Benefiting from this argument, the major societal problem in an independent Zimbabwe is that of a fractured nation and distorted identity. Hence this chapter through audience perspectives examines how streamed South African soap operas propel hybrid cultures and how they influence nation building.

Scholars such as Dutta (1999) and Aufderheie (2000) observe that soap operas are seen as alternative media carrying views, identities and issues of the subaltern and minority societies. In a multicultural, multi-ethnic and multilingual country as Zimbabwe, issues of exclusion of minority groups such as the Ndebele people from the socio-economic and political discourses has been attributed as a major stumbling block towards nation building. Thus, this study therefore hopes to examine how the inclusion of the minority in South African soap operas help inspire inclusive and pluralistic process of nation building in Zimbabwe.

## Cultural Imperialism, Nation Building and the Media

There is mature scholarship on issues to do with cultural imperialism and nation building as two independent research areas. Cultural imperialism has been studied within confines of public relations practice and education, advertising (see Friedman, 1992), development (see Ugochukwu, 2008) native cultures and identities (see Asemah, 2012) neo-colonialism and the media (Gatsheni-Ndlovu, 2016) among other research areas. Nationalism has also been studied in relation to how the media enhances and also destabilises the process of nation building (see Ndlovu, 2017). This brief literature review exposes the fact that there is dearth in scholarship on research that interrogates cultural imperialism and nationalism as related issues within the realm of nation building particularly in Zimbabwe. What further brings out the novelty of this chapter is that it also brings over the top media in the form of streamed South African soap operas to the discussion, something prior studies have turned a blind eye on. Methodology wise, this chapter uses the views of journalists unearthed through in-depth interviews. It should be emphasised that journalists' views on the intersection of South African soap operas, cultural imperialism and nation building have never been harvested, instead past scholarship has largely been clutched by textual determinism as it focused on analysing discourses carried by soap opera and documentaries. Therefore, this chapter is integral in growing literature on over the top media, hybrid cultures and nation building in Zimbabwe.

Asemah (2010) and Baran (2002) conceptualize culture as the beliefs, way of life, values, episteme, and practices of a given people which govern their behaviour and thought process. Consequently, cultural imperialism denotes the dominance of episteme, value systems and beliefs of capitalistic, enlightened and civilised societies in this instance Europe over the less enlightened Global South against their consent (Goonatilake, 2005:33; Ang 2001:29). In sum, cultural imperialism is seen as the free flow of knowledge and practices from the Global West to the South (Schiller, 1973). This unrestricted flow of knowledge is attributed to the fast rise and rampant use of digital technologies and media by Carter (2014) and McAlister and Taylor (2007) cited in Chibike and Nkomo (2022). In the same vein, Fuchs (2010) avers that the media has become a pipeline for the transfer of cultures, traditions, thoughts and behaviours between two mutually different people. This chapter benefits from this argument as a foundation for its attempt at interrogating the hybrid cultures birthed by streamed South African soap operas and their influence on the nation building project in Zimbabwe.

According to Ugochuwu (2008:36) cultural imperialism gives rise to universal cultures that are different to native cultures. Lee (1980:31) posits that the universal cultures are capitalistic in nature and bent at disadvantaging the deprived and eroding their social, cultural and historical fabric. Drawing from these postulations, the chapter is poised at examining the hybrid cultures that emerge in Zimbabwe after Zimbabwean audiences consume streamed South African soap operas. By doing, so the chapter exhibits that cultural imperialism is not cast in stone as a European ghost haunting African but can originate from fellow African countries.

The United States of America has been at the fore front of global media imperialism issues due to its robust creation, marketing and distribution of television content pregnant with American pop culture (Jin, 2007; Hesmondhalgh, 2002:178; Osama, 2001). Consequently, audiences of such television content get to adopt the American culture thus, clearly showing that cultural imperialism is anchored on media imperialism. This exhibits that cultural exchange leads to the emergence of unique cultures and political ideologies (Colakoglu, 2009:56). Against this background, this study examines the hybrid cultures (Marwan 2008) that emerge in Zimbabwe after Zimbabwean audiences consume South African soap operas and the role of these cultures in the nation building process. The assumption here is that native cultures are being altered, battered and forced into extinction thanks to foreign media programs such as South African soap operas.

Countries such as Brazil, Mexico and India according to Cater (2014) have been spared from the forces of cultural and media imperialism largely due to the fact that they produce their own quality content which is widely consumed by natives in the countries. Thus, there is little to no hybrid cultures, traditions and behaviours that are informed by Western media. In Nigeria, Asemah (2012) concludes that Western media has influenced the way Nigerians talk, eat and some of their mannerism a sign that cultural imperialism has taken its toll on the natives. According to Ronwire (2000) South Africa is a leader when it comes to media content thus, its media products are widely consumed in the African continent. However, Ndlovu (2011) recognises the influence of South African media but opts not to call it imperialism but instead avers that it breeds new sub cultures. Based on this, the chapter seeks to interrogate these cultures being produced after consuming South African soap operas in Zimbabwe and how they influence nationalism and nation building.

With regards to nationalism and nation building in Zimbabwe Sabelo Gatsheni-Ndlovu (2009) questions the existence of a Zimbabwean nation. In the same trajectory, Mlambo (2013:54) argues that there is no Zimbabwean nation instead a mixture of people dominated by the Shona, Ndebele, Coloured and Indian people who detested British colonial rule. Therefore, they are multiple cultures, identities and traditions in Zimbabwe.

According to Chiumbu (2004:32) state media in Zimbabwe has been largely used to push forward nation building ideas which propel national identity. However, the mediated narratives about nation building and national identity are not pluralistic as they carry the views, values, historiographies of majority Shona people (Chiumbu, 2004). Mpofu (2014) posits that state newspaper *Herald*'s discourses on nationalism are founded on ZANU PF interests. This is further explained by Richarson (2007:36) and Ranger (2006) who state that *Herald* practises patriotic journalism which subsequently forces it to publish national projects and values within the ideologies of the ruling party and government.

The advent of digital media have brought about avenues for some to challenge the hegemonic narrative of nation building and this is seen on *New Zimbabwe.com*, a diasporic online media outlet. *New Zimbabwe.com* has challenged the nationalist discourses carried by *Herald* on issues of Zimbabweanhood (Mpofu, 2014). This shows that issues of nation building in Zimbabwe has been studied with the contexts of mainstream and digital media with focus on media outlets primarily for Zimbabweans by Zimbabweans. What therefore is interesting is to examine the influence of foreign media particularly that from South Africa, created for South Africans by South Africans on nation building in Zimbabwe.

To do this, this study adopts an audience reception trajectory to examine views of audiences of streamed South African soap operas on how they give rise to hybrid cultures and how this impacts the nation building exercise in Zimbabwe.

## Theoretical Framework: Reception Theory and Uses and Gratifications Theory

The chapter draws from the Uses and Gratifications theory conceived in the Leicester Centre for Mass Communication (Mabweazara and Strelitz, 2009) and located within the functionalist tradition as a theoretical toolkit. The mainstay of the theory being that audiences select media content they want to consume for a specific purpose, based on their needs informed by the socio-cultural environment they are situated in (McQuil, 2010; Rubin, 2009; Katz et al.,1973). Thus, it therefore offers analytical lenses permitting the interrogation of how audiences of streamed South African soap operas understand and use the soap operas in relation to issues of creation of hybrid cultures, cultural imperialism and nationalism in Zimbabwe, consequently, helping to answer the objectives of the study. According to and Rubin (2009), Ang (1990:159) and Katz et al (1973) audiences of media products use the media for reinforcement of identity and values, to divert themselves from stressing issues, entertainment purposes, strengthening social ties, for status and confidence building and for widening one's knowledge. In light of this, the chapter therefore tests the practicability and applicability of these issues through examining audiences' perspectives of the role of over the top media in cultural imperialism and nation building in Zimbabwe. McQuil (2010) cited in Chen (2016) criticises the Uses and Gratifications theory by lamenting that it is weak and lacks depth. In the same vein, Mabweazara and Strelitz (2009) state that the theory though it recognises the background of consumers of media texts, its major emphasis is on what they use the media for. To counter the issues raised by McQuil (2010) and Mabweazara (2011) this chapter mixes Uses and Gratifications theory with the Reception Theory which gives emphasis on the cultural backgrounds in which audiences are situated in, and how this influences their understanding and ultimate use of media.

Originating from critical cultural studies, the reception theory pays attention to the audiences' reaction, interpretation, understanding and appropriation of meaning to mediated discourses based on their socio-cultural backgrounds (Iversen, 2015; Strelitz, 2000). Reception theory which Sins (2018) argues has not attracted wide spread academic use and interrogation is integral in understanding how audience critic rhetoric which they consume. Consequently, this explains Ang's (1996) submission that reception theory agrees that meaning making of texts occurs at the locus of reception. This locus of reception is embroiled in various societal discourses, thought processes and background (Herms, 1996 cited in Chibike, 2021; Hall 1980). The theory has been widely celebrated for recognizing the semiotic intelligence of audiences as active units that are capable of synthesizing and making meaning to texts when compared to other theories which see audiences as passive and only get the producer of the text's intended meaning (Hall, 1992). This chapter ultimately benefits from Sins (2018)'s argument that though Kjeldsen (2016) attest to the importance of reception theory in study audiences of print media there is need for the application of this theory in examining audiences of digital media. Thus, this study grabs the opportunity to fill this scholarly void identified by Sins (2018) through utilizing the reception theoretical heuristics as breeding ground for the examination of the views and perspectives of Zimbabwean audience of over the top media, specifically streamed South African soap operas.

## METHODOLOGY

This interpretive study is purely qualitative in nature and makes use of in-depth interviews with journalists who are viewers of streamed South African soap operas. By interpretive study, the researcher alludes to the nature of the study whose findings are based on audiences of streamed soap operas' views and perspectives of the soap operas within discourses of imperialism and nation building in Zimbabwe. Consequently, this perfectly makes this chapter qualify to be qualitative in the sense that Merriam (2002:3) argues that qualitative studies are concerned with examining how people make sense of the social world, something which is in line with interpretive studies. Thahn and Thahn (2015) posit that interpretive studies are concerned with the interrogation of how people intelligently and subjectively understand the universe and issues therein. The people, concerned in this study are journalists who watch streamed South African soap operas. Zimbabwean journalists were opted for this chapter because they have a better understanding of media products, what they entail and they role they play in the society. In the actual selection of journalists the researcher first purposively identified one journalist who was watching a South African soap opera online. In identifying the other participants snowball sampling was used as the first journalist refereed the researcher to more journalists who enjoyed watching South African soap operas online. All in all, 15 journalists (10 female and 5 male) were identified and participated in this study. The journalists were subjected to in-depth interviews which Ncube (2018) defined as serious dialogues that enable the researcher to draw important information making up the topic of discussion from an interviewee. In-depth interviews were ideal for this chapter because they help unearth audiences of South African soap operas' views on the role the soap opera in the creation of hybrid cultures and also in the project of nation building in Zimbabwe. The findings will be discussed under two themes informed by the objectives of the study which are;

- Hybrid cultures are birthed by consumption of streamed South African soap opera
- Influence of hybrid cultures on the nation building project in Zimbabwe

## DISCUSSION AND ANALYSIS OF FINDINGS

### Overview of Findings

The participants in the study had it that there are many ways in which South African soap operas alter and cement already existing cultures, beliefs and traditions. The participants state that the soap operas have introduced a culture of using new Bantu languages to audiences, introduced plural and diverse dress patterns, the importance of embracing multiculturalism, trusting ancestral spirits, appreciating gays and lesbians in Zimbabwe. These have led to the alteration of existing culture and the introduction of hybrid cultures in Zimbabwe. With regards to the chapter's second objective, the emergence of these hybrid cultures and traditions has according to the majority of the participants in the study positively influenced nation building in Zimbabwe. This is to say that these cultures have cemented and glued fissures that have stood in the way of nation building in Zimbabwe. The next section discusses the findings in greater detail.

## Hybrid Cultures Birthed by Consumption of Streamed South African Soap Opera

There were diverse interpretations of the ways in which streamed South African soap operas contribute to the emergence and creation of hybrid cultures in Zimbabwe. The participants reiterated that the consumption of these soap operas led to the emergence of new languages in Zimbabwe. In light of this participant 7 said;

*We watch South African soap operas in which the actors and actresses speak multiple languages. For example in House of Zwide and Scandal actors and actresses speak multiple languages such as Tswana, Sotho, Xhosa and Zulu. People here in Zimbabwe have started using some words from these languages which are minority and marginalised languages when communicating.*

Participant 13 added that

*The youth have created their own new languages which is made up of different words from different languages when communicating with each other. If you do not watch theses soap operas you will never understand their "private" language.*

This means that people in Zimbabwe after consuming South African soap operas now perform new cultures through the use of language. They speak what in Zimbabwe are minority languages and in some cases they mix the different languages to create one hybrid language. In line with Ugochuwu's (2008:36) argument that cultural imperialism gives rise to new cultures that are different from native cultures, the consumption of South African soap operas has birthed hybrid languages which are unique. The consumption of the soap operas has encouraged the use of minority languages in Zimbabwe which is unique to a country which largely identifies with English, Ndebele and Shona as noted by Hadebe (2005). This consequently confirm the Uses and Gratification Theory's notion that audiences use the media for multiple purposes and in this instance audiences use streamed South African soap operas to revive minority languages at the same time to create new private languages. The creation of private languages which embolden private cultures exhibit the semiotic intelligence which Hall (1992) alludes to in the Reception theory when he states that audiences are capable of appropriating meaning to media texts based on their socio cultural background. For the creation of private languages to be a success, audiences ought to have first understood the meanings of different words in different languages (Hall, 1992) and then use it in the creation of unique languages which also are a conduit for the creation of hybrid linguistic cultures in Zimbabwe.

In South African soap operas there is rampant reference to ATR. This ATR has however been reconfigured and entangled with modernity as the rich and famous subscribe to it and attribute their success to it. This was emphasised by participant 3 who stated that

*We watch soap operas online such as Nikiwe, Isitha and Smoke & Mirrors were the rich people such as Jaxon, Caesar, Bhungane were given powers by traditional healers to sustain their power and control over their multi-million business ventures.*

Participant 15 added that

*In these Soap operas it is perfectly normal and acceptable for one to consult traditional healers. The traditional healers look modern, drive decent cars and stay in decent houses. In Zimbabwe we are slowly having people who are coming out clean about them practising ATR. You now even hear others say "asbonge amadlozi" (thank the ancestors).*

This means that South African soap operas have helped Zimbabwean audiences normalise the practice of ATR which has long been demonized and classified as Satanic by the British colonial masters. Instead, the colonisers replaced ATR with Christianity hence contemporary inhabitants of the geographic area called Zimbabwe call themselves Christians and their country a Christian nation (Ruzivo, 2018). Speaking to Geraghty (2005) and Buckingham (1987)'s assertion that soap operas churn out societal problems and ways to fix them, this chapter is of the notion that South African soap operas teach audiences to lean towards ATR as a panacea to their problems especially those involving riches. Thus, Zimbabweans have normalized the culture of consulting and speaking freely about ATR something that was never accepted. ATR has been de-demonized, decolonialized, de-demonized, purified and has been accepted as a normal cultural practice. Thus, this exhibits the fact that Zimbabwean audiences are now using the media to cement certain values in this instance ATR enshrined values and practices. This is in line with Rubin (2009), Ang (1990:159) and Katz et al (1973)'s argument that the media is used ascertain and strengthen the values and beliefs of a people which in turn also influences audiences understanding of media products (Iversen, 2015; Strelitz, 2000). All in all, the consumption of streamed South African soap operas has led to the creation of a new culture among Zimbabwean audiences, that of appreciating ATR which has all along been demonised and relegated to the margins.

Participant 2 notes the cross-cultural dressing and attributes the mixing of European and African dress as fashionable to the consumption of South African soap operas. The participant states that;

*In Soap operas such as Muvhango and Isitha actors combine African traditional attires with Eurocentric suits and styles and this dressing style is now being seen in Zimbabwe. You find this in many ceremonies such as wedding and funerals were people have African traditional attires complemented by European attires.*

In addition participant 13 also states that

*In the Muvhango soap opera, actors and actress sometimes completely wear traditional gear (Venda attire) while in House of Zwide other completely wear their Zulu regalia and this has led to Zimbabweans especially those from Southern Zimbabwe being comfortable with wearing traditional attires.*

This therefore means that people in Southern parts of Zimbabwe such as Bulawayo, Gwanda, Beitbridge among others have embraced the culture of wearing traditional outfit regardless of the event they are attending. Thus, the culture of wearing traditional regalia has been accepted as opposed to the past times were traditional regalia was seen as primitive and in some instances barbaric. This new culture that has been adopted allows different traditions to flourish without being blanketed by Western styles and opinions which detect dress code of what they call a civilised people. This hybrid culture in Zimbabwe sees people from different ethnic groups now also wearing traditional attires belonging to the other groups. This validates the argument by Fuchs (2010) that the media leads the transfer of cultures and beliefs among different groups. This is facilitated by the fact that the media carries various identities and cultures (Aufderheie, 2000) which audiences consume and use to create new cultures and in other instances used to cement their already existing cultures.

There has also been a shift in the way Zimbabweans entertain themselves particularly with reference to dance and music. They have adopted a major dance style associated with contemporary *amapiano* music. Most South African soap operas showcase scenes of dance and the dances will be revolving around the contemporary *amapiano* genre. Also, it is generally said that South Africa is a nation synonymous with dance (van Heerden, 2019)thus, Zimbabwe has also joined this bandwagon of dance with more of its dances now learning towards the amapiano genre. Looking at social media the youth have normalised taking videos of themselves dancing to amapiano music. Not only is the youth the only people who have been caught up in the dance storm but also the elderly something synonymous with South African life. This shows a rift from the traditional Zimbabwean music and other music genres that are indigenous to the country.

Zimbabwe since its attainment of independence has prominently and bravely identified itself as homophobic. This homophobia was cemented by the late President Robert Mugabe who declared that there is no space for pigs and dogs in Zimbabwe (Johwa, 2004). By pigs and dogs Mugabe refereed to homosexuals, gays and lesbians (Johwa, 2004). However, though homophobia is still strong in Zimbabwe, of late they are some sections of the society have been embracing the homosexual, gays and lesbians. This was said by participant 11 who lamented that;

*Soap operas have cultivated in us a culture of tolerance. In South Africa there is tolerance particularly when it comes to issues of gays and lesbians. We have slowly started to embrace these people who in the past we were not ready to accept within our communities. We have started to understand the world through their perspectives.*

This shows that the consumption of South African soap operas has widened the corridors of social acceptance. The minority in particular gays and lesbians are now slowly being accepted into the community. As such this validates the postulation that soap operas are a pipeline for the promulgation of minority people's identities and issues by Dutta (1999).

Having discussed the hybrid cultures and behaviours among Zimbabwean audiences of streamed South African soap operas the next section discusses the influence of these hybrid cultures on the nation building exercise in Zimbabwe.

## Influence of Hybrid Cultures on the Nation Building Project in Zimbabwe

The fusion of cultures exhibited in streamed South African soap operas into the cultures of Zimbabwean audiences has largely been seen as important in aiding the ailing project of nation building in Zimbabwe. According to participant 4 the fact that Zimbabweans have also adopted the use of multiple languages concurrently in conversing, enables the nation building project. In light of this the participant said:

*We have also normalised using multiple languages. This on its on shows that we are now in the nation building trajectory as we are recognising our linguistic differences at the same time appreciating each other's languages without putting hierarchies on language. This is good for nation building, no language must be seen as more important than the other.*

The participant here implies that the use of all languages is important to nation building exercise as it dismantles linguistic hierarchies which favour other languages at the expense of other languages which are seen as less equal and inferior. This therefore validates the assertion that soap operas are instrumental in arresting societal problems (Geraghty, 2005; Buckingham, 1987) in the sense that soap operas help to challenge linguistic hierarchies in Zimbabwe and help fight off colonial vices undermined linguistic plurality and diversity in the process which flagrantly disregard efforts bent at nation building. Thus, it can be argued that the consumption of South African soap operas le to linguistic pluralism which culminate to efforts of nationalism at the same time doing away with linguistic hierarchies and the love for English language which collectively widen fissures within ploy racial and multi-cultural nations such as Zimbabwe. This consequently challenges Lee(1980)'s assertion that cultural imperialism leas to the erosion of local languages, cultures and tradition in the sense that South African soap operas have brought renaissance of thought and communication in various indigenous languages which pivots diverse local cultures and traditions. Since McQuil (2010) is of the notion that audiences use the media for their needs the argument here is that even though audiences initially consumed and used South African soap operas for entertainment they consequently used them for self-exorcism and repentance from colonial vices which had no-nation building motives.

Participant 1 is of the notion that South African soap operas are important in nation building process in Zimbabwe and this is through helping Zimbabweans appreciate ATR. In light of this the participant said;

*These soap operas have taught us to appreciate African Traditional Religion which is the fountain of our religion, the religion of our fore fathers. ATR is a religion which ties us together. I feel ATR binds us together as this is something we have in common dating back to our ancestors.*

These views of the participant aptly indicate the importance of religion in nation building. This therefore exhibits that South African soap operas have taught Zimbabweans to appreciate ATR, a religion which is a common denominator among the natives. The act of adopting a common religion therefore eradicates partisan religious incongruities which dissuade natives from embracing each other as argued by Agbiji and Swart (2015). This validates Ang (1990:159) and Katz et al (1973)'s assertions that the media is capable of amplifying identities, values and traditions of a people. This chapter extends this assertion by positing that the amplification of medieval traditions, religions, beliefs and practices is watering down split patriotic notions which incubate tribalism and divisions. Hence this amplification is important in the gluing together of a people and constructing the idea of nationalism and belonging. The fact that the audience singled out issues of ATR in relation to nation building is a testimony of Iversen (2015) and Strelitz's (2000) supposition that texts are understood by audiences in relation to their background. Zimbabwe is a country which is in the murky waters of religious oblivion as it is slowly abandoning ATR for Christianity. So, the participant here read ATR narrative in the soap operas and located them in the issues of nation building in the sense that ATR can be a viable tool for nation building against the background that Zimbabwe now calls itself a Christian nation.

In South African soap operas actors and actresses are seen clad in their own traditional attires and clothes resembling their ration and beliefs. This has been copied in Zimbabwe as people are now taking pride in their different traditional attires. This according to participant 11 is important in nation building. In respect of this the participant states that

*I think the fact that we have adopted the traditional attires like South Africans shows that we tolerate each other and this builds a strong nation. For example the Shona people are seen wearing Ndebele attires, the Ndebele wearing Kalanga attires the Tonga wearing Sotho attires. I think this is good for nation building.*

The participant here speaks to inter-culturalism as a panacea to nation building challenges in Zimbabwe. The fact that ethnic groups recognise each other and share cultural dressing and values is a sign of acceptance. Thus, it is invaluable to reiterate that South African soap operas have cultivated an irrigated a culture of acceptance in Zimbabwe which is the engineering hub for nation building. On the contrary participant 8 feels that the wearing of traditional attires further pushes away natives thus injuring attempts at nation building. In light if this the participant submits that:

*The wearing of different tradition outfits divides us against nation building because we are emphasising more n our difference. This then evokes a spirit of difference and the ethnic groups which dominate will obviously be seen as the majority that matters at the expense of others.*

Here the participant is of the notion that the fact that people will be forced to wear their different traditional outfits will then highlight the societal fissure anchored on ethnicity in Zimbabwe. Thus, soap operas are seen as a pipeline of discourses that further rift apart the different Zimbabwean ethnic groups thus hindering the project of nationalism. Since meaning is made at the locus of reception an informed by one's background (Ang, 1996) the contradictions here bear testimony to that. Participant 8 and 11 gave views on the same issues a sign that they are situated within different background thus their construction of meaning is different.

A hybrid culture that has slowly been seeping into the veins of Zimbabweans is that of accepting the LGBT community. This is attested to by participant 12 who states that:

*After watching these soap operas we now embrace LGBT community and I feel this is important for the nation building exercise. These people are usually left out of important issues and are not treated as people. So embracing them gives a sense of unity and belonging for them*

Here the participant is of the notion that South African soap operas have facilitated a humane heart which has allowed Zimbabweans to accept those that they have in the past marginalised. Thus, the notions of acceptance are a benchmark for the realisation of a unified nation. The media in this instance is a sight of the proliferation of minority citizens and the recognition of these minorities is seen as a new culture as argue by Colakoglu (2009:56) particularly in Zimbabwe.

## CONCLUSION

The chapter has revealed the practicability of Uses and Gratifications theory and the Reception theory as theoretical toolkits enabling the investigation of the hybrid cultures birthed by the consumption of streamed South African soap operas and how these cultures play a role in the nation building exercise in Zimbabwe. With regards to the first objective the chapter unearthed that the sampled participants who are viewers of streamed South African soap operas are of the belief that the soap operas have influence in the creation of hybrid cultures in Zimbabwe. These cultures rover around issues of acceptance of ATR, gays and lesbians, using hybrid languages and adopting foreign ways of dance and singing. Secondly, the chapter unravels the fact that these hybrid cultures are very instrumental in aiding the nation building exercise in Zimbabwe. All in all, over the top media is instrumental in the diffusion of foreign cultures and offers cues to help important national issues such as that of nation building. There is need however for more studies with different sets of audiences to verify or even improve the insights offered in this chapter. Besides being audience reception studies, future scholarship can delve into the narratives carried by soap operas and these influence the mushrooming of hybrid cultures.

## REFERENCES


Agbiji, O. M., & Swart, I. (2015). Religion and social transformation in Africa: A critical and appreciative perspective. *Scriptura, 114*, 1–20. doi:10.7833/114-0-1115

Ang, I. (1990). Culture and Communication: Toward an Ethnographic Critique of Media Consumption in the Transnational Media System. *European Journal of Communication, 5*(5), 239–266. doi:10.1177/0267323190005002006

Ang, I. (1996). *Living Room Wars: Rethinking Media Audiences for a Postmodern World*. Routledge. doi:10.4324/9780203289549

Ang, I. (2001). Desperately guarding border: media globalization, "cultural imperialism", and the rise of "Asia". In Y. Soucho (Ed.), *House of Glass: culture, modernity, and the state in Southeast Asia* (pp. 27–45). Seng Lee Press.

Asemah, E. (2011). *Selected Mass Media Themes*. Jos University Press.

Asemah, E. S. (2012). Media imperialism: The cultural implication for Nigeria. In M. Mboho & H. Batta (Eds.), *Communication approaches to development: Essays in Honour of Professor Des Wilson*. BSM Resources Limited.

Aufderheide, P. (2000). *The Daily Planet*. University of Minnesota Press.

Baran, S. (2002). *Introduction to Mass Communication, Media and Culture* (2nd ed.). McGraw Hill Higher Education.

Barker, C. (1999). *Television, Globalization and Cultural Identities*. Open University Press.

Buckingham, D. (1987). *Public Secrets*. BFI.

Carter, L. (2014). The cultural imperialism model in international communication.

Chen, L. (2016). *Talking about health and health related issues: an inquiry into the social media use of Chinese celebrity physicians and their fans*. [PhD Thesis, University of Iowa].

Chibike, J. J. (2021). *Photography in a human rights crisis: Readers' perspectives and views of BBC's representation of the 2018 post-election violence in Zimbabwe*. [Unpublished Master's thesis, National University of Science and Technology].

Chibike, J. J., & Nkomo, T. (2022). Public relations education and cultural sensitivity in the digital epoch: A case of National University of Science and Technology's Journalism and Media Studies Department. *Journal of African Education, 3*(3), 55–78. doi:10.31920/2633-2930/2022/v3n3a3

Chiumbu, S. (2004). Redefining the National Agenda, Media and Identity-Challenges of Building a New Zimbabwe. In H. Melber (Ed.), *Media, Public Discourse and Political Contestation in Zimbabwe*. Nordiska AfrikaInstitutet.

Colakoglu, B.E. (2009). Perceptional difference of international news: Western media influence on non-Western media. *China Media Research,* 55-63

Dutta, M. B. (1999). Taming the victim: Rape in soap opera. *The Journal of Popular Film and Television, 27*(1), 34–39. doi:10.1080/01956059909602795

Friedman, T. (1992). The World of The World of Coca-Cola. *Communication Research, 19*(5), 642–662. doi:10.1177/009365092019005005

Fuchs, C. (2010). New Imperialism: Information and media imperialism? *Global Media and Communication, 6*(1), 33–60. doi:10.1177/1742766510362018

Geraghty, C. (2005). The study of the soap opera. In J. Wasko (Ed.), *Companion to Television* (pp. 308–323). Blackwell Publishing. doi:10.1002/9780470997130.ch17

Goonatilake, S. (2005). Cultural Imperialism: A Short History, Future, and a Postscript from the Present. In B. Hamm & R. Smandych (Eds.), *Cultural Imperialism: Essays on the Political Economy of Cultural Domination* (pp. 33–51). Broadview Press.

Hadebe, S. (2005). Reflections on the proposed Ndebele/Shona-Shona/Ndebele dictionary. *Lexikos, 15*, 265–275.

Hall, S. (1980). Encoding/decoding. In S. Hall, D. Hobson, A. Lowe, & P. Willis (Eds.), *Culture, media, language*. Hutchinson.

Hall, S. (1992). The question of cultural identity. In S. Hall, D. Held, & T. McGrew (Eds.), *Modernity and its futures*. Polity Press.

Hayward, J. (1997). *Consuming Pleasures Active Audiences and Serial Fictions from Dickens to Soap Opera*. University Press of Kentucky.

Hesmondhalgh, D. (2002). *The Cultural Industries*. Sage.

Iversen, M. H. (2015). Employing a rhetorical approach to the practice of audience research on political communication. In Kramp, L., Carpentier, N., Hepp, A. Nieminen, H and Olsson, T. (Eds). *Journalism, Representation and the Public Sphere*. Lumiere.

Katz, E., Gurevitch, M., & Haas, H. (1973). On the use of the mass media for important things. *American Sociological Review, 38*(2), 164–181. doi:10.2307/2094393

Khalid, M. Z., & Ahmed, A. (2014). Entertainment –education strategies for social change: Opportunities and emerging trends. *Review of Journalism and Mass Communication, 2*(1), 69–89.

Kjeldsen, J. E. (2016). Studying Rhetorical Audiences–a Call for Qualitative Reception Studies in Argumentation and Rhetoric. *Informal Logic, 36*(2), 136–158. doi:10.22329/il.v36i2.4672

Lee, C. (1980). *Economic Dependency and cultural imperialism: Theoretical perspectives Media imperialism reconsidered: The homogenizing of television culture*. Sage.

Mabweazara, H., & Strelitz, L. (2009). Investigating the popularity of the Zimbabwean tabloid newspaper UMthunywa: A reception study of Bulawayo, *African. Journalism Studies, 30*(2), 113–133.

Mankekar, P. (2000). National texts and gendered lives: An ethnography of television viewers in a North Indian city. *American Ethnologist, 20*(3), 299–322.

Marwan, K. (2008). *Hybridity: The Cultural Logic of Globalization*. Temple University Press.

McDonald, M. (1995). *Representing Women*. Arnold.

McQuail, D. (2010). The Uses and Gratification Approach. In D. McQuail (Ed.), *McQuail's Mass Communication Theory* (6th ed.). Sage.

Merriam, S. (2002). *Qualitative research in practice: Examples for discussion and analysis*. Jossey-Bass.

Mlambo, A. (2013). Becoming Zimbabwe or becoming Zimbabwean: Identity, nationalism and state-making. *Africa Spectrum, 48*(1), 49–70. doi:10.1177/000203971304800103

Mpofu, S. (2014). *Public and Diasporic Online Media in the Discursive Construction of National Identity: A Case of Zimbabwe*. [Unpublished PhD thesis, University of the Witwatersrand].

Ncube, B. J. (2018). *Radio Mthwakazi and the construction of Ndebele Ethnic Identity*. PhD thesis. University of Johannesburg.

Ndlovu, M. (2017). *Site of remembrance and forgetting: New media reconstructions of distinct Ndebele collective memory and history in the context of hegemonic Zimbabwean nationalism*. [Unpublished PhD thesis, Stellenbosch University].

Ndlovu-Gatsheni, S. (2009). *Do 'Zimbabweans' exist? Trajectories of nationalism, national identity formation and crisis in a postcolonial state*. Peter Lang.

Osama, E. (2001). *Media imperialism in the global era*. Middle East News Online.

Ranger, T. (2004). Historiography, Patriotic History and the History of the Nation: The Struggle over the Past in Zimbabwe. *Journal of Southern African Studies, 30*(2), 215–234. doi:10.1080/0305707042000215338

Richardson, J. (2007). *Analysing Newspapers: An Approach from Critical Discourse Analysis*. Palgrave Macmillan. doi:10.1007/978-0-230-20968-8

Rubin, A. M. (2009). Uses-and-gratifications perspective on media effects. In J. Bryant & M. B. Oliver (Eds.), *Media effects: Advances in theory and research* (3rd ed.). Routledge.

Schiller, H. (1976). *Cultural Domination: Sources, Context and Current Styles. Communication and Cultural Domination*. International Arts and Science Press.

Sins, R. K. (2018). The commenting persona: Reception theory and the digital rhetorical audience. *The Journal of Medical Research*, *11*(1), 55–70.

Strelitz, L. (2000). Approaches to Understanding the Relationship between Texts and Audiences. *Communication*, *26*(2), 37–51.

Thanh, T., & Thahn, N. (2015). The Interconnection Between Interpretivist Paradigm and Qualitative Methods in Education. *American Journal of Education Science*, *1*(2), 24–27.

Van Heerden, M. (2019). A little kind of community: South African students dance for self, other and society. In M. Marx (Ed.), *Dance and the quality of life* (pp. 383–397). Springer Publishers.

# Chapter 7
# Uses and Gratifications of "The Last Dance" and the Viewing Intention for a Chicago Bulls Game:
## A Sports Documentary Series as a Promotional Tool

**Sebastiano Mereu**
https://orcid.org/0000-0003-2607-1782
*Sports Business Research Academy, Switzerland*

## ABSTRACT

*This chapter explores uses and gratifications of watching "The Last Dance," a sports documentary series that premiered in 2020 on ESPN and Netflix about Michael Jordan and the 1990s Chicago Bulls basketball team. The study examines the influence of uses and gratifications on brand-user imagery fit and the intention to watch a Chicago Bulls game after having seen the show. Primary data was collected through a self-administered online questionnaire from people that had been watching "The Last Dance." Results were computed through exploratory and confirmatory factor analysis and structural equation modeling. Results suggest that the realistic nature of the story and characters portrayed in "The Last Dance," the relaxation experienced from watching the show, and social interaction with other viewers influence the intention to watch a Chicago Bulls game when mediated through brand-user imagery fit. These are new findings to the respective field of study and contribute towards the understanding for producing similar documentaries.*

## INTRODUCTION

Documentary films are generally understood to creatively represent and depict real events in order to entertain, inform, and influence viewers (Hughes, 1996; Juel, 2006; Reiss, 2019). Documenting events on film or video as close as possible to reality can be exciting for viewers and boost the effect that the

DOI: 10.4018/979-8-3693-3526-0.ch007

conveyed message has on its recipients (Pettersson, 1993). In the second-half of the twentieth century, television (TV) sets became more advanced and gained widespread acceptance around the world, leading to the adoption in households, businesses, and institutions (Diggs-Brown, 2011). With TV broadcasters able to reach larger audiences with a wider array of shows, they became further able to provide more financial support to documentary films producers in order to produce better quality shows (Hughes, 1996). Eventually, TV grew to become one of the most important media channels to communicate historical events through documentaries, which led to the adoption of specific aesthetic and narrative forms and elements, like the one's applied in feature films (Ebbrecht, 2007). This would enhance the entertainment value of documentary films, making them even more appealing to a wider, general audience. The technological advancements in consumer electronics, internet, and broadband connectivity of the early twenty-first century, enabled companies to build media and entertainment channels that are directly available through the internet, i.e. in an online browser or mobile app, hence, disrupting the way people consume films and videos (Balbi & Magaudda, 2018). This has allowed consumers to access audiovisual media products faster and more conveniently through linear-delayed or subscription-based viewing services, as compared to linear viewing services on traditional TV or in movie theatres; in many cases, people are now able to watch a film or an episode of a show at their convenience, without having to be in front of a specified screen at a time set by a broadcaster or a movie theater (Afilipoaie et al, 2021).

Looking at this opportunity from the perspective of a sports organization, it becomes legitimate to argue that offering sports fans non-fictional footage, such as behind-the-scenes, interviews or expert commentary of or about their favorite team and athletes, along a certain narrative, for example through social media videos, may strengthen their association with the franchise or club and, thus, bolster fans' interest in the sports product, which may be the game, championship, tournament or similar (Goldsmith, 2013; Hipke & Hachtmann, 2014). Videos on social media are often produced as short-form videos to quickly document and share a real-life moment from the respective sports brand with fans and followers (Newman et al., 2013). Nevertheless, despite the wide reach social media videos can offer to sports organizations, they lack the narrative depth and aesthetic brilliance that professionally-produced long-form sports documentaries can offer (Vogan, 2014).

Over-the-top (OTT) platforms and streaming services like Amazon Prime, Netflix, and DAZN understood the perceived value that unscripted sports films about a team or athlete can add to the main sports product and invested in the production of original sports documentary series (Iordache et al., 2023); popular examples include "All or Nothing: Manchester City" (Amazon Prime; football/soccer), "Formula 1: Drive to Survive" (Netflix; car racing), and "The Making of Canelo" (DAZN; boxing), to name a few. The entertaining and captivating characteristics of shows based on the real life of athletes or sports teams can attract large viewing audiences, because people direct their focus towards stimuli that pertain to the fulfilment of their fundamental desires, which include curiosity, idealism, romance, social contact and status, among other motives (Reiss & Wiltz, 2004). This supports the business case for streaming services and other networks to invest in the production of original sports documentaries. Looking at documentary content from the perspective of sports organizations, NFL Films, the film and television production company of the National Football League (NFL), a professional American football league and one of America's most popular sports leagues, offers a compelling case: Founded in 1962, the company has proven that a consistent, glamorized, and well-constructed recount can be used to create a mythological aura around a story to promote a specific sports product or brand, i.e. the NFL (Vogan, 2014).

"The Last Dance", a sports documentary series that premiered in 2020 on ESPN and Netflix, can be considered a similarly captivating case in regard to uses and gratifications of watching a sports documentary. The series garnered widespread attention and became a cultural phenomenon upon its release by focusing on the career of the iconic basketball player Michael Jordan, generally considered one of the greatest athletes of all time, and providing an in-depth exploration of the Chicago Bulls' championship-winning era in the 1990s (ESPN, 2023). The series generated significant viewer interest and sparked conversations worldwide, making it a relevant and widely recognized subject for studying how popular culture can transport a brand-story and influence consumer behavior (Weinish & Lev, 2021). Furthermore, by displaying Nike and Jordan logos numerous times and discussing Chicago Bulls games in every episode, it can be argued that the series can indirectly function as a promotional vehicle for those products (Adweek, 2020). Considering the ubiquity of video content in consumers' lives resulting from the omnipresence of digital devices, unrestricted connectivity, and the ease of access to a variety of social media and other content channels, sports brands are presented with the opportunity to exploit the situation and use this type of communication and marketing tool to convey a more affective brand-story and strengthen the association with existing and new fans (Fetchko et al., 2018). Based upon the discussion above, "The Last Dance" is hereby considered an appropriate and interesting case to be explored in an academic and professional context.

The aim of this study is to explore the impact and potential of sports documentaries, particularly focusing on the evolving landscape of media consumption, the role of OTT platforms, and the influence of sports documentaries on consumer behavior. Using "The Last Dance" as subject, the study aims to investigate how such documentaries can function as promotional vehicles for sports products and brands, examining their potential to strengthen the association between sports organizations and fans. The overarching research question for this research is defined as follows: "How do uses and gratifications of watching the docuseries "The Last Dance" influence brand-user imagery fit of its viewers and the intention to watch a Chicago Bulls game?" The individual elements of the research question will be defined and discussed in the next section of this chapter.

## THEORETICAL BACKGROUND

This research employs an adaptation of theoretical frameworks suggested by Papacharissi and Mendelson (2007) and Rubin (1983) to investigate the extent to which viewers engage and derive satisfaction from watching a sports documentary series such as "The Last Dance". The applied conceptual model includes seven uses and gratifications for the following elements: reality versus fiction, entertainment, relaxation, habitual pass time, companionship, social interaction, and voyeurism. In order to assess the influence of uses and gratifications on brand-user imagery fit and the intention to watch a Chicago Bulls game, the model was extended with elements from Lee and Watkins (2016).

### Uses and Gratifications

Traditionally, research on mass media concentrated on the question, "What do the media do to people?"; with a wider range of mass communication channels available, and the rise of popular culture, Katz (1959) recognized the need for bridging mass communication research into the field of social science and popular culture with focus on the question, "What do the people do with the media?". In the context of watching televised sports using the framework of uses and gratifications, Gantz (1981) proposed an examination of the respective reasons, which revealed four distinct dimensions: to thrill in victory, to let loose, to learn, and to pass time. However, Gantz (1981) acknowledged the limitations of the study and cautioned that the provided list may not be exhaustive. With that in mind, this study expands the respective framework with further elements proposed by Papacharissi and Mendelson (2007), Trail (2012), Nabi et al. (2003), and Lee and Watkins (2016). The following paragraphs will define the uses and gratifications as applied in this study and discuss them with regard to a possible influence on a viewer's general evaluation of how well they relate or connect with other viewers of "The Last Dance" and on a possible effect on the viewing intention of a related media product, in this case, Chicago Bulls games.

## Reality vs. Fiction

Papacharissi and Mendelson (2007) compile questionnaire items from their uses and gratifications construct that refer to "reality versus fiction" and "entertainment" into one factor they call "reality entertainment". However, this research splits that factor into two factors, as the six questionnaire items load onto two different factors in this study's exploratory factor analysis (EFA). Thus, reality versus fiction represents a unique feature of reality-based programming and is hereby defined as the "realistic and not fictional nature of [a show's] story and characters" (Papacharissi & Mendelson, 2007, p. 363). It offers the foundation to question, if the portrayed real-life characters are more interesting than characters interpreted in fictional programs, and, subsequently, if a reality-based program is more enjoyable than a program based upon a fictional story. Sports documentaries like "The Last Dance" feature prominent athletes who serve directly or indirectly as ambassadors for the brands and products presented in the show, i.e. the National Basketball Association (NBA) championship, Chicago Bulls games, Nike sneakers, other merchandise, etc. However, such connections need to be presented with authenticity in order to strengthen brand-user associations and possible purchase intentions (Arya et al., 2019; Kucharska et al., 2020). Furthermore, placing a sports brand in a program about sports increases perceptions of congruency for that sports brand as compared to placing it in a non-sports program (Cheng & Kim, 2016). This leads to the first hypothesis:

H1: Reality versus fiction has a positive effect on (a) brand-user imagery fit and (b) the intention to watch a Chicago Bulls game.

## Entertainment

Entertainment enables viewers to passively absorb what is shown in a program and allows them to enjoy the program, because it is amusing or exciting and, therefore, arouses certain emotions (Pine & Gilmore, 2020). Schmitt (1999) adds that "people deliver feelings at point of consumption and use communication to "frame" the consumption experience" (p. 219). In "The Last Dance", the protagonists, i.e. athletes, managers, and other staff members, arouse feelings of the viewing audience by telling compelling stories, while the producers of the show frame the narrative accordingly. Sports documentaries often evoke strong emotions in viewers by capturing the triumphs, challenges, and personal stories of athletes and others involved (Vogan, 2014). This leads to emotional engagement that plays a crucial role in creating a strong brand association (Schmitt, 1999). When viewers connect emotionally with the athletes and their achievements, this emotional engagement can transfer to the brands and products associated with them (Smilansky, 2017). With the above in mind, the next hypothesis is defined as follows:

H2: Entertainment has a positive effect on (a) brand-user imagery fit and (b) the intention to watch a Chicago Bulls game.

## Relaxation

Relaxation refers to the activity of calming down while watching a TV program or streaming a show after a possibly stressful day; an activity that Zillmann (1982) describes as "an antidote to the rousing "fight for survival" in society" (p. 53). Relaxing while watching a docuseries like "The Last Dance" can therefore serve as a form of escape, providing fans with an opportunity to temporarily detach from their daily routines and immerse themselves in the world of their beloved team and athletes (Trail, 2012). Viewers are less concerned with the outcome of the show, especially since it documents historical sporting events widely known to the public, than with the general entertainment quality (Ebbrecht, 2007; Vogan, 2014). Highlighting the emotional properties of the show might foster the intention of viewers to engage in more escapism and, thus, watch further episodes of the show (Hall et al., 2010). This would then strengthen their commitment and association to the show and its protagonists and possibly increase their intention to purchase a product directly or indirectly associated with the show (Sokolova & Kefi, 2020). Thus, the subsequent hypothesis is offered:

H3: Relaxation has a positive effect on (a) brand-user imagery fit and (b) the intention to watch a Chicago Bulls game.

## Habitual Pass Time

Habitual pass time can be defined as an activity that occupies a viewer's time when there is arguably nothing better to watch or do (Papacharissi & Mendelson, 2007). In the age of social media and online streaming services, the amount of online content about one's favorite team or athletes, especially if popular, can be perceived as close to infinite; moreover, access to their content is oftentimes free and easily available (Nielsen.com, 2022). Because of that, an occasional checking-in with their favorite club, athlete, or league on any online platform can contribute towards the fan experience and strengthen a fan's respective association with that sports brand; this may then reinforce the intention to watch a specific sports media product (Sutera, 2013). Anghelcev et al. (2021) offer an additional viewpoint by maintaining that "watching more episodes [of a show] per session makes the experience more enjoyable and facilitates the development of parasocial interactions with the shows' characters" (p. 142), possibly strengthening the association between viewers and the characters. Arguably, Netflix supported that notion by releasing the ten episodes of "The Last Dance" in installments of two episodes per week over the timespan of five weeks (ESPN Press Room, 2020). Therefore, the following can be hypothesized:

H4: Habitual pass time has a positive effect on (a) brand-user imagery fit and (b) the intention to watch a Chicago Bulls game.

## Companionship

Companionship is defined as the passive use of watching a show for the benefit of not feeling lonely despite the physical absence of other people, instead of watching it for its content (Perse & Ferguson,

1993). Papacharissi and Mendelson (2007) noted that viewers who perceive reality-based shows as more realistic than fictional and have a general affinity for such shows, are more inclined to watch them in order to satisfy their desire for companionship. From the perspective of OTT platforms, companionship can be a driving force behind the decision to abandon one media channel and switch to an alternative platform (Menon, 2022). Hence, it can be reasoned that social presence is an additional element to be considered in this regard, as it may offer viewers a convincing "degree to which users perceive one or many others as being present via [a] mediated interface" (Lim et al., 2015, p. 160), although their presence is solely imaginary. The immediacy and intimacy of such an imaginary presence can contribute to the sense of social connection by reducing the perceived psychological gap between viewers (Rettie, 2003). Arguably, this notion can also be applied to the viewing experience of a sports documentary series like "The Last Dance" (Sutera, 2013). Based upon this, the next hypothesis is defined as follows:

H5: Companionship has a positive effect on (a) brand-user imagery fit and (b) the intention to watch a Chicago Bulls game.

## Social Interaction

Trail (2012) defines social interaction as "the need to interact and socialize with others of like interests to achieve feelings that one is part of a group" (p. 4). Various studies have found social interaction to have a significant effect on the decision to watch a TV series (see Flayelle et al., 2018). Leslie-Walker and Mulvenna (2022) emphasize that engaging in social interactions during a sports media product cultivates favorable sentiments towards the overall experience. It could also be understood as to add social utility to the viewing experience by talking and spending time with others that watch the same show (Nabi et al., 2003). A further perspective can be observed with communal engagement through social TV, i.e. live interaction and communication on social media that takes place concurrently with a broadcasted program, where engaging with people who watch the same show is a significant motive behind continuing to watch that show (Lim et al., 2015). Thus, the following is hypothesized:

H6: Social interaction has a positive effect on (a) brand-user imagery fit and (b) the intention to watch a Chicago Bulls game.

## Voyeurism

A possible definition for voyeurism is the enjoyment derived from secretly observing other people engaged in sexual encounters or, in a broader sense, finding satisfaction in observing the private lives of others (Cambridge Dictionary, 2023). However, this research presents an explicit understanding of voyeurism that departs from the psychiatric definition focusing on sexual arousal from observing others during intimate moments. Instead, contemporary perspectives highlight voyeurism as a common pleasure derived from having figurative access to other people's otherwise inaccessible private lives, not necessarily of a sexual nature; furthermore, the prevalence of electronic media has contributed to a pervasive curiosity about the private lives of others, shaping contemporary society (Baruh, 2009). In the context of reality TV, Nabi et al. (2003) use the term voyeurism in a more colloquial way and consider viewers of such programs "to feel connected to others, to be surprised, to gain personal insight" (p. 312). With that in mind, the subsequent hypothesis is proposed:

H7: Voyeurism has a positive effect on (a) brand-user imagery fit and (b) the intention to watch a Chicago Bulls game.

## Brand-User Imagery Fit

The establishment of brand image and brand equity relies significantly on a collection of associations that shape the perception and significance of the brand for consumers, enabling product differentiation within the market (Keller, 1993). This sort of alignment and similarity between the conventional brand-user image and a consumer's self-concept can have a positive effect on their attitudes and purchasing behavior (Lee & Watkins, 2016). Miller and Mills (2012) define brand-user imagery fit as "a consumer's overall assessment of the compatibility or match between themselves and users of the brand" (p. 1475). This leads to query how the alignment between the interests of the viewers of "The Last Dance" and the sports brands portrayed in the show, e.g. the Chicago Bulls, the NBA, Michael Jordan, etc., can be influenced by uses and gratifications, as well as how that alignment can then motivate people to purchase a product from those portrayed brands, which, in the case of this study, is a Chicago Bulls game. Hence, the last hypothesis of this research is defined as follows:

H8: Brand-user imagery fit has a positive effect on the intention to watch a Chicago Bulls game.

## Conceptual Model

Building upon the preceding theoretical discussion, this study aims to analyze the impact that the uses and gratifications of watching the docuseries "The Last Dance" can have on the fit between brand imagery and user perception, as well as the intention to watch a Chicago Bulls game. To reiterate the guiding research question for this study: "How do uses and gratifications of watching the docuseries "The Last Dance" influence brand-user imagery fit of its viewers and the intention to watch a Chicago Bulls game?" A visualization of the hypothesized causal relationships is offered in Figure 1.

*Figure 1. Conceptual model*

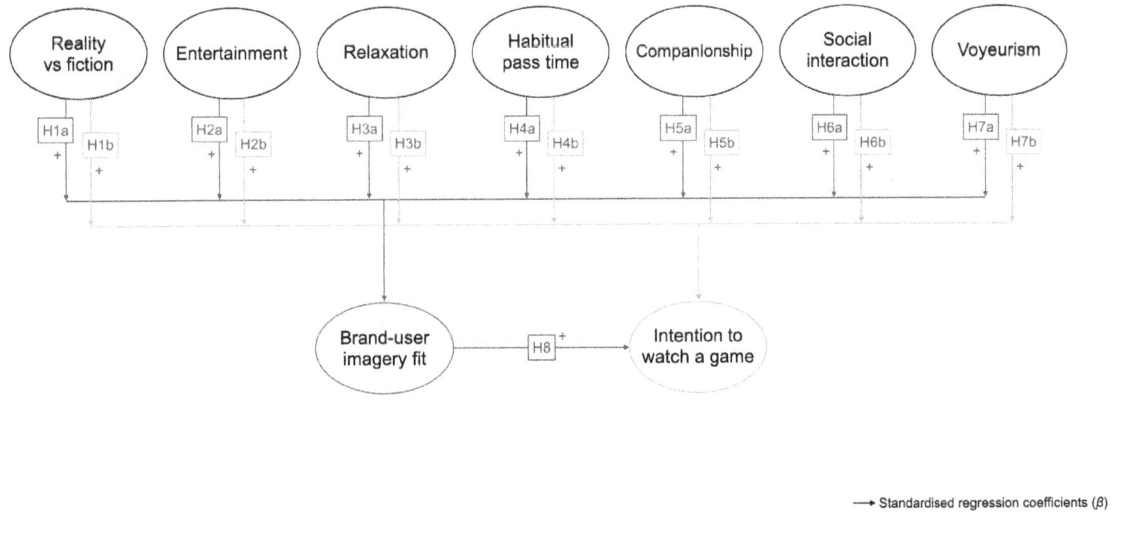

## METHODOLOGY

In order to examine causal relationships between uses and gratifications, brand-user imagery fit, and the intention to watch a Chicago Bulls game, as portrayed in Figure 1, this research follows a positivist approach. Primary data was collected from a convenience sample through a self-administered online questionnaire that was distributed across various digital communication channels. The target audience included people that had been watching "The Last Dance". Responses were collected between Wednesday, 6 May and Friday, 8 May 2020. This time period was chosen as it coincided with the release of the seventh episode of the show, which is the approximate middle of the season. "The Last Dance" had been running for about 3 weeks and had not yet reached its finale. Hence, feelings about watching the show were still fresh in the minds of respective respondents; also, viewers would still not know what would be shown in the final episode.

### Measurement Tool

The self-administered online questionnaire included 26 closed questionnaire items to be answered through a 5-point Likert-scale. The items utilized to assess the uses and gratifications of watching "The Last Dance" and their impact on brand-user imagery fit and viewing intention for a Chicago Bulls game were adapted to the context at hand from Papacharissi and Mendelson (2007), Trail (2012), Nabi et al. (2003), and Lee and Watkins (2016). Table 1 provides an overview of the utilized questionnaire items. The questionnaire included four more questions, asking for respondents' gender, age, location, and most-used streaming/OTT platform for sports-related content. The collected data was then analyzed via EFA, confirmatory factor analysis (CFA), and structural equation modeling (SEM) in IBM SPSS and Amos.

## RESULTS

### Respondents of the Study

The sample consists of a total of 442 respondents, split into 399 male respondents (90.3%) and 43 female respondents (9.7%). The largest age-group is represented in the age-bracket 35 to 44 with 178 respondents (40.3%); 97 respondents were recorded in the age-bracket 25 to 34 (21.9%), followed by 83 respondents in the age-bracket 45 to 54 (18.8%), 56 participants in the age-bracket 18 to 24 (12.6%), and 18 people in the age-bracket 55 to 64 (4.1%); 7 respondents are 17 years of age or younger (1.6%), and 3 respondents are older than 65 (0.7%). 400 respondents represent viewers from Western Europe (90.5%) with an adequate dispersion across the region; 16 respondents live in Eastern Europe (3.6%); the rest of the sample, 26 respondents, are spread across the rest of the world (5.9%). When asked which streaming/OTT platform they spend most time on watching sports-related content, almost half of the sample (219 respondents; 49.5%) said Netflix, followed by YouTube with 74 respondents (16.7%). The rest of the sample recorded a variety of other platforms as their most-used for sports-related video content. In summary, this convenience sample represents a male perspective, with Millennials and Generation Xers from Western Europe representing a great majority of the sample. Table 2 provides a respective overview of the characteristics of the sample.

*Table 1. Questionnaire items*

| I watch "The Last Dance" (because)... |
|---|
| **Factor 1: Reality versus fiction** |
| Characters are more interesting than those in fiction programs. |
| It is more enjoyable than fiction programming. |
| **Factor 2: Entertainment** |
| I just like to watch. |
| It amuses me. |
| It entertains me. |
| It is exciting. |
| **Factor 3: Relaxation** |
| It relaxes me. |
| It allows me to unwind. |
| It is a pleasant rest. |
| **Factor 4: Habitual pass time** |
| Just because it's there. |
| When I have nothing better to do. |
| It gives me something to do to occupy my time. |
| **Factor 5: Companionship** |
| It makes me feel less lonely. |
| So I won't have to be alone. |
| When there's no one else to talk or be with. |
| **Factor 6: Social interaction** |
| I enjoy socializing with others online or offline while watching the show. |
| I enjoy interacting with others online or offline while watching the show. |
| I enjoy talking to other people online or offline about the show. |
| **Factor 7: Voyeurism** |
| I feel like I'm getting a peek into other people's lives. |
| I get to see a side of people that I wouldn't normally get to see. |
| You never know what you might see. |
| **Factor 8: Brand-user imagery fit** |
| I am very much like the typical Chicago Bulls fan. |
| I am similar to people who like the Chicago Bulls. |
| I identify with people who prefer the Chicago Bulls to other teams. |
| **Factor 9: Viewing intention** |
| My willingness to watch a Chicago Bulls game would be high, if I were looking to watch an NBA game. |
| If I were to buy tickets or pay-per-view for an NBA game, the likelihood I would pay to watch the Chicago Bulls is high. |

*Table 2. Characteristics of the sample*

| | N | % |
|---|---|---|
| **Gender** | | |
| Female | 43 | 9.7 |
| Male | 399 | 90.3 |
| **Age** | | |
| 0-17 | 7 | 1.6 |
| 18-24 | 56 | 12.6 |
| 25-34 | 97 | 21.9 |
| 35-44 | 178 | 40.3 |
| 45-54 | 83 | 18.8 |
| 55-64 | 18 | 4.1 |
| 65+ | 3 | 0.7 |
| **Geographical areas** | | |
| Western Europe | 400 | 90.5 |
| Eastern Europe | 16 | 3.6 |
| Rest of the world | 26 | 5.9 |
| **Most-used streaming/OTT platform** | | |
| Netflix | 219 | 49.5 |
| YouTube | 74 | 16.7 |
| Other | 46 | 10.4 |
| Sky Sports | 42 | 9.5 |
| NBA App | 23 | 5.2 |
| ESPN+ | 14 | 3.2 |
| DAZN | 10 | 2.3 |
| Amazon Prime | 8 | 1.8 |
| Rest of the list | 6 | 1.4 |

*Note.* N = 442.

## Reliability, Validity, and Model Fit

Internal consistency of the applied scales was measured through Cronbach's Alpha (α) with construct reliability (CR) and average variance extracted (AVE) in support (Saunders et al., 2007). Despite a few values falling slightly below the recommended thresholds, the constructs are deemed acceptable. A CFA measured the fit of the collected data with the defined conceptual model, resulting in a good model fit with $\chi2/df = 1.860$, RMSEA = .044, CFI = .959, NFI = .916, TLI = .948, and GFI = .927 (de Mesquita & Kostelijk, 2021). The computed coefficient of multiple determination ($R^2$) values support that suggestion (Cohen, 1992; Sarstedt & Mooi, 2014). An overview of the above-mentioned metrics is provided in Table 3.

*Table 3. Reliability and validity*

| | α | CR | AVE | $R^2$ | M | SD |
|---|---|---|---|---|---|---|
| **RVF** | .742 | .724 | .570 | - | 4.062 | .869 |
| **ENT** | .635 | .644 | .324 | - | 4.201 | .649 |
| **REL** | .839 | .805 | .579 | - | 3.582 | .907 |
| **HPT** | .805 | .789 | .560 | - | 1.841 | .900 |
| **COM** | .871 | .837 | .635 | - | 1.694 | .919 |
| **SOI** | .854 | .859 | .680 | - | 2.666 | 1.211 |
| **VOY** | .799 | .793 | .564 | - | 3.477 | 1.041 |
| **BUI** | .858 | .803 | .576 | .151 | 2.373 | 1.134 |
| **INT** | .847 | .641 | .471 | .455 | 2.578 | 1.241 |

*Note 1.* α: Cronbach's Alpha, CR: Construct reliability, AVE: Average variance extracted, $R^2$: Coefficient of multiple determination, M: Mean, SD: Standard deviation, RVF: Reality versus fiction, ENT: Entertainment, REL: Relaxation, HPT: Habitual pass time, COM: Companionship, SOI: Social interaction, VOY: Voyeurism, BUI: Brand-user imagery fit, INT: Intention to watch a Chicago Bulls game.

*Note 2.* All means are based on a 5-point Likert-scale.

## Hypotheses Testing

Given the satisfactory reliability and validity of the primary data collected, along with an acceptable model fit, the study was able to proceed with hypotheses testing. This involved employing the respective SEM in order to examine the hypotheses outlined in the theory section. A summary of the tested hypotheses is provided in Table 4, including standardized regression weights ($\beta$), significance levels

*Table 4. Hypotheses testing results*

| Hypothesis no. | IV | DV | $\beta$ | $p$ | *result* |
|---|---|---|---|---|---|
| **H1a** | **Reality vs fiction** | **Brand-user imagery fit** | **.171** | **.007** | **supported** |
| H1b | Reality vs fiction | Intention to watch a Chicago Bulls game | .034 | .531 | not supported |
| H2a | Entertainment | Brand-user imagery fit | .063 | .387 | not supported |
| H2b | Entertainment | Intention to watch a Chicago Bulls game | -.024 | .697 | not supported |
| **H3a** | **Relaxation** | **Brand-user imagery fit** | **.134** | **.042** | **supported** |
| H3b | Relaxation | Intention to watch a Chicago Bulls game | .024 | .682 | not supported |
| H4a | Habitual pass time | Brand-user imagery fit | .078 | .227 | not supported |
| H4b | Habitual pass time | Intention to watch a Chicago Bulls game | .063 | .264 | not supported |
| H5a | Companionship | Brand-user imagery fit | .104 | .098 | not supported |
| H5b | Companionship | Intention to watch a Chicago Bulls game | .063 | .812 | not supported |
| **H6a** | **Social interaction** | **Brand-user imagery fit** | **.145** | **.006** | **supported** |
| H6b | Social interaction | Intention to watch a Chicago Bulls game | .055 | .228 | not supported |
| H7a | Voyeurism | Brand-user imagery fit | -.017 | .770 | not supported |
| H7b | Voyeurism | Intention to watch a Chicago Bulls game | .066 | .196 | not supported |
| **H8** | **Brand-user imagery fit** | **Intention to watch a Chicago Bulls game** | **.623** | *** | **supported** |

*Note.* ***$p < .001$. IV: Independent variable, DV: Dependent variable, $p$: Strength of statistical significance, $\beta$: Standardized regression coefficient.

($p$), and the final regression results. Results show that H1a, H3a, H6a, and H8 are supported; all other hypotheses are not supported. A comprehensive analysis and respective discussion on the findings will be offered in the next section of this chapter.

## DISCUSSION

As reported in Table 4 and displayed in Figure 2, the uses and gratifications variables reality versus fiction (H1a, $\beta = .171$, $p < .01$), relaxation (H3a, $\beta = .134$, $p < .05$), and social interaction (H6a, $\beta = .145$, $p < .01$) have a significant and positive influence on brand-user imagery fit; all other uses and gratifications variables record no significant influence on brand-user imagery fit. Furthermore, all uses and gratifications variables record no significant direct influence on the intention to watch a Chicago Bulls game after having seen the docuseries "The Last Dance". However, brand-user imagery fit shows a significant and positive influence on the intention to watch a Chicago Bulls game (H8, $\beta = .623$, $p < .001$). This suggests that brand-user imagery fit functions as a mediator between the significant uses and gratifications of watching the docuseries and the viewing intention for a Chicago Bulls game. The subsequent discourse expands upon the presented results by providing theoretical and managerial implications.

### Reality vs. Fiction

The realistic, non-fictional nature of the story and the characters presented in "The Last Dance" positively influence the alignment and similarity between a viewer's concept of self and the conventional brand-user image (H1a; see Figure 2). Furthermore, the documentary format of the show is the second-strongest reason for watching the show (see M in Table 3). This is supported by Vogan (2014), who explains that sports documentaries have the unique ability to capture real-life stories and characters, offering viewers

*Figure 2. Results of SEM*

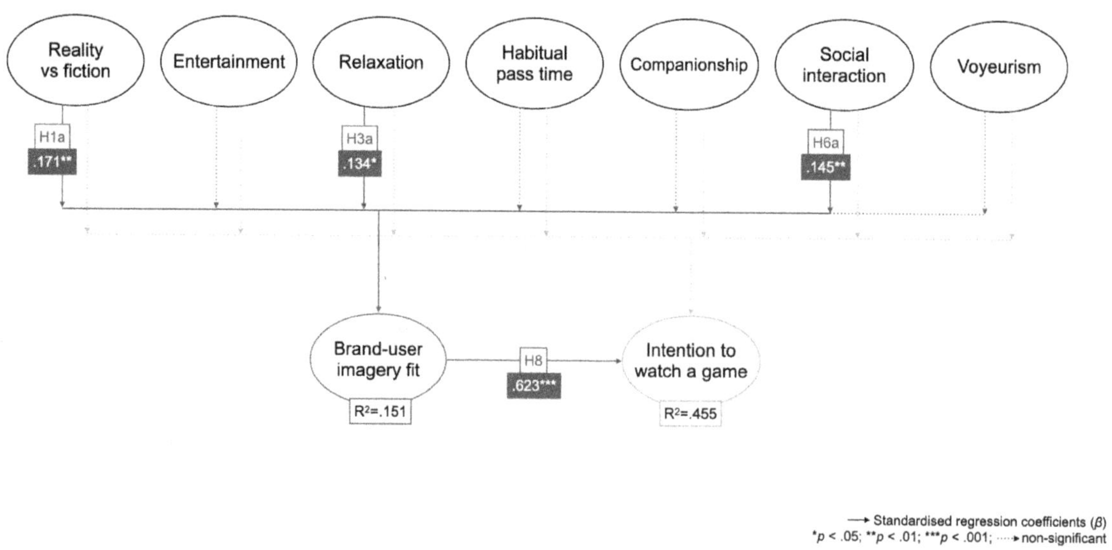

an authentic and unfiltered glimpse into the world of sports. It can further be asserted that building and strengthening the above-mentioned viewer-protagonist connection upon a holistic viewing experience can evoke emotions, inspire empathy, and foster a sense of relatability (Schmitt, 1999; Pine & Gilmore, 2020). The realistic nature of "The Last Dance" lends credibility and authenticity to the events and experiences depicted. When viewers know they are watching real struggles and real triumphs of Michael Jordan and his teammates, it adds a layer of authenticity that enhances their emotional investment in the story. Unlike fictional narratives, sports documentaries draw from actual events and real-life individuals, showcasing their challenges, sacrifices, and achievements (Paige, 2005). This realism allows viewers to witness the human drama unfold firsthand, creating a sense of relevance (Boyle, 2020).

The portrayal of authentic characters in sports documentaries can foster a connection with viewers (McDonald, 2007). "The Last Dance" delves deep into the lives and personalities of various protagonists of the Chicago Bulls organization, as well as other athletes and individuals shown in the docuseries. By presenting their vulnerabilities, aspirations, and personal journeys, the documentary humanizes these characters, making them relatable and multidimensional (Grainger et al., 2006). Viewers may see themselves reflected in the hopes, dreams, and struggles of these real-life individuals, forging a bond that transcends the screen (O'Connor & Boyle, 1993). This is further supported with the results of this study, which find a mediated connection to the viewing intention for watching a Chicago Bulls game (see Figure 2).

"The Last Dance" also explores universal themes found in other sports documentaries, such as determination and resilience (Sarkar & Soni, 2023). The realistic representation of these themes can resonate with viewers, as they can draw parallels between the challenges faced by the documentary's subjects and their own life experiences (O'Connor & Boyle, 1993). Watching Michael Jordan and his teammates push their limits, endure setbacks, and ultimately triumph serves as an inspiration, motivating viewers to persevere in their own pursuits in life. The raw emotions and real personal narratives presented in "The Last Dance" evoke empathy, allowing viewers to feel a genuine connection with the characters and their stories (McDonald, 2007). Thus, the apparent reality offered in the documentary can strengthen the parasocial relationship perceived as friendship between viewers and protagonists, which reinforces their credibility and affects the purchase intention for a directly or indirectly promoted product, e.g., a Chicago Bulls game (Chung & Cho, 2014; Sokolova & Kefi, 2020).

## Relaxation

This study finds that watching "The Last Dance" in order to unwind and escape from daily routines can positively affect brand-user imagery fit and, hence, foster the viewing intention for a Chicago Bulls game (H3a; see Figure 2). Similarly, Won and Kitamura (2006) found a positive effect from escapism on the consumption intention for sports media products. Based upon that, it can be contended that while relaxation is the third most important motive for watching "The Last Dance" (see M in Table 3), the immersive nature of the series, combined with its captivating storytelling, allows viewers to form a strong emotional bond with the story and its characters. One significant reason is the show's focus on the iconic Michael Jordan and the legendary 1990s Chicago Bulls teams. By focusing on one of the most popular athletes of all time and a basketball team that achieved unprecedented success, the series can capture its viewers' attention and offer relaxing entertainment (Tefertiller & Sheehan, 2019; Iordache et al., 2023). Viewers who are fans of basketball, the Chicago Bulls, or Michael Jordan, may already have a pre-existing connection or affinity, making it easier for them to become emotionally invested in the story and, therefore, the show (Weinish & Lev, 2021).

Analyzing "The Last Dance" from an aesthetic perspective, it is notable that the show uses a combination of archival footage, interviews, and narration to tell a cohesive and compelling narrative. These components add a specific aesthetic of nostalgia to the narrative that recreate and, arguably, fictionalize to a greater or lesser degree the story told in the show by combining visualizations found in documentaries and feature films; that explicit aesthetic style offers viewers warrants of authenticity in an affective cinematographic environment that seeks to foster empathy with the protagonists of the show (Ebbrecht, 2007). Building upon the importance of empathy in documentary films, it can be argued that film as a narrative art form places particular importance on the concept of empathy due to its inherent connection to the way individuals interact with art and gain insight into the inner emotions and experiences of others, i.e. the show's characters (Stadler, 2017).

Moreover, "The Last Dance" weaves together the past and the present, while playing with respective aesthetics and stories to build suspense and keep viewers engaged. As viewers immerse themselves in the story of a reality-based program, they become emotionally invested in the outcomes, eagerly anticipating the next episode and the resolution of various plotlines (Beck et al., 2012). This investment deepens the connection viewers feel with the characters and their journey, heightening the emotional impact of the documentary, which can further lead towards enhancing the connection and resemblance between the typical brand-user image and a viewer's self-perception and, ultimately, inciting a certain behavior (Lee & Watkins, 2016), like watching a Chicago Bulls game, as in the case at hand.

## Social Interaction

Watching "The Last Dance" as a means of social interaction with like-minded others can positively influence the degree of alignment or compatibility between the show's portrayed image or identity and the characteristics, values, and self-perception of its viewers (H6a; see Figure 2). Although findings suggest that social interaction is not as strong a motivator for watching the show as reality versus fiction or relaxation (see M in Table 3), this shared viewing experience fosters a sense of community, enhances emotional engagement, and deepens the connection between viewers and the media product (Lim et al., 2015). The viewing experience with a group of like-minded individuals, who share a passion for a specific sport or athlete, can create a sense of camaraderie (Trail, 2012). The collective excitement and enthusiasm during moments of triumph or suspenseful scenes in "The Last Dance" heighten the emotional impact of the documentary. Viewers feed off each other's energy, engaging in discussions, debates, and reactions, which enriches their overall viewing experience; this also applies for social interaction via digital communication channels like social media (Sutera, 2013). The shared social interaction amplifies the connection viewers feel with the story and characters, as they become part of a community that is invested in the narrative (Papacharissi & Mendelson, 2007).

Discussing the documentary with others allows viewers to gain different perspectives, insights, and interpretations. Engaging in conversations about the series deepens their understanding of the story and characters, leading to a more profound connection (Nabi et al., 2003). As viewers exchange thoughts and opinions, they further invest in the narrative, creating a collective bond that extends beyond the screen (Sutera, 2013). This social interaction provides an opportunity to explore various themes, motivations, and emotions portrayed in the documentary, thereby enhancing the connection viewers feel with the story, its characters, and with each other, motivating viewers to keep watching the series (Anghelcev et al., 2021).

Similarly, watching a show with peers, be it in the same room or remotely via social TV, can evoke nostalgia and spark memories related to the events depicted in the show (Hahm et al., 2021). Viewers of "The Last Dance" may recall their own experiences as fans, reliving moments of excitement, disappointment, or admiration for the players and teams. Sharing these personal anecdotes with others who have similar memories, enhances the connection with the story, as viewers draw parallels between their own lives and the experiences portrayed in the documentary (O'Connor & Boyle, 1993). This shared sense of nostalgia and reminiscence strengthens the emotional bond among viewers and deepens their connection to the characters and their journey, while acting as a marketing tool (Cho, 2023).

Social interactions surrounding "The Last Dance" extend beyond immediate viewing groups. Online communities, forums, and social media platforms provide platforms for viewers to engage with a wider audience of like-minded others (Newman et al., 2013). These digital spaces allow viewers to connect with people from diverse backgrounds and locations, sharing their thoughts, analyses, and emotions related to the show (Sutera, 2013). The ability to connect with a larger community of fans creates a sense of belonging and further solidifies the connection viewers feel with the story and its characters, making it a memorable and meaningful viewing experience (Lim et al., 2015).

## Brand-User Imagery Fit

The level of agreement between what is depicted in "The Last Dance" and the qualities, values, and self-image of its viewers has a positive influence on their intention to watch a Chicago Bulls game (H8; see Figure 2). Similarly, Lee and Watkins (2016) report a significant and positive effect between brand-user imagery fit and the purchase intention of YouTube viewers for luxury brands promoted by vloggers. It can be reasoned that the causal relationship between brand-user imagery fit and the viewing intention for a Chicago Bulls game may find its significance in "The Last Dance's" intimate and behind-the-scenes look of athletes and sports brands that viewers are familiar with and interested in knowing more about while being entertained (Nabi et al., 2003; Vogan, 2014). The documentary highlights the unwavering commitment to success of Michael Jordan and his teammates, their fierce competitiveness, and their ability to overcome challenges. Viewers who value determination, perseverance, and a winning mindset are, therefore, likely to find a strong alignment with the qualities portrayed in "The Last Dance". Consequently, their intention to watch a Chicago Bulls game can be heightened as they might seek to witness those same qualities displayed on the court (cf. Mancini, 2013).

Moreover, "The Last Dance" offers an in-depth analysis of the cultural impact that Michael Jordan and the Chicago Bulls had on a global audience and examines the widespread fan following they garnered during their dominant years (NPR, 2020). This is highlighted by the show's showcase of the protagonists' ability to captivate and unite people from diverse backgrounds and walks of life. Viewers who identify themselves as fans of Michael Jordan or the Chicago Bulls or resonate with the sense of community, camaraderie, and pride associated with them, are likely to be influenced by the documentary (Lee & Watkins, 2016). These viewers may also feel a sense of nostalgia (Weinish & Lev, 2021) or a desire to be part of that collective experience, motivating them to watch a Chicago Bulls game.

Lastly, "The Last Dance" explores the broader social and historical context of the characters' success (Bryant, 2020). Delving into the intense rivalry with other NBA teams, such as the Detroit Pistons and the Utah Jazz, adds an element of drama and excitement to the plot. Viewers who enjoy compelling narratives, rivalries, and the emotional rollercoaster of sports may find the portrayal of these aspects captivating (Vogan, 2014). Subsequently, this kind of engagement with the storyline can enhance their intention to watch a Chicago Bulls game, as viewers might anticipate the thrill and intensity of the rivalry.

## CONCLUSION

This chapter answers the question, "How do uses and gratifications of watching the docuseries "The Last Dance" influence brand-user imagery fit of its viewers and the intention to watch a Chicago Bulls game?" The findings of this research indicate that the realistic depiction of the story and its characters, the sense of relaxation derived from watching the show, and social interaction with other viewers have an impact on the viewing intention for a Chicago Bulls game when mediated by the association and similarity between the established brand-user image and a consumer's self-concept. The study found no direct influence from uses and gratification on said viewing intention.

In order to enhance the intention to watch a related sports media product, such as a Chicago Bulls game, the company and people responsible for influencing that behavior should focus on strengthening the above-mentioned elements and highlight them in promotional material and communication. For instance, the realistic portrayal of events and experiences in "The Last Dance", as well as the authentic depiction of the lives and personalities of the Chicago Bulls organization and other athletes, fosters relatability and empathy (Pine & Gilmore, 2020). By showcasing real struggles, triumphs, and universal themes like determination and resilience, the documentary inspires viewers and allows them to draw parallels with their own lives (Paige, 2005).

Moreover, the escape from daily routines and immersion in the entertaining and inspiring world of Michael Jordan and the Chicago Bulls allows viewers to relax while being entertained, which fosters their emotional investment in the journey and accomplishments of the protagonists (Weinish & Lev, 2021). Additionally, combining visualizations from documentaries (e.g. archival footage and present-day interviews) and feature film (e.g. dramatic screenplays and bombastic soundscapes), in order to recreate and fictionalize the narrative, can foster credibility in an emotional cinematographic aesthetic (Ebbrecht, 2007) and, thus, strengthen a desired behavioral engagement with the show or with a related media product.

From a social interaction perspective, it can be concluded that watching "The Last Dance" with like-minded others can enhance the connection viewers feel with the story and characters portrayed (Vogan, 2014). By sharing a common interest in sports and a passion for basketball, viewers can engage in lively discussions, debates, and conversations about the series, deepening their emotional investment in the narrative (Sutera, 2013). This collective experience fosters a sense of camaraderie, belonging, and shared understanding, which further strengthens the connection viewers have with the story and its protagonists (Trail, 2012; Iordache et al., 2023). Consequently, it can influence the intention to keep watching the show or start watching a related media product (Lim et al., 2015; Anghelcev et al., 2021).

Finally, it can be maintained that "The Last Dance's" reality-based story and characters, the desire to escape from daily routines and relax, and the need for social interaction with like-minded other viewers should be enhanced with qualities, values, and self-image that match the perception of the individual viewer and other viewers. This could be implemented by portraying the determination and winning mindset of the protagonists. Additionally, the cultural impact of Michael Jordan and his teammates, their ability to unite diverse audiences, and the sense of community associated with them can affect viewers who identify as fans or seek a nostalgic collective experience (NPR, 2020; Weinish & Lev, 2021). This, ultimately, influences the intention to watch a Chicago Bulls game.

## LIMITATIONS AND FUTURE RESEARCH

A main limitation of the presented study includes the sample reflecting a predominantly male perspective (399 respondents; 90.3%). Future studies should try to gather data from a more balanced sample in regard to gender, as female and non-binary viewers might draw satisfaction from different uses and gratifications in comparison to male viewers. Moreover, the sample reflects the perception of Western European viewers (400 respondents; 90.5%). Although this might not necessarily be a significant limitation, the findings of this study cannot be generalized across other geographical regions. A crucial consideration of this limitation is the fact that the Chicago Bulls are an American sports brand and might be differently appealing to American audiences as compared to European audiences. Future research could assess the proposed conceptual with samples from different geographical regions and compare resulting findings. Furthermore, a specific examination of the social relevance of "The Last Dance" on different age groups or social classes would be beneficial.

## REFERENCES

Afilipoaie, A., Iordache, C., & Raats, T. (2021). The 'Netflix Original' and what it means for the production of European television content. *Critical Studies in Television*, *16*(3), 304–325. doi:10.1177/17496020211023318

Anghelcev, G., Sar, S., Martin, J. D., & Moultrie, J. L. (2021). Binge-watching serial video content: Exploring the subjective phenomenology of the binge-watching experience. *Mass Communication & Society*, *24*(1), 130–154. doi:10.1080/15205436.2020.1811346

Arya, V., Verma, H., Sethi, D., & Agarwal, R. (2019). Brand authenticity and brand attachment: How online communities built on social networking vehicles moderate the consumers' brand attachment. *IIM Kozhikode Society & Management Review*, *8*(2), 87–103. doi:10.1177/2277975219825508

Balbi, G., & Magaudda, P. (2018). *A history of digital media: An intermedia and global perspective*. Routledge., doi:10.4324/9781315209630

Baruh, L. (2009). Publicized intimacies on reality television: An analysis of voyeuristic content and its contribution to the appeal of reality programming. *Journal of Broadcasting & Electronic Media*, *53*(2), 190–210. doi:10.1080/08838150902907678

Beck, D., Hellmueller, L. C., & Aeschbacher, N. (2012). Factual entertainment and reality TV. *Communication Research Trends*, *31*(2), 4–27.

Boyle, R. (2020). A new Golden Age? In R. Steen, J. Novick, & H. Richards (Eds.), *Routledge Handbook of Sports Journalism* (pp. 355–358). Routledge., doi:10.4324/9781315616704-33

Bryant, H. (2020). What 'The Last Dance' reveals about Michael Jordan's legacy. https://www.espn.com/nba/story/_/id/29365615/what-last-dance-reveals-michael-jordan-legacy

Cambridge Dictionary. (2023). *Voyeurism*. Cambridge Press. https://dictionary.cambridge.org/dictionary/english/voyeurism

Cheng, L., & Kim, S. (2016). The relationship between brand-program congruency and purchase intention. *International Journal of Applied Sports Sciences : IJASS*, *28*(2), 123–136. doi:10.24985/ijass.2016.28.2.123

Cho, H. (2023). Nostalgia in sport and leisure. *Current Opinion in Psychology*, *49*, 101551. Advance online publication. doi:10.1016/j.copsyc.2022.101551 PMID:36702009

Cho, H., Pyun, D. Y., & Lim, S. H. (2019). Enhancing the stability of the leisure nostalgia scale: Assessment of the measurement invariance between two different leisure populations. *Leisure Studies*, *39*(2), 294–303. doi:10.1080/02614367.2019.1653356

Chung, S., & Cho, H. (2014). Parasocial relationship via reality TV and social media: its implications for celebrity endorsement. In *Proceedings of the ACM International Conference on Interactive Experiences for TV and Online Video* (pp. 47–54). ACM. 10.1145/2602299.2602306

Diggs-Brown, B. (2011). *Strategic public relations: An audience-focused approach*. Cengage Learning.

Ebbrecht, T. (2007). Docudramatizing history on TV: German and British docudrama and historical event television in the memorial year 2005. *European Journal of Cultural Studies*, *10*(1), 35–53. doi:10.1177/1367549407072969

ESPN. (2023). *The Last Dance: The untold story of Michael Jordan's Chicago Bulls*. ESPN. https://www.espn.com/nba/story/_/id/28973557/the-last-dance-untold-story-michael-jordan-chicago-bulls

ESPN Press Room. (2020). *ESPN and Netflix set new April 19 premiere date for highly anticipated documentary series "The Last Dance"*. ESPN. https://espnpressroom.com/us/press-releases/2020/03/espn-and-netflix-set-new-april-19-premiere-date-for-highly-anticipated-documentary-series-the-last-dance/

Fetchko, M. J., Roy, D. P., & Clow, K. E. (2018). *Sports marketing*. Routledge. doi:10.4324/9781315175614

Flayelle, M., Canale, N., Vögele, C., Karila, L., Maurage, P., & Billieux, J. (2019). Assessing binge-watching behaviors: Development and validation of the "Watching TV Series Motives" and "Binge-watching Engagement and Symptoms" questionnaires. *Computers in Human Behavior*, *90*, 26–36. doi:10.1016/j.chb.2018.08.022

Gantz, W. (1981). An exploration of viewing motives and behaviors associated with television sports. *Journal of Broadcasting*, *25*(3), 263–275. doi:10.1080/08838158109386450

Goldsmith, B. (2013). "SportTV": The legacies and power of television. In B. Hutchins & D. Rowe (Eds.), *Digital media sport: Technology, power and culture in the network society* (pp. 52–65). Routledge. doi:10.4324/9780203382851

Grainger, A., Newman, J. I., & Andrews, D. L. (2006). Sport, the media, and the construction of race. In A. A. Raney & J. Bryant (Eds.), *Handbook of Sports and Media* (pp. 482–505). Lawrence Erlbaum Associates. doi:10.4324/9780203873670.ch27

Gupta, G., & Singharia, K. (2021). Consumption of OTT media streaming in COVID-19 lockdown: Insights from PLS analysis. *Vision (Basel)*, *25*(1), 36–46. doi:10.1177/0972262921989118

Hahm, J., Kang, T. A., & Matsuoka, H. (2021). Understanding the relationship between past experience of a sports mega-event and current spectatorship: The mediating role of nostalgia. *Sustainability (Basel)*, *13*(6), 3504. doi:10.3390/su13063504

Hall, J., O'Mahony, B., & Vieceli, J. (2010). An empirical model of attendance factors at major sporting events. *International Journal of Hospitality Management*, *29*(2), 328–334. doi:10.1016/j.ijhm.2009.10.011

Hipke, M., & Hachtmann, F. (2014). Game changer: A case study of social-media strategy in Big Ten athletic departments. *International Journal of Sport Communication*, *7*(4), 516–532. doi:10.1123/IJSC.2014-0022

Iordache, C., Raats, T., & Mombaerts, S. (2023). The Netflix original documentary, explained: Global investment patterns in documentary films and series. *Studies in Documentary Film*, *17*(2), 151–171. doi:10.1080/17503280.2022.2109099

Juel, H. (2006). Defining documentary film. *POV A Danish Journal of Film Studies, 22*, 5–15.

Keller, K. L. (1993). Conceptualizing, measuring, and managing customer-based brand equity. *Journal of Marketing*, *57*(1), 1–22. doi:10.1177/002224299305700101

Kucharska, W., Confente, I., & Brunetti, F. (2020). The power of personal brand authenticity and identification: Top celebrity players' contribution to loyalty toward football. *Journal of Product and Brand Management*, *29*(6), 815–830. doi:10.1108/JPBM-02-2019-2241

Lee, J., & Kang, J. (2022). Sport brand authenticity: Scale development and validation. *Korean Journal of Sport Science*, *33*(4), 659–673. doi:10.24985/kjss.2022.33.4.659

Lee, J. E., & Watkins, B. (2016). YouTube vloggers' influence on consumer luxury brand perceptions and intentions. *Journal of Business Research*, *69*(12), 5753–5760. doi:10.1016/j.jbusres.2016.04.171

Leslie-Walker, A., & Mulvenna, C. (2022). The Football Association's Women's Super League and female soccer fans: Fan engagement and the importance of supporter clubs. *Soccer and Society*, *23*(3), 314–327. doi:10.1080/14660970.2022.2037218

Lim, J. S., Hwang, Y. C., Kim, S., & Biocca, F. A. (2015). How social media engagement leads to sports channel loyalty: Mediating roles of social presence and channel commitment. *Computers in Human Behavior*, *46*, 158–167. doi:10.1016/j.chb.2015.01.013

Mancini, G. A. (2013). Can the effects of brand user imagery congruence be important if the brand is not as well known? *Journal of International Management Studies*, *13*(4), 85–90. doi:10.18374/JIMS-13-4.10

Menon, D. (2022). Purchase and continuation intentions of over-the-top (OTT) video streaming platform subscriptions: A uses and gratification theory perspective. *Telematics and Informatics Reports*, *5*, 100006. Advance online publication. doi:10.1016/j.teler.2022.100006

Nabi, R. L., Biely, E. N., Morgan, S. J., & Stitt, C. R. (2003). Reality-based television programming and the psychology of its appeal. *Media Psychology*, *5*(4), 303–330. doi:10.1207/S1532785XMEP0504_01

Newman, T., Peck, J., Harris, C., & Wilhide, B. (2013). *Social media in sport marketing*. Holcomb Hathaway Publishers., doi:10.4324/9781315212975

Nielsen.com. (2022). *Sports fandom is increasing, powered by new digital platforms, global report finds*. Nielsen. https://www.nielsen.com/news-center/2022/sports-fandom-is-increasing-powered-by-new-digital-platforms-global-report-finds/

NPR. (2020). *'The Last Dance' tries to cut through all the Bulls*. NPR. https://www.npr.org/2020/04/30/848229079/the-last-dance-tries-to-cut-through-all-the-bulls

O'Connor, B., & Boyle, R. (1993). Dallas with balls: Televized sport, soap opera and male and female pleasures. *Leisure Studies*, *12*(2), 107–119. doi:10.1080/02614369300390101

Paige, T. (2005). Golden age is now for sports documentaries: Genre gains popularity among programmers. *Multichannel News*, *26*(33), 16.

Papacharissi, Z., & Mendelson, A. L. (2007). An exploratory study of reality appeal: Uses and gratifications of reality TV shows. *Journal of Broadcasting & Electronic Media*, *51*(2), 355–370. doi:10.1080/08838150701307152

Perse, E. M., & Ferguson, D. A. (1993). The impact of the newer television technologies on television satisfaction. *The Journalism Quarterly*, *70*(4), 843–853. doi:10.1177/107769909307000410

Pettersson, R. (1993). Visual information. *Educational Technology*.

Pew Research Center. (2019). *Defining generations: Where Millennials end and Generation Z begins*. Pew Research. https://www.pewresearch.org/short-reads/2019/01/17/where-millennials-end-and-generation-z-begins/

Pine, B. J., & Gilmore, J. H. (2020). *The experience economy: Competing for customer time, attention, and money*. Harvard Business Review Press.

Reams, M. (2020). *Nike and Jordan logos appear over 100 times in a single episode of The Last Dance*. AdWeek. https://www.adweek.com/brand-marketing/nike-and-jordan-logos-appeared-over-100-times-in-one-episode-of-the-last-dance/

Rein, I., Kotler, P., & Shields, B. (2006). *The elusive fan*. McGraw-Hill.

Reiss, S. (2019). *Sport, Representation and culture in the modern world, 1920-2020*. Faculty Research and Creative Activities Symposium, Northeastern Illinois University. https://neiudc.neiu.edu/cgi/viewcontent.cgi?article=1095&context=frcas

Reiss, S., & Wiltz, J. (2004). Why people watch reality TV. *Media Psychology*, *6*(4), 363–378. doi:10.1207/s1532785xmep0604_3

Rettie, R. (2003). *Connectedness, awareness, and social presence*. Paper presented at the 6th International Presence Workshop, Aalborg, Denmark.

Rubin, A. (1983). Television uses and gratifications: The interactions of viewing patterns and motivations. *Journal of Broadcasting*, *27*(1), 37–51. doi:10.1080/08838158309386471

Sarkar, S., & Soni, M. J. (2023). Is Reality Less Real? Proliferation of Sports Documentaries in a Locked-Down World. In B. Basu, M. Desbordes, & S. Sarkar (Eds.), *Sports Management in an Uncertain Environment* (pp. 221–253). Springer. doi:10.1007/978-981-19-7010-8_10

Schmitt, B. (1999). *Experiential marketing: How to get customers to sense, feel, think, act, relate to your company and brands*. Free Press. doi:10.1362/026725799784870496

Smilansky, S. (2017). *Experiential marketing: A practical guide to interactive brand experiences*. Kogan Page Publishers.

Sokolova, K., & Kefi, H. (2020). Instagram and YouTube bloggers promote it, why should I buy? How credibility and parasocial interaction influence purchase intentions. *Journal of Retailing and Consumer Services*, *53*, 101742. doi:10.1016/j.jretconser.2019.01.011

Stadler, J. (2017). Empathy in Film. In H. Maibom (Ed.), *The Routledge Handbook of Philosophy of Empathy* (pp. 317–326). Routledge. doi:10.4324/9781315282015-29

Sutera, D. (2013). *Sports Fans 2.0: How Fans Are Using Social Media to Get Closer to the Game*. Scarecrow Press.

Tefertiller, A., & Sheehan, K. (2019). TV in the streaming age: Motivations, behaviors, and satisfaction of post-network television. *Journal of Broadcasting & Electronic Media*, *63*(4), 595616. doi:10.1080/08838151.2019.1698233

Trail, G. (2012). *Manual for the Motivation Scale for Sport Consumption (MSSC)*. Sport Consumer Research Consultants LLC.

Vogan, T. (2014). *Keepers of the flame: NFL Films and the rise of sports media*. University of Illinois Press. doi:10.5406/illinois/9780252038389.001.0001

Wann, D. L., & James, J. D. (2019). *Sport fans: The psychology and social impact of fandom* (2nd ed.). Routledge.

Weinish, S., & Lev, A. (2021). Who leads The Last Dance? *International Review for the Sociology of Sport*, *56*(3), 436–441. doi:10.1177/1012690220969345

Won, J. U., & Kitamura, K. (2006). Motivational factors affecting sports consumption behavior of K-league and J-league spectators. *International Journal of Sport and Health Science*, *4*, 233–251. doi:10.5432/ijshs.4.233

## KEY TERMS AND DEFINITIONS

**Brand-User Imagery Fit:** The alignment and similarity between a viewer's concept of self and the conventional brand-user image.

**Documentary Film:** A non-fictional film that presents real events, people, or issues, typically with the purpose of informing, educating, or raising awareness.

**"The Last Dance":** A 10-episode sports documentary series released in 2020 on ESPN and Netflix that chronicles the final championship-winning season in 1997-1998 of the Chicago Bulls, an NBA franchise, and provides an in-depth look at the team's journey and the impact of Michael Jordan.

**Michael Jordan:** A retired professional basketball player widely regarded as one of the greatest athletes of all time, whose illustrious career includes winning six NBA championships with the Chicago Bulls and his groundbreaking partnership with Nike, resulting in the iconic Air Jordan sneaker line.

**Sports Documentary Series:** A serialized collection of non-fictional episodes that explore and chronicle various aspects of the world of sports, including athletes, teams, competitions, and related stories.

**Uses and Gratifications:** A communication theory that examines how individuals actively seek out and use media to fulfill specific needs, desires, and gratifications.

# Section 3

# Psychological and Social Implications

# Chapter 8
# Streaming Success:
## Examining the Impact of OTT Media on Content Production and Consumption of New Malayalam Cinema

**Sanjna Sebastian T.**
https://orcid.org/0000-0002-1722-5754
*Christ University, India*

## ABSTRACT

*Exploring the prosperous union of New Malayalam Cinema and OTT platforms, this chapter scrutinises the circumstances under which the former thrived on OTTs during the COVID-19 pandemic. It delves into Malayalee market trends, the content diversification incited by OTTs, and the expansion of New Malayalam Cinema's market. As it explores changing audience consumption patterns and the prospects of India's OTT landscape, it particularly accentuates Malayalam regional cinema. By providing examples of films that debuted on OTT platforms, such as C U Soon and Joji, the chapter offers an in-depth analysis, highlighting the success of this symbiotic relationship.*

## INTRODUCTION

The Over-the-top (OTT) media services, specifically the subscription video-on-demand (SVoD) services, serve as a reserve for the spontaneous requirements of media and entertainment. The function of supply met by the plethora of streaming platforms for the demand created by the evolving nature of the audiences and their new viewing practices has become an increasingly sought-after field of academic enquiry in recent years (VS, 2022). The introduction of 'new cinemas' from the regional film industries, tapping into the pan-Indian film market due to the advancement of digital technology and its reach, has established a cause-and-effect relationship between OTT platforms and the filmic content made available for subscription-based viewership.

In the case of post-2010s, New Malayalam cinema, often memorialised for its 'generational shift', the proliferation of video services on OTT platforms alongside its development marks a combination of

DOI: 10.4018/979-8-3693-3526-0.ch008

disruptions that serve as interventions in the markets for audiovisual content on a global scale (Fitzgerald, 2019). This also has had a direct relation to the exponential rise in regional cinemas in India due to the boost provided by OTT media as they have conquered new territories and expanded beyond their language markets by claiming a share of popular culture.[1]

Considering the COVID-19-induced nationwide lockdown in India as a watershed moment that marked the heightened visibility of the OTT media service in the Indian mediascape (Nayak & Biswal, 2021; Srivastav & Rai, 2021) and its correlation with the advances that New Malayalam cinema as the alternate popular gives visibility to overlaps and the essay seizes this opportunity to engage with the subject matter academically. These are visible in the newer and contemporary themes that came to light in the narratives (from the 2010s onwards and seen an accelerated growth rate during the pandemic) and in the kind of stylistic endeavours that have surfaced. Malayalam films available on OTT platforms necessitate a comprehensive examination of their themes, style and the revised dimensions of production, elucidating the platforms' transformative impact on filmmakers' creative agency, the filmmaking landscape and the ever-evolving dynamics between OTT media and New Malayalam Cinema.

The steady movement towards rural locations contradicts how this contemporary shift has been termed a new 'generational' shift.[2] It becomes post-Travancore in its treatment and retirement from the established Capital 'M' star-studded Malayalam film industry.[3] The generational shift of New Malayalam cinema, as reported by media, works on two levels: (1) it reflects upon the new crop of directors, which includes Dileesh Pothan, Mahesh C. Narayanan, Lijo Jose Pellissery, Zakariya Mohammed, Aashiq Abu, to name a few, and (2) the 'modern' nature themes that are explicitly highlighted within urban geographies.

Limiting the potential of New Malayalam cinema solely by understanding it as a generational shift would warrant the engagement in an incomplete exercise of understanding the motive of this project and essentialising it to a grand narrative of the loss of traditional structures of the established Malayalam cinema. Films in New Malayalam cinema have narrativised rural spaces, as seen in films such as *Angamaly Diaries* (2017), *Ee. Ma. Yau* (2018) and *Jallikattu* (2019) by Lijo Jose Pellissery, *Kammattipadam* (2018) by Rajeev Ravi, *Kumbhlangi Nights* (2019) and *Halal Love Story* (2020) by Zakariya Mohammed. "For the longest time, rural landscapes in Malayalam cinema have been sites of feudal revivalism and nostalgia, with stories often told exclusively from the perspectives of elite upper-caste feudal landlord figures" (P T, 2020). The capability of New Malayalam cinema beyond the generational shift, a shift that attempts to draw one's attention solely to a defamatory manner of a sudden jolting from the rural and traditional sensibilities of Kerala state, has to be evaluated as not limited to their intimate relationship to city spaces but also a construction of a region that retains its socio-politico-religious markers, steering away from the homogenous (P T, 2020).

Analysing the impact of OTT media on content production and consumption of New Malayalam cinema, one is made aware of the evolving dynamics between the two, one of which is a technological boom, and the latter is an ideological shift in the treatment of its themes, style and spaces. These become complementary in understanding the off-centre of new Malayalam cinema, synchronously greeted by the innovation of OTT and SVoD platforms.

Following this dominant strand, the present essay has explored the various streams of revenue available along with opportunities for monetisation available within the Malayalam OTT industry. It has attempted to study how diverse content creators within the filmic setup, such as the producers, writers, scriptwriters, actors, cinematographers, editors and other professionals actively producing audio-visual content can generate income through newer models. These include the subscription-based model and advertising (including pre-roll, mid-roll and post-roll ads and tactics of viewer engagement with ad-blocking provisions).

By understanding its working within a regional cinematic industry setting such as the Malayalam film industry, this essay has weighed the considerations of exclusivity between the two ends of the spectrum of reach and revenue. It casts light on the changing business landscape in the Malayalam film industry due to the advent of OTT by explaining the other avenues for generating income, including brand partnerships, sponsorship and offering premium, exclusive content.

Additionally, the essay has discussed the issue of market expansion and global reach. It focuses on how OTT platforms have made Malayalam content available to audiences worldwide, specifically the Malayali diaspora (Pillai, 2021). It studies the increasing appeal and admiration of Malayalam films internationally and discusses the marketing strategies utilised to promote and distribute the films. It does so by answering two anchoring research questions which address both the form and content of films from New Malayalam Cinema on OTT platforms:

1. How have the conditions within which New Malayalam Cinema's position as the contemporary alternative popular proven to be favourable in alignment with Malayalam cinema's success in the movement towards OTT platforms during the COVID-19 pandemic?
2. What thematic and stylistic trends have emerged due to the complementary advantages in terms of cost of production and budget and non-adherence to (previously conceived as normative) traditional structures of the Malayalam cinematic industry?

## LITERATURE REVIEW

Studying the impact of OTT platforms and their subsequent changes in media consumption and production raises questions of spectatorship, viewing experience, and the revised digital means of production, which become the preoccupations of both those engaged in generating the 'portable' filmic experience and scholars engaging with the material academically. The impact of OTT media on Malayalam cinema's content production and consumption can be examined from various perspectives. On the one hand, OTT media has enabled filmmakers to experiment with new genres, themes, styles, and formats, breaking away from the conventional norms and expectations of the mainstream industry. On the other hand, OTT media has also influenced the tastes and preferences of the audiences, who have become more discerning and demanding in their consumption patterns. In critically examining the implications and consequences of OTT media alongside the rise of New Malayalam cinema, a few themes come to the fore from the existing scholarly literature:

### Revised Viewing Experience of the Cinephile Audiences

Devadas Menon's "Purchase and continuation intentions of over-the-top (OTT) video streaming platform subscriptions: a uses and gratification theory perspective" (2022) explores the factors influencing OTT platform subscription and continuation intentions in India using the U&G theory. The study found that convenient navigability, binge-watching, and relaxation U&G are significant predictors of subscription intentions, while entertainment U&G influences continuation intentions. Following the same, Nayak and Biswal (2021) explore the impact of the COVID-19 lockdown on the consumption patterns and preferences of OTT media streaming platforms in India. The authors surveyed 300 respondents from different age groups, genders, and occupations to understand their OTT media streaming behaviour during the lockdown.

Dahiya, S. and Nijhawan, G. S.', "Role of COVID as a catalyst in increasing adoption of OTTs in India: A study of evolving consumer consumption patterns and future business scope" (2020) further examines the impact of the COVID-19 pandemic on the growth and consumption of over-the-top (OTT) platforms in India. The authors argue that the pandemic has accelerated the adoption of OTTs by offering consumers various content, ease of access, and choice of devices. Their article explores the changing preferences and behaviours of different audience segments, such as children, adults, and the elderly, and the implications for the OTT industry. It also discusses the challenges and opportunities for the OTT players regarding content creation, regulation, and competition.

## OTT Platforms and Changing Cinematic Cultural Industries

Bouquillion, P. and Ithurbide, C.'s "Audio-visual industry and digital platforms in India: A contribution from the political economy of communication" (2022)" studies how digital platforms affect the audio-visual industry in India using a political economy lens. They analyse the interactions between old and new media players, the foreign ownership trends, and the localisation tactics of global platforms. By explaining how digital platforms show new forms of economic and cultural power in the cultural industries and the growth and diversification of Indian capital in the digital and cultural fields, they consider the impacts of digital platforms on the diversity and quality of audio-visual content and the policy and regulation issues and prospects.

Fitzgerald, S.'s "Over-the-Top Video Services in India: Media Imperialism after Globalization" (2019) also explores how OTT video services affect the international communication order in the context of India, the second-largest internet market and a major content producer. The article examines the 'platformisation process' in India, which involves the development of digital infrastructures, the competition of local and global OTT players, and the state regulation of digital media. The article uses the media imperialism framework to critique the impact of OTT video services on the Indian media industry and culture and to expose the conflicts and dichotomies of the global and the local, the dominant and the alternative, and the old and the new in the platform era.

Investigating the adoption and use of OTT video streaming platforms in India during the COVID-19 pandemic, Malewar and Bajaj (2020) surveyed 277 users. They found that performance expectancy, price value, habit and content availability influenced their behavioural intention and usage. Establishing the relevance of their model, the authors suggested that the managers of OTT video streaming platforms should adopt a three-pronged approach of (a) enhancing quality and variety of content, (b) creating user-friendly interfaces to attract and retain customers, and (c) offer competitive price points.

Srivastav and Rai's article titled "Transforming the Entertainment Industry: Video on Demand Services in India" examines the emergence, functions, and impact of the video-on-demand (VOD) industry in India, rapidly changing the dynamics of the entertainment sector. The authors argue that the VOD industry offers alternative storytelling from various paradigms, catering to the new age audience who are not passive consumers but active participants in content creation. The article also explores the challenges faced by the practitioners in the VOD industry, such as censorship, regulation, competition, and piracy, and how they differ from the existing television and film industries.

The entry on 'marketing and promotion' by Sarah Berry in the *Critical Dictionary of Film and Television Theory* (2005), edited by Roberta E. Pearson and Philip Simpson, is a concise overview of the history, theory and practice of film and television marketing and promotion. Berry's entry is a useful introduction to marketing and promotion for film and television students, scholars and practitioners. She traces the development of marketing and promotion from the early days of cinema and broadcasting to the contemporary era of digital media and globalisation. She also discusses the key concepts and debates in the field, such as the role of audiences, the impact of technology, the ethics of persuasion, the politics of representation, and the cultural implications of branding.

"Shift Focus: Women Shaping the Narrative in Media and Entertainment" is a report published by the Women in Cinema Collective (WCC), a group of women working in the Malayalam film industry, in collaboration with Sakhi Women's Resource Centre, a feminist organisation based in Kerala, India. The report, supported by the United States Consulate General in Chennai, aims to document the experiences, challenges, and aspirations of women working in various film industries across India and to provide recommendations for creating more gender-inclusive and safe workplaces. The report is based on the findings of two conferences held in Kochi and Mumbai in 2022, where women from different film industry sectors, such as actors, directors, writers, producers, technicians, and journalists, shared their stories and perspectives. This report is relevant for studying OTT platforms, as it provides insights into the potential and challenges of digital media for women in cinema. It notes that OTT platforms have emerged as a new avenue for content creation and consumption, offering creative freedom, flexibility, and reach for filmmakers and audiences. The report also acknowledges that OTT platforms have enabled more women to enter and excel in the film industry and explore diverse and unconventional genres, themes, and characters. However, the report also cautions that OTT platforms are not immune to the existing biases, barriers, and power dynamics that affect women in cinema and that there is a need for more research, regulation, and dialogue to ensure that OTT platforms are genuinely inclusive, accessible, and empowering for women in media and entertainment.

The present essay contributes to the topic by expanding the scope of New Malayalam cinema by studying the exponential rise of OTT media and streaming platforms, partly because of the COVID-19 pandemic, in a cause and effect relationship, wherein one of the 'causes' is the latter, which has an effect on the former. It studies the Malayalam cinema industry in particular to study the shifts in narrative and production that have made it a generational shift and also narrows its focus further to two films which are commercial success and examples of the catalyst-like function of OTT platforms.

## Methods and Methodology

This chapter undertakes a scholarly enquiry into the nature of the content created and released on OTT platforms, how it contributes to the growth of New Malayalam cinema, and its distribution and the patterns of its consumption and reception. Subsequently, it qualitatively analyses the themes and content of select films via a close textual reading of the films *C U Soon* (2020) and *Joji* (2021).

The rationale for choosing the two films are a few; firstly, the chosen films are grouped within the contemporary era of New Malayalam cinema and, simultaneously, are films that have been commercially viable after their release on OTT platforms. Due to the direct release of Mahesh Narayan's *C U Soon* and Dileesh Pothan's *Joji's* on Amazon Prime Video, an on-demand, OTT streaming platform, these filmic narratives have been nudged to directly enter a global market and be recognised for their distinctive style and methods of treatment of the themes.

Identifying the changing landscape of the aspects mentioned above and their implications, OTT platforms have dislodged the traditional models of film-making and provide creative space and license to directors to segue into robust diversity of content. The experimental and unconventionally intertwining storytelling techniques employed by independent, novae filmmakers, as seen in the films released on OTT platforms, explore themes and initiate discourse circling regional narratives unconceivable within the traditional structures of the Malayalam cinematic industry. This essay, thus, employs a qualitative methodology to study the form and content of these films by using the methods of close textual reading and thematic analysis. By analysing these films as texts, the essay studies them as a 'subject' that makes shifts in the industry through the reimagination of spatialities, themes, characters and renewed means of production.

## Discussion

### Exploration of OTT Media and its Malayalee Market Trends

OTT media has had a short yet fruitful endeavour in the Indian markets, which has opened up content from all over the world across languages, available for immediate consumption. There is also a rapid acceleration of homegrown OTT players mushrooming across the country, with an estimate that the OTT app business would be worth $200 billion and by 2027, the global OTT industry is expected to reach nearly $1039.03 billion (Mudaliar, 2023). Within this dynamic ecosystem, the OTT platforms acquire content through two means: (1) self-production and (2) by purchasing broadcasting rights. In the former, original content like web series, films, documentaries, reality shows, and so on are produced explicitly for the respective streaming platform.

In this context, understanding contemporary Malayalee market trends that have evolved alongside the growth of OTT cinema requires a two-pronged approach; firstly, examining the concept of internet film promotion is an upcoming idea that threads along with new Malayalam cinema. As a regional film industry that has begun to attempt to present ideas from the fringes con, the contemporary popular replete with the realist aesthetic called New Malayalam cinema also offers a constant tussle with tradition and modern methods and modes of storytelling. The impact of social media on Malayalam film promotion through teasers and curated posters is a new phenomenon factor like social media marketing and its tendency to calculate films as commodities for sale (Berry, 2005).

Secondly, the OTT platforms facilitate an investigation of the revenue streams within this ecosystem. According to a report published by the Confederation of Indian Industry (CII), the revenues of the South Indian film industry have doubled. Adding earnings from multiple streams, the film industry's revenue includes the funds generated through digital or OTT rights and other language dubbing rights. Reflecting on the report, Tamil cinema dominated the OTT and Satellite rights, generating revenue of 580 crores (*South Indian film industry,* 2023).

Interestingly, the conditions under which OTT media garnered interest and proved a financial success were backed by the unique potential tapped by the New Malayalam cinema. The aforementioned adaptive nature of Malayalam cinema was not only a result of its history of having been developed by independent studios (Raghavendra & Pillai, 2017) but also from the low-budget nature of its production. Considering Malayalam cinema as a regional cinema industry, avoiding its branding under the umbrella of 'Mollywood', New Malayalam cinema becomes its recent experimentation with its already established visual grammar.[4] Beyond a venture, the project of New Malayalam cinema primarily disengages with the traditional structures of meaning-making in its particularities and media production, consumption and distribution on a macro level.

## Content Diversity

Popular culture is a porous phenomenon that initiates and provides a platform for dialogue that either perpetuates ideology or representation or presents it as a matter of content. At its heart, it is driven by those with access to it. As part of this culture, OTT platforms contribute to the large repository of visual culture but are not porous regarding access. Due to the digital divide in India combined with the mobile gender gap, the question of access becomes highly evident in the conversations that circle the nature of the audience, the means of access and the narratives available on these streaming platforms.

Often bestowing the characters of a functioning character to a 'home', New Malayalam cinema has undertaken a reinvention of the idea and toyed with its creative possibilities. Some films from this list include two discussed in this chapter at length: Mahesh Narayanan's *C U Soon* (2020) and Dileesh Pothan's *Joji* (2021).

## The Medium and the Message in *C U Soon* (2019)

Written and directed by Mahesh Narayanan, *C U Soon is* a testament to the adaptive nature of Malayalam cinema. A narrative delivered through various screens, *C U Soon* sheds light on a socially relevant issue through new mediums. It expands the scope of the deliberately paradoxical statement heralded by Marshal McLuhan, 'the medium is the message', in his *Understanding Media: The Extensions of Man* (1964). As part of the New Malayalam Cinema era, it checks all the boxes of the contemporary tendencies and shifts in the represented themes. Emphasising the mode adopted to deliver the narrative, the film *C U Soon* becomes a stylistic endeavour that also ties together the social importance of the narrative. Although the social consequence of this medium of narrativising is vastly different from those that existed before or after the film's release, the film's content cannot be discounted.

The film is one of the few Indian thriller films that stylistically adhere to the screen-life genre of visual narrativising. The plot of the film centres on the character of Anumol Sebastian (Anu) and her mysterious introduction into the life of Jimmy Kurian, a client executive at Union National Bank in the UAE. The film utilises the ambiguous nature of online dating to spin the two characters into a complex web of human trafficking, which draws in Jimmy's cousin Kevin, a cyber security specialist, along with a host of supporting characters in unravelling the true identity of Anu. Experimental in nature due to the COVID-19-induced nationwide lockdown and the strict protocols in public gatherings, initiating a 'new normal' becoming a term of common parlance. The film is a circumstance of this renewal of narrative space and exploration of the potential provided by limiting the narrative to domestic interiors.

The film feels similar to an episode of *Black Mirror* due to its exploration of the topics of invasion of privacy, data leaks and the dynamics of the present-day situation due to the excess of the internet. Still, at the same time, it gives a hopeful ending to the audience who were a part of the journey. It takes up an issue before the pandemic but naturally ties in the sensibilities of the lockdown within the budgetary constraints. The film was primarily produced with an OTT release agenda and on a minimal budget, and the experience generated from watching it, on the level of audience reception, is unique.

The effort in *C U Soon* effectively encapsulates the extent to which New Malayalam cinema consciously attempts to rewrite the pre-established narratives and character tropes. The film also accentuates the Foucauldian theory of the panopticon in its reflection on a surveillance society. New Malayalam cinema,

*...it is to be noted that despite being accused of banality, triteness and superficiality, 'New Generation' movies attempt to faithfully record the changing ethos of Malayalam cinema in the new millennium. Though rootedly middle class and dealing primarily with urban alienation, it creates a cinematic subculture that reflects the anxieties of a generation who are the children of liberalisation, the satellite television boom and globalisation. Many of these movies sought to show ordinary men and women. Unlike the paradigms that confined earlier on-screen feminine representations as paragons of virtue, attempts were afoot in these kinds of films to portray realistic women, cast in a more liberal and progressive idiom and premised on the hope that there would be a rupture in the mould of the ideal feminine in Malayalam cinema (Raghavendra & Pillai, 2017).*

*Joji* (2021)*: A Tale of Masculinities*

Restricting the narrative within domestic interiors due to the restrictions posed by the COVID-19 lockdown, *Joji* became synonymous with commercial success as one of the films that put Malayalam cinema on the world map. The film successfully ties in the postmodern elements of contemporary visual storytelling. *Joji* is known for several filmic components and finds itself in a long list of Shakespearean adaptations, specifically of the playwright's *Macbeth.*

*The postmodern incredulities that deconstruct the metanarrative structural authority of Shakespeare are the new defining parameters of [recent] cinematic endeavours. The utilisation of contemporary postmodern aesthetics through new critical and rhetorical paradigms helped a postmodernist Shakespeare universe that replaced the previous idealisations of the Shakespearian drama and its different adaptations to emerge. The postmodern adaptations of Shakespeare in Malayalam tend to show a form of simplicity in character formation and [are] aided with a realistic plot in which they are embedded in a modernised contemporary space of Kerala. For example, Annayum Rasoolum is set in the urban outskirts of Kochi, Iyyobinte Pusthakam in the hilltops of Munnar, Eeda in North Malabar, and Joji within the rubber plantations of Kottayam. In all these films, the tragic element is deeply associated with the perception of the modern regional space of the land and its many social, cultural, and political complexities (Jalarajan & Suresh, 2021).*

In an article published in The New Yorker, titled "Joji," Reviewed: The First Major Film of the COVID-19 Pandemic", Brody explains the 'graceful integration' of the situation posed by the pandemic within the film's plot (Brody, 2021). With the use of aerial cinematography for the establishing shots and the showcasing of the fleeting exterior juxtaposed with the dynamic nature of the interior, the narrative not only incorporates symbols and visuals of the pandemic but, at the same time, utilises household objects to their utmost potential to build the main narrative of the film.

From the alphabet stickers that spell out 'Joji's Palace' whenever the titular character closes the door of his room and bursts out into an almost rehearsed 'silent rebellion' to drawing the viewers' attention to the granite used to construct the makeshift kitchen island, *Joji* is as much a story of the domestic and its spaces as it is a discussion about human nature, patriarchy, hypermasculinity in a highland spatiality.

The film revolves around Joji, one of the three sons of the hypermasculine patriarch, Kuttapan Panachel. Joji's frustration stems from his 'self' and his surroundings—a significant part of which is his father, Kuttapan. The other crucial character in this constellation of power is Bincy, Joji's sister-in-law. The spaces she occupies and the functions she carries out are gendered and serve as a social commentary about the role of women within the intimate space of a family. This trope has been eloquently represented and discussed in another film from New Malayalam cinema, *The Great Indian Kitchen* (2021).

The narrative accelerates moving the wheels of (mis)fortune with the tragedy that struck upon the Panachel family when Kuttapan suffers from a stroke that requires immediate medical attention. At the hope of his death dangling as bait, Joji and Bincy convey in silent glances the opportune moment before them and the need to wait to savour the inheritance they will receive. As the camera lingers on interconnected gazes shuttling from one character to the other, the narrative shifts to that of building a partnership between the two characters to expedite the dream that the two have seen together and parts in isolation; a dream which requires an actor to go on a killing spree.[5]

The Malayalam film industry has been known for its increased reliance on realism and realist aesthetics (Pillai, 2017). Following the same vein, the screenplay by Syam Pushkaran in *Joji* 'mirrors' the other side of the patriarchal tyranny, often the centre of some other Malayalam films.[6] It does not offer an ending in a cathartic sense. However, it places the narrative in a series of films from the contemporary era of New Malayalam Cinema that interrogates pre-established norms of the society.

As a postmodern Shakespearean adaptation, *Joji* ties in the everyday, regional and socio-cultural politics to accentuate the theme of patriarchal violence. The film is a narrative about patriarchy and hypermasculinity and, at the same time, is a story of different masculinities in the setting of a home and the restrictions posed by the pandemic. Although masculinities within the context of masculinity studies have been identified into four broad types: hegemonic, complicit, subordinate and marginalised, *Joji,* on the surface, a film about four men, explores masculinities as a reaction to the hypermasculine Kuttapan Panachel. Repressed under the patriarch and his power in both his presence and absence, the male characters are in a constellation of power relations wherein each is more powerful than the other, with the titular character Joji at the receiving end. Interestingly, despite Joji's position and his typified performance of gender, he exerts power on Bincy, a female character who appears powerless but earmarks the silence in the murder of the patriarch.

## Market Expansion of New Malayalam Cinema

The position of New Malayalam cinema at a global market level emphasises its market expansion due to the rise and success of OTT platforms, specifically due to the COVID-19 pandemic and the lockdown. Due to this, regional cinemas, like of that Malayalam, have reached a global audience. Although the Malayalam film industry resists the grouping within the 'pan-Indian' film, the film's realist aesthetics have invited audiences beyond regional borders to participate in a meaning-making process rooted in the local but transmitted through the universalising practice of subtitling.

Films such as *Joji* and *C U Soon,* as discussed in this chapter, are intimately scripted around an intimate reflection of the contemporary ethos of Kerala. Still, at the same time, subtitling efforts have allowed a global audience to appreciate the work of New Malayalam cinema. This space provided by the rise of OTT platforms to the regional Malayalam cinema is proven by the spike in subscriptions during the COVID-19 pandemic, up to 55-60 per cent (*Paid OTT subscriptions,* 2020).

Playing a significant role in the market expansion of New Malayalam cinema, OTT platforms such as Netflix, Amazon Prime, Disney+ Hotstar, and regional OTT platforms like Zee5 and Aha have been widely recognised as critical players in the promotion of the same on an international level. Diversifying the viewership base, accompanied by the surge in content consumption during the pandemic, allowed filmmakers to turn to these platforms for their releases to recoup their investments. Following the same vein, newer artists found avenues of recognition through this digital mode of market expansion as well.

As a means of survival and growth, the resuscitation provided by OTT platforms for other film industries was not met with the same gesture in the Malayalam film industry as it proved to be a boost to the already existing, low-budget production projects undertaken by the directors paving the way for New Malayalam cinema. Take the example of the film *C U Soon;* the film was allegedly produced at the expense of less than one crore rupees and was sold to Amazon for around eight crore rupees, which led to the development of a trading formula called a 'table profit' which ensured that producers could earn money to cover their expenses through OTT sales and did not have to wait for the box office collection after its theatrical release (*Mollywood suffers as OTT platforms,*2023).

Playing a pivotal role in the globalisation of New Malayalam cinema, OTT platforms allowed the bypass of traditional distribution chains. This segued into providing the ground for directors, especially those who directly released their films on the abovementioned platforms, experimenting with diverse narratives, complex themes and narratives that detoured from the mainstream. As discussed earlier, *Joji* and *C U Soon* are filmic successes immediately available to audiences in SVoDs. At the same time, they contribute to the larger project of New Malayalam cinema and add to its expansion of definition and diversity.

## Changing Content Consumption Patterns of (Cinephile) Audiences

On-demand content delivered by the OTT platforms during the home-bound period of the pandemic changed the patterns of content consumption in two ways; firstly, it created an audience that was readily accepting and participating in composing the viewership of a large group or culture of cinephiles who were watching and critiquing the world cinema. Secondly, it allowed the audience to have immediate responses to what they were watching in the cinema through social media.

The shift from multiplex screens to smaller screens was jolted into action due to the unexpected nature of the pandemic and led to technological advancements in both the production and the consumption of filmic content, which led to an excess of choices for cinephile audiences. The individualistic viewing experience of viewing films from SVoDs differs from the communal experience of a multiplex setting. However, due to the peculiar formation of fan bases on social media and related applications such as Telegram, there is a more extensive, like-minded formation of fans that discuss their interests and films together. Through this, the OTT experience becomes more intimate due to its personalisation and access to larger fan groups working on a global scale. Although New Malayalam cinema essentially is post-star cinema, there has been increased discourse and discussions about the symbolism and contexts in which the films have been situated on websites like YouTube. In other words, the parallel growth of OTT platforms and New Malayalam cinema have been working complementarily to shift the focus from the film's actors to the film's narrative as a complete cultural artefact of its own.

This shift in consumption patterns from the end of the audience has dramatically impacted viewers' expectations and the critical reception of films. Scouting for quality content has created a need for narratives with thematic depths to the film, made apparent through artistry and the script. This, once again, has proven to be conducive to developing new Malayalam cinema.

The interactive nature of content consumption due to the rise of OTT platforms makes viewers active participants in a global cinematic experience. This is evident in the reviews that the audience leaves about films as either Google reviews or comments on posts published on social media. The engagement in discussions, alternate fan theories and so on provide the space for the tech-savvy cinephile individual to be a part of a greater whole. Furthermore, the creation of social media trends and the recent replication of the 60-second video concept across the platforms of TikTok, YouTube, Instagram, and so on promote films through the integration of songs or popular dialogues. In more ways than one, it points towards a contested yet believable situation of a cultural singularity.

The fear of a certain homogenisation of content available on OTT platforms is inevitable in a discussion that addresses consumption patterns as it poses a challenge to traditional cinema. Interestingly, this claim does not apply to the Malayalam film industry due to its shift to New Malayalam cinema in the 2010s. As the debunking idea of the 'traditional' was already set in motion, OTT cinema's challenge to the traditional was not experienced as a crisis to the industry. It was instead an opportunity to adapt and innovate.

The changing consumption patterns of the audience have also nudged filmmakers across film industries to experiment with different formats of visual storytelling, such as anthologies, which further challenge the traditional means of storytelling and create an evident diversion between the conventional and unconventional. Although a hybrid model of release is not sanctioned post the OTT boom, there have been steps taken by the film exhibitor's body to raise concerns regarding the impact of OTT releases on the traditional theatrical business model (*Film exhibitor's body,* 2023).

## CONCLUSION

### Future Prospects in the OTT Mediascape Alongside the Propagation of New Malayalam Cinema

Discussing the success of OTT platforms within the Indian context and estimating their reach and prospects is a question about the audience viewing patterns of those with access. The mobile gender gap within the broader domain of the digital divide is crucial in this discussion. However, an optimistic deliberation on the prospects of the OTT mediascape as a distribution medium should consider its continuous state of flux. The reason for this is its growing database and the dynamic ability of OTT platforms to spark an interest in regional narratives on a global scale. The ease of choice goes along with the question of access to lead to a democratised viewing experience across the masses.

The nature of the composition of the audience is also an important point of discussion within this segment as it finds itself at the intersections of gender, language, region, and access (Malewar & Bajaj, 2020; Menon, 2022). The latter is further divided into two more critical and complementary discussions: access to the devices to access it and the funds to purchase subscriptions for SVoDs. Recently, streaming platforms like Netflix have deliberated about ending password sharing of their accounts, a habit adopted by several subscribers to make it cost-effective (*Netflix ends password sharing in 'big market' India,* 2023). In this case, the viewer gets discouraged from using OTT platforms and returns to the theatres or pirating films.

The possibility of cross-cultural collaborations becomes more approachable through the globalisation of New Malayalam cinema, which was already primed for these endeavours. These efforts can increase the industry's visibility and incorporate global perspectives in its narratives and filming techniques. Additionally, the prospect of the proliferation of the OTT mediascape is a delve into the question that poses two experiences at the opposite ends of a spectrum: OTT versus cinema. It is a question of equating the filmic content produced on OTT streaming platforms as cinema and not as a separate experience, which requires a logistical change in the location of viewing the films and the overall ambience of it. Considering OTT as against cinema runs the risk of it as an option concerning the Malayalam cinematic industry.

The risk of content homogenisation becomes glaringly apparent while factoring in audience reception to understand their consumption patterns. The dilemma that would make itself apparent would be the intentional creation of a formula or formulaic film, once again, that dilutes cultural specificities and geographical peculiarities and creates an appeal for a 'relatable' film narrative.

Although New Malayalam cinema has significantly benefitted from OTT platforms, one segment of society solely dependent on the traditional film distribution model is the cinema workers or cine workers. They form an integral part of the ecosystem comprising two significant elements: film production and distribution. They include a range of professionals like camera crews, set designers, make-up artists, costume designers, and sound designers. As an extension, they also have the cinema hall staff, projectionists and ticket sales personnel. With the OTT model of film distribution, a different set of skills is needed that may make them obsolete. For instance, due to digital cinematography, specifically in a film such as *C U Soon*, the impact may have been felt in jobs related to film-based camera work.

Additionally, shooting on locations has resulted in a decrease in the demand for studios with certain types of set design. However, in defence of OTT platforms, with the advent of New Malayalam cinema, films had moved out of studios to real locations post-2010s. This also contributed to the easy assimilation of this regional film industry to adapt to the challenges posed by the pandemic, the ripples of which were felt across the globe in the context of the entertainment sector. To add to it, the role of the intermediaries who formed the link from producers to theatres, like the distributors, sales agents, sub-distributors, and so on, become invisible in the OTT model as the process becomes direct with the producers selling the rights of the content to the OTT platforms.

Despite the shift to digital platforms placing the position of the Malayalam film industry and India on the whole higher on the linear scale of measuring development, the loss of cineworkers' jobs is one of the most pressing issues which remained secure in the traditional cinema ecosystem. This was observed after the release of *Joji* when the director and lead actor, Fahad Faasil, donated ten lakh rupees from the profit generated by the release of their film to the Film Employees Federation of Kerala (FEFKA) (Sidhardhan, 2020; Praveen, 2022).

While this presents a bleak picture for those who have intricately related jobs to the traditional cinema ecosystem, an objective study of the impact of OTT platforms should reflect upon the newer job opportunities in digital marketing and data analysis domains on the same level. With the OTT model, there is also an effort to create original content restricted to these platforms, as mentioned before. This offers openings for scriptwriters, directors and actors who can contribute to this growing medium and its exclusive audiences.

In conclusion, studying digital streaming platforms offers insights into new forms of economic and cultural hegemony in the cultural industries (Bouquillion, 2022). Although the rise of OTT platforms has affected the job market specifically for cineworkers, at the same time, it ushers a new era of opportunities and skills. The challenge, which then remains at the end, is ensuring that the workforce is equipped to adapt to the changing demands of the industry. At the same time, there is an ongoing effort to sustain traditional cinema and its experience alongside the OTT model.

## REFERENCES

Anand, S. N. (2021). We let 'Joji's' plot grow organically: Dileesh Pothan. *The Hindu*. https://www.thehindu.com/entertainment/movies/dileesh-pothan-speaks-about-his-new-filmjoji-and-the-inspiration-from-shakespeares-macbeth/article34264470.ece

Baby, J. (2021). *The Great Indian Kitchen*. Mankind Cinemas.

BBC. (2023, July 20). *Netflix ends password sharing in "big market" India*. BBC News. https://www.bbc.com/news/world-asia-india-66253656

Berry, S. (2005). Marketing and promotion. In R. E. Pearson & P. Simpson (Eds.), *The Critical Dictionary of Film and Television Theory* (pp. 371–375). Routledge.

Bouquillion, P., & Ithurbide, C. (2022). Audiovisual industry and digital platforms in India: A contribution from the political economy of communication. *Global Media and Communication, 18*(3), 345–364. doi:10.1177/17427665221125548

Brody, R. (2021, June 1). "Joji," reviewed: The first major film of the COVID-19 pandemic. *The New Yorker.* https://www.newyorker.com/culture/the-front-row/joji-reviewed-the-first-major-film-of-the-covid-19-pandemic

Dahiya, S., & Nijhawan, G. S. (2020). Role of Covid as a Catalyst in Increasing Adoptions of OTT in India: A Study of Evolving Consumer Consumption Pattern and Future Business Scope. *Journal of Content, Community & Communication., 12*(6), 298–311. doi:10.31620/JCCC.12.20/28

Fitzgerald, S. (2019). Over-the-top video services in India: Media imperialism after globalisation. *Media Industries Journal, 6*(2), 89–115. doi:10.3998/mij.15031809.0006.206

Malewar, S., & Bajaj, S. (2020). Acceptance of OTT video streaming platforms in India during COVID-19: Extending UTAUT2 with content availability. *Journal of Content. Community and Communication, 12*(6), 89–106. doi:10.31620/JCCC.12.20/09

McLuhan, M. (1964). *The extensions of man.*

Menon, D. (2022). Purchase and continuation intentions of over-the-top (OTT) video streaming platform subscriptions: a uses and gratification theory perspective. *Telematics and Informatics Reports, 5*, 01–15.

Mohammed, Z. (2020). *Halal Love Story*. Papaya Films.

Mudaliar, A. (2023, February 13). The evolution of content in India's Ott Ecosystem. *Times of India Blog*. https://timesofindia.indiatimes.com/blogs/voices/the-evolution-of-content-in-indias-ott-ecosystem/

Narayanan, M. (2019). *Kumbhlangi Nights*. Fahadh Faasil and Friends.

Narayanan, M. (2020). *C U Soon*. Fahadh Faasil and Friends.

Nath, P. J. (2021, August 26). *Streaming a new reality: Malayalam movies in the Ott World*. BusinessLine. https://www.thehindubusinessline.com/blink/cover/ott-lifeline-a-moment-of-reckoning-for-malayalam-cinema/article36111055.ece#:~:text=C%20U%20Soon%2C%20Joji%2C%20Sara's%2C,knees%20by%20a%20rampaging%20virus

Nayak, S. C., & Biswal, S. K. (2021). OTT Media streaming in COVID-19 lockdown: The Indian experience. *Media Watch*, *12*(3), 440–450. doi:10.15655/mw_2021_v12i3_166019

P T, Y. (2020, October 18). *Dileesh Pothan and reframing the rural landscape in Malayalam cinema • reading room co*. Reading Room Co. https://readingroomco.com/2020/10/18/dileesh-pothan-and-the-reframing-of-the-rural-landscape-in-malayalam-cinema/

Hindu Business Line. (2020, December 16). Paid ott subscriptions up by nearly 60%: CII-BCG Report. *The Hindu BusinessLine*. https://www.thehindubusinessline.com/info-tech/paid-ott-subscriptions-up-by-nearly-60-cii-bcg-report/article33346493.ece

Pellissery, L. J. (2017). *Angamaly Diaries*. Friday Film House.

Pellissery, L. J. (2018). *Ee. Ma. Yau*. OPM Cinemas.

Pellissery, L. J. (2019). *Jallikattu*. Opus Penta.

Pillai, S. (2021, October 21). *Ott revolution: A boon for filmmakers*. Forbes India. https://www.forbesindia.com/article/south-celebrity-special-2021/ott-revolution-a-boon-for-filmmakers/71057/1

Pothan, D. (2021). *Joji*. Bhavana Studios.

Praveen, S. R. (2022, August 13). Explained: Tussle between Malayalam cinema exhibitors and producers over Ott releases. *The Hindu*. https://www.thehindu.com/news/national/kerala/malayalam-cinema-exhibitors-producers-clash-over-ott-releases-explained/article65764558.ece

Raghavendra, M. K., & Pillai, M. (2017). Bearing Witness: Malayalam Cinema and the Making of Keralam. In Beyond Bollywood: The Cinemas of South India (pp. 267–330). HarperCollins Publishers India.

Raj, S. J., & Suresh, A. K. (2021). *Postmodern transpositions of Shakespeare in Malayalam cinema–a transformative discourse of regional tragedy*.

Ravi, R. (2016). *Kammatipaadam*. Global United Media.

Sidhardhan, S. (2020, October 6). Fahadh, Mahesh Narayanan donated rs 10 lakh to fefka: Malayalam movie News - Times of India. *The Times of India*. https://timesofindia.indiatimes.com/entertainment/malayalam/movies/news/fahadh-mahesh-narayanan-donate-rs-10-lakh-to-fefka/articleshow/78507178.cms

Srivastav, S., & Rai, S. (2021). Transforming the entertainment industry: Video on Demand services in India. *Media Asia*, *49*(1), 44–57. doi:10.1080/01296612.2021.1977499

TOI. (2023, June 6). Film exhibitor's body calls for theatre shutdown in protest against '2018's OTT release: Malayalam Movie News - Times of India. *The Times of India*. https://m.timesofindia.com/entertainment/malayalam/movies/news/film-exhibitors-body-calls-for-theatre-shutdown-in-protest-against-2018s-ott-release/articleshow/100797557.cms

VS. G. (2022, February 8). The "homely" shift in Malayalam cinema. *The Indian Express*. https://indianexpress.com/article/entertainment/malayalam/the-homely-shift-in-malayalam-cinema-7761953/

## ENDNOTES

1 This has been made due to the advancement in digital technology alongside translation practices in films, which include dubbing and subtitling that increase the reach of the film beyond their regional contexts.

2 The movement towards rural and semi-urban locations is not only a deviation from the traditional model of using sets and exotic locations wherein the narrative would unfold (specifically in the case of Bollywood cinema) but also a shift in the narrative. Locations and spatialities become indispensable parts of the narrative, giving them the status of an equal, contributing character.

3 Capital 'M' stands for the M's in the names of the superstars, Mamooty and Mohanlal and their massive fan base within the Malayalam film industry. Other stars include Suresh Gopi and Jayaram, who featured in the films with their flexed muscles and churned out hegemonic muscle characters (Raghavendra & Pillai, 2017).

4 As a portmanteau originally derived from 'Hollywood', 'Mollywood' becomes a part of a larger essentialization project in which the unique characteristics of the Malayalam cinema industry is lost due to the packing and rebranding of its narrative. This essay refrains from using the term 'Mollywood' to avoid its grouping under a similar kind of cinematic cultural industry, available for scholarly enquiry using the same parameters such as Hollywood and Bollywood.

5 The director Dileesh Pothan, in an interview with The Hindu, stated that *Joji* was shot in the 185 format instead of the Cinemascope, as the viewing experience was different and individual compared to the cinema hall setup (Anand, 2021).

6 The term 'mirrors' here refers to the use of mirrors in the film and the manner in which they act as truth windows for the characters.

# Chapter 9
# Analyzing the Narrative Structure of a Tent-Pole Bollywood Film vs. Exclusive Indian OTT Cinema

**Kanika Arya**
https://orcid.org/0000-0002-6460-0174
*D.Y. Patil University, India*

**Manish Verma**
*Chitkara University, India*

**Anisha Jain**
*Independent Researcher, India*

## ABSTRACT

*The chapter is an attempt to compare and contrast the narrative structure of a tent-pole popular Indian cinema film and an exclusive to OTT Indian popular cinema release. Indian popular cinema has followed some conventions of storytelling since its inception, which gives it unique storytelling character compared to global cinema. Though popular Indian cinema has been severely criticized by critics and filmmakers around the world, it has survived the criticism and the competition by international film industries for survival in the ever-growing Indian market, and also has been able to garner a faithful reception. OTT and streaming brought a plethora of entertainment options to the audience besides mainstream cinema, one very successful among these is exclusive to OTT Indian cinema. Analysis of the storytelling technique of both cinemas has been done using case of two films Jab We Met and Meenakshi Sundareshwar which helps us understand how cinematic storytelling has been and is gradually changing in India with interesting twists and turns.*

DOI: 10.4018/979-8-3693-3526-0.ch009

## INTRODUCTION

Storytelling is important to humans because it helps them make sense out of the world around them. Listing to poetry recitals brings catharsis, watching play performances makes one a *rasic*[1]; novels are detailed and can create complex discourses, can engulf qualities of other genres and have the same stature in storytelling as epics had in olden times, novel is a complex space and makes one feel intricately involved (Mambrol, 2018). Mythology and folklore bring one close to rustic art and entertainment. Intertextuality and adaptations make these worlds mingle, crossover and still stand distant but creating varied perceptions in the readers' or audiences' mind. Cinema emerged much later, it is contentious whether you call it art or entertainment but it is a young-old form of storytelling which has changed its form and style suiting the filmmaker's needs or the needs of a certain culture and audiences sensibilities.

Sometimes the auteur takes over and presents art in the form of cinema; like the cinema of *Avant Gard-ists* or *Neo-realists* or those belonging to art or parallel cinema (*margi*) as we call it in India (Gupta, 2008). With the onset of digital storytelling the story-world of the audiences has grown by leaps and bounds, stories from all around the world are just a click away, stories by artists, entertainers or amateurs all ready to reveal in fraction of a second, created by the latest and the best technological innovations in moviemaking, but not constricted to that. As Mc Luhan said medium is the message, thus, plethora of mediums merged on this new multifarious medium to give stories in seconds like a short reel or in a few minutes (short film); realistic and experimental narrative (documentary) and feature film whose timespan can be anything from 1.5 to 3 hours. The emergence of various formats and genres on new medium, especially OTT video streaming platforms makes stories and their rendition even more interesting and dynamic. There is no dearth of stories on streaming platforms (whether ad-based or subscription based), maybe they just miss the time, pondering and a sense of authenticity to do with conventional television and film production. Truffaut (qtd. In *Film Art* (2012)) says:

*There are two kinds of directors, those who have the public in mind when they conceive and make their films and those who don't consider the public at all. For the former, cinema is an art of spectacle; for the latter, it is an individual adventure. There is nothing intrinsically better about one or the other; it's simply a matter of different approaches.*

This chapter analyses narrative structure of Popular Indian cinema i.e. Hindi cinema popularly known as Bollywood post 1990's (*Bollywoodisation*[2]) versus exclusive to OTT (video streaming platforms) Indian cinema. Our contention is that the narrative strategy for cinema exclusive to OTT (in India) has changed drastically from the conventional norm. Narrative is the path that story follows; through time and space; philosophers, literary critics, and researchers have discussed various models and theories of narrative to make sense out of a story's structure and storytelling methods followed by various storytellers. Cutting (2016) elaborates on the concepts of *fabula* and *syuzhet,* the terms used earliest by Russian formalists to describe story and storytelling . *Fabula* is simply the story 'in all its semantic, chronological and causal detail', while as *syuzhet* is the 'dramaturgy' of the film, the organised set of cues that are used to assemble the story for us in a way that we comprehend it. Bordwell (2007) delineates on Narrative Structures, highlights the importance of understanding the structures that underpin film narratives. He emphasizes that narratives aren't merely stories but intricate constructions involving plot, character development, and various cinematic elements. Especially for classical Hollywood Cinema and its narrative conventions, he emphasizes its reliance on cause-and-effect logic, clear motivations for characters, and the use of specific narrative devices like setups and payoffs. He also delves into the techniques employed by filmmakers to shape narratives, such as manipulating time (e.g., flashbacks, non-linear storytelling) and point of view (e.g., subjective or unreliable narration), all of which significantly impact how stories are perceived and understood.

Indian cinema is unique especially Hindi cinema which has audiences not only in India but globally, those who enjoy it, try to emulate fashion trends from it, ingrain its music and dance performance in important events in their lives and are aspirational about it (Dwyer, 2014; Banaji, 2006). The narrative structure of a Bollywood tent-pole film is what mostly represents a *masala*[3] film- a patchwork quilt of genres (Gupta, 2008); it cannot be decoded completely through western models of narrative like Aristotle's or Freytags' or Todorovs' or contemporarily what Field or Thompson suggest for cinematic narrative; but also needs knowledge of Indian storytelling methods to be deciphered completely; also a look at its verisimilitude becomes vital which, in turn, effects its storytelling (Prasad, 1998; Mishra, 2002; Lutgendorf, 2006; Khandekar, 2022). Contrastingly, digital is unconventional, challenging and democratic; from an amateur to an expert everyone gets an opportunity to release their work and let audiences see it and decide whether to watch another film by the maker or not. This platform also known as new medium is huge, it has humongous content to sift through, it's effervescent, interactive and niche. Due to fragmentation of audiences, people in the same time and space can watch different content; also audience is exposed to variety of content- some that is made exclusively for this medium (Drennan & Baranovsky, 2018; McQuail and Dueze, 2020; Khandekar, 2022).

Partho Dasgupta (Business Standard, 2022), former CEO BARC India explains the formidable challenge faced by foreign businesses in India due to the nation's diverse consumer preferences and values. Despite global triumphs, Netflix struggled in the Indian market, finding its subscription model at odds with local expectations. The discussion delves into various OTT business models, ranging from ad-based (AVOD), subscription-based (SVOD), pay-per-view (TVOD), to hybrid models, highlighting their respective strengths and weaknesses. Central to success is the overarching importance of compelling content, albeit amidst controversies, as seen in hit series like *Sacred Games* (Kashyap,2018) and *Mirzapur* (Anshuman, Singh and Desai, 2018). Platforms flourish by consistently delivering engaging content, as evidenced by SonyLiv's substantial growth propelled by shows like *Scam 1992* (Mehta, 2022). There is a necessity for ongoing innovation and consistent creation of appealing content to sustain audience interest and competitive edge in India's dynamic OTT industry.

A hybrid video streaming platform like Youtube has everything from an amateur shot, podcast, to a documentary to a well-made fiction film. Audiences who can spend some money on subscriptions usually prefer watching ad free streaming platforms which have content from all over the world- though slightly different from each other based on TG (target group or target audience) base. Examples include Netflix, Prime, Hotstar etc. These platforms are also known as OTT[4] video streaming platforms– these comprise varied content from documentaries to feature films to docudramas to shorts to web-series and serials and podcasts and it goes on and on and the newest content that has entered the Indian market is exclusive to OTT Indian cinema. Examples include *Freddie* (Ghosh, 2022), *Qala* (Guptan, 2022) and *Gulmohar* (Chittella, 2023).

Films in India that are exclusive to OTT, that have no theatrical release but are released directly on OTT have a narrative structure that breaks the unique story construction and storytelling norms of popular Hindi cinema. Such films can have niche topics, i.e. they are made with a selected few in mind. Also they might not hold digressions in storytelling like popular Hindi films have. Also the character and story arc are created differently compared to popular cinema in India. Thus, we review the narrative structure of a tent pole Bollywood film and an exclusive to OTT film in India through a discussion/critique of two films *Jab We Met* (Ali, 2007) and *Meenakshi Sundareshwar* (Soni, 2021)

This chapter is a critical essay which problematizes and makes an attempt at deciphering storytelling in films, especially films exclusively for the OTT medium in Indian entertainment space. We begin with a modest look at storytelling through literature and then films, especially films in India. The essay looks at story and its rendition as is expressed by various philosophers and researchers; especially storytelling in films. India storytelling is unique, whether it is the epic verses *Ramayana* or *Mahabharata* or the various stories that emerged out of it and were adapted in TV series or films based on these stories (Booth,1995) or stories that were created, inspired by Arabic or Persian literature (*Hatim Tai*,1990) or even British Literature and Art (*Rang Rasiya*, 2014) each of these adaptations were created with a very strong sense of Indian storytelling traditions. Thus, chronologically starting with narrative theories and models and then understanding Indian cinematic storytelling and finally moving to how OTT has digressed from typical Indian cinematic storytelling to suit the medium, audiences and the global competition in the OTT space. We finally transcend to analysing two films *Jab We Met* (Ali,2007) and *Meenakshi Sundareshwar* (Soni, 2021), the former a typical masala film which had a tentpole release as OTT hadn't emerged in such a big way at that time, but a critical and cult film as it shifted rom-com genre conventions in Bollywood (Lallantop, October 2023), and *Meenakshi Sundareshwar* because it falls in the same genre category (rom-com) but the story and its presentation changes as it had an OTT release; besides the film had a mixed reception and critical reception. It also released on one of the most significant OTT platforms globally Netflix,which had its own struggle and success story in India (Netflix's Monika Shergill on Understanding Audiences, Stories and the Storytelling Journey, 2021); (Netflix Next, n.d.)).

Of course, there is a significant time gap also between the two films, thus, that has also played its part, but when it comes to filmic storytelling Imtiaz Ali is an unconventional storyteller whose made films deviating from typical *masala* rom-coms; in-fact he is someone after Raj Kapoor and Yash Chopra who redrew the lines for rom-com genre in Indian popular films category (Arya, 2017; Newsroom Lallantop, 2023); thus, comparing his film and the new-age OTT romcom gives us a critical view at how filmic storytelling for romcom genre especially has changed on the new medium. Conclusion gives a synoptic view of the chapter besides revising the significant observations that we made through the analysis of the two films . This is a qualitative study, the films chosen for analysis are a purposive sample based on the critical appeal of the films; the choice of the subscription based platform Netflix is again based on the global appeal and dominance of the platform, the amazing story it has till date in India (Netflix's Monika Shergill on Understanding Audiences, Stories and the Storytelling Journey, 2021). We use narrative theory as prescribed by Bordwell(2007) -the observations on *fabula, syuzhet* and *film style* on chosen films in both the platforms will be made, model of screenplay suggested by Field (2005) and Indian Filmic storytelling as suggested by Hogan (2009) will be used further to understand the narrative structure of Indian popular films - tentpole and OTT; besides Baranosky and Drennan's(2018) model of new media storytelling will also be considered while analysing the OTT film.

## Story and Storytelling

Stories are as old as is humankind, poetry and oral storytelling are considered to be the oldest forms of literature; poems like *Beowulf* in the tradition of English Literature; *Ramayana* and *Mahabharata* as longest and oldest written verses in India and still very relevant in storytelling traditions here; the *Iliad* and *Odyssey* in Greeks; *Jataka Tales* or *One Thousand and One Nights* in South Asia and *Aesop's Fables* in west. Stories have stayed with us and kept us engaged and quenching for more!

*Figure 1. Four representations of narrative theories of literature and films*[5]

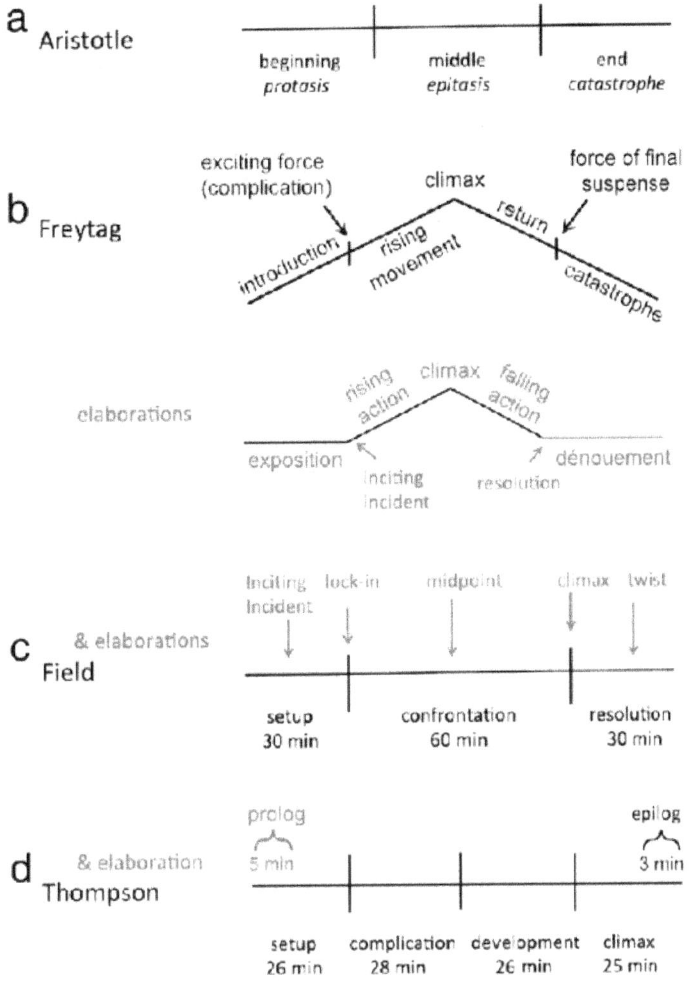


Elements of story include theme, plot, setting, characters, POV, symbolism and tone. Plot is the outline or structure a story follows, setting is the time and place of the story, characters include the protagonists and other characters which make the story - these can be round, static or stereotypes; POV is the perspective from which story is told, symbolism is the recurrent motifs and theme that an author or a storyteller repeats to emphasise something vital in the story and tone is the feel with which a storyteller narrates the story. Short story or novel or drama are literature but film is a different medium altogether where technical and aesthetics have to come together to narrate a story (Rudolf Arnheim's Formalist Film Theory, 2021; Hugo Munsterberg's The Photoplay, 2021). Of all the narrative structures mentioned above in the diagram the one that suits most to cinema is Field's (2005) structure of narrative or Thompson's (2003) especially w.r.t Hollywood narratives.

Field (2005) calls out for dividing the screenplay into three significant parts the beginning, middle and the end and in-between the beginning and middle we have the first plot point and in-between the middle and end we have the second plot point, these need not be in the same chronology but one look at the story should clearly carve out these for us, plot points one and two are those vital incidents in the story which makes the story transcend from one part to the other; the most important entity in a narrative film that drives the story forward is the protagonist - this character should be well etched out; the best character example that he gives of and rightly so is of Charles Foster Kane from the film *Citizen Kane* (Welles, 1941). Orson Welles' masterpiece revolves around this character, it is a character based story which looks at the protagonist in a very *Roshomonic*[6] (fluidly looking at the character through various personality and relationship based lenses) way, then the use of visual elements in a unique way-use of deep focus and close ups to enhance character and emotional details as well as the context in an intricate way. (camera movements and *misc-en-scene*). The use of dialogues is the crux of this story, that one word from the dying media magnet, '*rosebud*' evokes a time and space that moves beyond the filmic space into the audiences' most sensitive moments and back to the delineation of the character story. We will use Fields model of storytelling in films to understand the cinematic storytelling and its newest form but with the cinema being discussed being Indian, inclusion of the *rasa* theory elements becomes imperative along with the cognitive culturalist approach and because storytelling in films is overshadowed by director's perception (Ray, 1976) we will take the auteur's streak also in consideration while analysing the films. Also Bordwell's (2007) narrative theory elements will also be used while analysing the intricate text, that is, cinema; Drennan and Baranovsky's storytelling model for OTT will also be considered; it will be a multipronged analysis.

Bordwell delves into the multifaceted nature of narrative in films, positing that it can be approached as either a representation (semantic aspect of narrative), a structure (syntactic approach), or an act. He emphasizes the importance of a robust narrative theory, highlighting the need for internal coherence, empirical breadth, discriminating power, and an awareness of historical evolution. He distinguishes between Diegetic (telling) and Mimetic (presentation of spectacle) theories of narration.Three pivotal principles connect *syuzhet* to *fabula*—narrative logic, spatial arrangement, and temporal unfolding. Eisenstein's perspective on narration as scenography, framing cinema as a meticulously calculated spectacle for the spectator, is acknowledged. In the realm of fiction film, narration becomes the dynamic interplay between the film's syuzhet and style, working in tandem to shape cues and channels that guide the spectator's construction of fabula. Furthermore, the syuzhet actively molds our perception of fabula by regulating our access to information, determining the relevance we assign to presented details, and establishing a formal correspondence between syuzhet presentation and fabula data.

Narrative is recognized as a trans-medium phenomenon, capable of transcending language to manifest in various forms such as pantomime, dance, images, and even music. When delving into narrative poetics, the focus shifts from ontological considerations to a more psychological exploration. The principles, processes, and practices uncovered are more likely psychological insights than models of rigorous reasoning. A story unfolds across three dimensions: the story world, the plot structure, and the narration. In the cinematic realm, there is a notable encouragement of inferential elaboration. The fine-grain details of the medium play a pivotal role in shaping our interpretation of events within a film. Therefore, narration extends beyond mere storytelling to encompass the intricate patterning of a film's surface texture, including its audio and visual style. In essence, narration within film is the dynamic process through which the viewer is prompted to construct the ongoing fabula based on the organizational framework of syuzhet and the stylistic patterns inherent in the medium.

## Indian Popular Cinema Storytelling Norms

Indian films since their inception have been unique, when put on a global stage the films that earliest garnered most respect and immediate recognition were the art films of Satyajit Ray; of course from popular cinema films of Bimal Roy, Raj Kapoor and Guru Dutt garnered appreciation at that time but as Rajadhakshya (2016) says Ray's film were the second big bang moment of Indian cinema (First Big bang moment of Indian cinema was release of Phalke's movies in theatres), because of their subtle and realistic storytelling and following the form and style of neo-realists; cinema lovers world over could identify with those films; when it came to Indian popular cinema especially Hindi popular cinema; critics and analysts have had been sceptical about it.

Somewhere in late 1980s the tradition of criticism of looking afresh at Hindi popular cinema began, some significant works over the years which have looked at Indian cinema afresh include –Nandy (1981) and Kakar (1981) looking at it as necessary new folk, Dwyer (1998, 2000, 2012, 2014) looking at it as a cultural product and as an aspiration to the new middle class, Hogan (2008, 2009) delineating on its narrative and roots of its storytelling tradition, Roy (2017) deciphering it through *Natyashastra*[7] elements, Mishra (2002) looking at it as a spectacle as a grand syntagm, Ganti (2013) dwelling into it to understand its production ethnography, Lutgendorf (2005,2016), Booth (1995) and Sarrazin (2009) looking at its unique narrative structure and where it takes roots from. Indian popular films were financed by petty Indian businessmen to begin with; the visual structure of it was inspired by the images of calendar art of Raja Ravi Verma; actors initiated with the standard Indian theatre acting, the storytelling structure of folktales, Parsi theatre and mythological scriptures were used to form stories in the tradition of this cinema (Rajadhakshya, 2016)

With the advent of the Golden age in Indian popular cinema (1950s) - directors like K.A. Abbas, Raj Kapoor, Guru Dutt, and Bimal Roy initiated Hollywood editing style, acting stylistics of Stanislavski school (method acting), costumes, set design, *mise-en-scene* and camera movements and techniques inspired by filmmakers worldwide especially Hollywood and Italian Neorealism in Hindi popular cinema. But looking at its story structure, Aristotle's three act structure or a Field's or Thompson's screenplay structure cannot completely analyse the storytelling traditions here. Like mythological scriptures here have multiple stories within a story –a labyrinth of stories, numerous characters and each characters backstory might bring a new connotation to the story present within a story; experimental travel through time and space; existence of multiverses, concept of the wheel of *karma* and the cycle of births and rebirths and the immensity of life cycles; the infusion of digressive elements like song and dance which are though departures but important tools of storytelling . Filmic storytelling is affected due to storytelling traditions here - mixing of different genre elements, dialogue heavy narratives, ahistorical stories and caricature characters and a didactic streak in the stories (Hogan, 2009) - thus these stories cannot be completely and satisfactorily analysed through western models (Pattanaik, 2010). Everything from stereotypical characters to dialogue heavy script to melodramatic acting to infusion of multiple stories in one story, the nature of story being ahistorical (Zankar, 2013), inclusion of song and dance as a story element, grand sets and costumes and a folklorist style of storytelling which is symbol laden and metaphorical yet stereotypical at the surface to give it a mass appeal (Schechner, 2001; Schneider, 2009).Hogan's cognitive culturalist approach, rooted in cognitive theory, offers a distinctive lens for evaluating Indian films. By examining the cognitive processes of perception, attention, and memory, this approach delves into the intricate ways in which cultural nuances shape the cinematic experience. Hogan's perspective proves

*Figure 2. Graphical representation of Indian popular cinema storytelling elements*

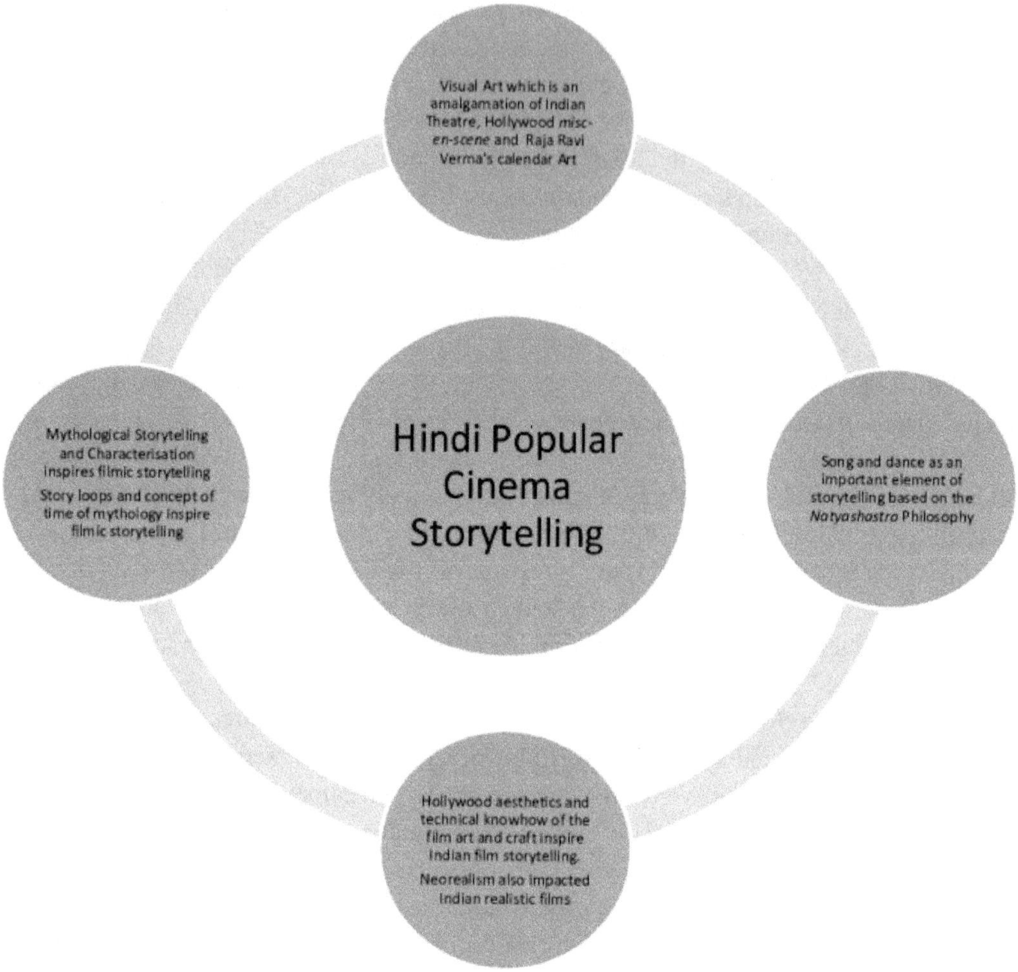


invaluable in dissecting the complex layers inherent in Indian films, facilitating a nuanced critique that carefully navigates the interplay between culture and cognition.

Indian popular cinema has followed the storytelling paradigms as mentioned above but with the onset of digital cinema storytelling there have been detours from the popular cinema storytelling norm. The 1950s Golden age of Indian popular cinema (Gupta,2018) and the *hatke*[8] Bollywood cinema of post 2000 (Wright, 2015) besides the 1970s new wave Hindi cinema are the diversions in filmmaking that challenged the *fabula, syuzhet* and film style parameters of Indian popular cinema

## Exclusive to OTT Indian Cinema Storytelling Norms

During pandemic the Indian cinema business lost two thirds of the revenue and thus, had to become real, they understood that theatrical release is not going to be possible for some time and consequently in order to save themselves from complete shutdown many film producers, financers and distributors decided to release films online on the OTT medium for the first time- some such films include *Gulabo Sitabo* (Sircar, 2020), *Shakuntala Devi* (Menon, 2020) and *Dil Bechara* (Chabbra, 2020). Khandekar (2022) in her book on Indian Media Business painstakingly gives details of how the loss in profit, loss of jobs in film industry and shutting of theatres impacted this industry a lot. OTT could never give back the industry its financial glory – a well-established theatre industry, a strong studio business which can churn out big and small budget films and a pandemic free country, which are very important for cinema business to survive in India (FICCI FRAMES 2023 - Industry Reports, n.d.).

It was post pandemic that a new avenue for small budget and niche films opened up; but this was not the first time; in 1970s in sync with the prevalent sentiments popular cinema had counterparts like middle and parallel cinema; with the onset of consumerism post 1990s Hindi popular cinema came up with many new types of films like domestic drama films (Malhotra and Alag, 2014), the city of ruin films (Majumdar, 2008) and *hatke* films post 2000 (Wright, 2015). But we would say that the pandemic liberated Indian popular cinema once again from its standard practice; as big production houses started giving chances to new directors and filmmakers who have taken fresh and unconventional topics and not only that revamped the storytelling structure as well. Streaming platforms have also provided a platform and financial investment according to business arrangements to the new and budding filmmakers. This seems like the third big bang moment in Indian film industry as audience is exposed to plethora of films based on different subjects, especially made on small budget and in a sleek and organised way in short time span; especially films with narrative structures which are way different from conventional popular cinema narrative structure (Drenn and Baranovsky, 2018; Khandekar, 2022).

New medium is innovative, interactive, effervescent, experimental, democratic, liberal, fragmented and private. It provides the opportunity to watch in our confined and comfortable private spaces at our own leisure. Because of its flexible viewing, films like *Freddy* (Ghosh, 2022), *Qala* (Guptan, 2022) and *Gulmohur* (Chittela, 2023) get the perfect platform to release. Freddy (Kartik Aryan as Dr Freddy Ginwala) is about a nerdy doctor who avenges mistreatment by his girlfriend (Alia F) in the most gruesome way – the cinematic story element that sets it apart from conventional popular cinema storytelling is its lead character Freddy, how the character arc never fails to surprise us! It's a psychological thriller, the cinematography adds up to the creation of spooky ambience especially when we see Aryan alone and in close-ups in frames contemplating to avenge his betrayal. The character is remorseless, there are no moral grounds to look at that is so typical of an Indian lead character, also inflections of Arthur Fleck (Joker,2019) in this character makes the character more dark, intense and with a dead conscience. The flashbacks of a horrible past memory emerging when Freddy is most vulnerable give the film a bit of the melodrama it needs.

*Qala* is poignant cinema; complex and layered; Dimri's character Qala's internal world has been visually created to bring a sense of tangibility through the art design of Meenal Agarwal and cinematography of Siddharth Divan; the dark, desperate, spooky, frustrated and guilt ridden world comes to life with chiaroscuro aesthetics. Qala's character arc is phenomenal starting from a love deprived child to a determined and dreamy youngster, to a vile and ambitious woman who can go to any limits to achieve her goal and prove herself to her mother; and ultimately an insane and frightened adult; which brings strong Freudian and Lacanian interpretative bent to this cinema's visual world. Female characters - the caricatures that have been created predominantly in Indian popular cinema have been criticised and rightly so (Virdi, 1999; Sharma, 2014; Kaur and Sharma, 2017); but OTT has been though at a minuscule scale but been able to show us some real women of mettle; for e.g.- Suchitra's (Priyamani) character from

*Figure 3. Tripti Dimri in a still from Qala*[9]

*Family Man* (Nidhimoru &D.K., 2019); Sandhya (Sanya Malhotra's) character in *Pagglait* (Bist, 2021) and Nimisha Sajayan's character as wife in *The Great Indian Kitchen* (Baby,2021); strong, unapologetic and real characters with depth, complexity and intricacy.

Qala's challenged relationship with her mother (Swastika Mukherjee); short-lived but jealousy ridden relationship with Jagan Batwal (Babil) and her guilt eating away her mental peace; portrayal of yester-year's film industry, position of woman singers, the kind of compromises they did to remain popular and rich. This is a psychological thriller which has a *mise-en-scene* intricately working on creating bygone era of glitz and glamour and psychological battles of a leading star; also due emphasis has been given to troubled relationship between a mother and daughter and the character of mother has been shown in numerous shades which breaks the 'mother image' stereotype of Indian cinema.

*Figure 4. Typical narrative structure of a web-series*[10]

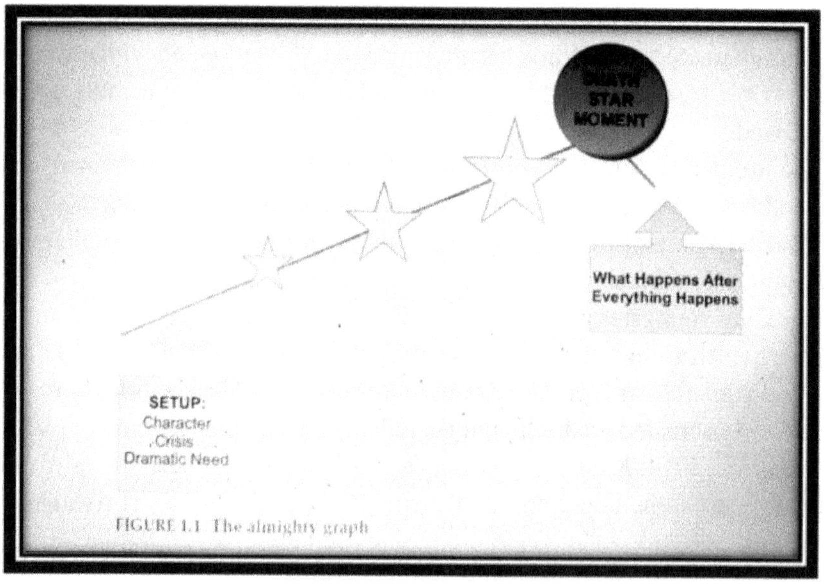

*Figure 5. Narrative structure of epic inspired stories on digital medium*[11]

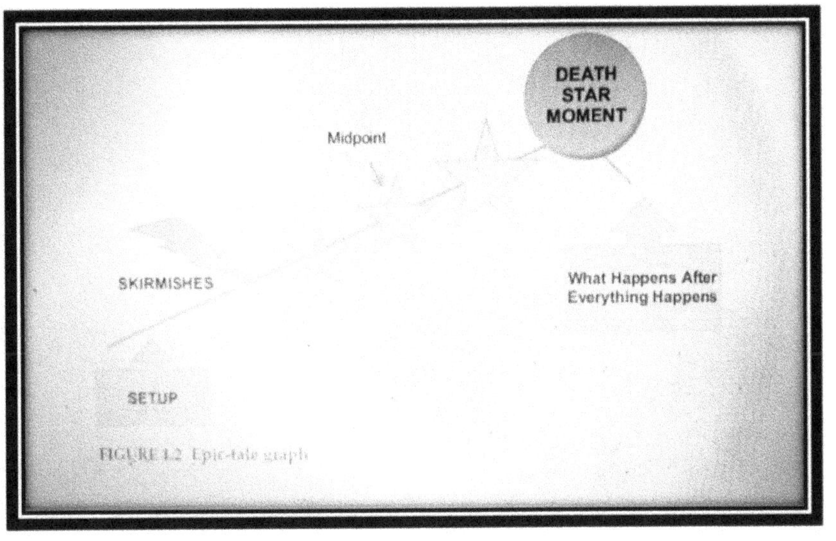


A web series never starts with exposition; rather it starts with the character in a crisis situation and then later on introduces the character and his world in the midst of crisis situations that are thrown at him/her/them. A battery of crisis situations which lead to one major life bursting crisis connoted by the 'death star moment' after which if the character survives and is ready to fight another battle or situation it is described by 'WHAEH'(What Happens After everything Happens) (Baranosky and Drennan,2018). So if visually we try to show Indian popular cinema (tent-pole film) narrative prompts and an exclusive to OTT Indian film, it will somewhat be depicted like this-

*Fabula* in exclusive to India OTT cinema has an amalgamation of *syuzhet* and film style which is inspired by Hollywood narrative strategy i.e. films are made with genre specifications in mind, characters are no more stereotypical, are more realistic and contemporary; song and dance is not a necessary inclusion, these films are also not necessary new folk but intricate and subtle stories with perceptive *syuzhet*. Narrative in these films which can be well understood by Field's (2005) model but Drennan and Baranovsky's (2018) insights need to considered as well where exposition happens somewhere after the plot point two and introductions are character already immersed in problems and on the go; the conflicts are never ending and there are multiple sub climaxes before the main climax and then as the resolution arrives before that we have the WHAEH moment ; which is the most important moment for protagonist; life changing moment and then an open ended conclusion which leaves the audience with a lot to think on. The film style has also contributed in a big way to experiment with the film narrative; in a film like *Qala,* chiaroscuro lighting in *misc en scene* throughout the film makes the mental tormented world of the protagonist very prominent and is also a first when it comes to mainstream cinema, as such stylistic experiments are not a norm. Gulmohar normalising LGBTQIA+ characters, imperfect parenting, ugly family secrets and obscure marriages is the OTT cinema revolutionising screenplay elements for mainstream popular cinema.

*Figure 6. Narrative elements of a tent-pole Hindi film*

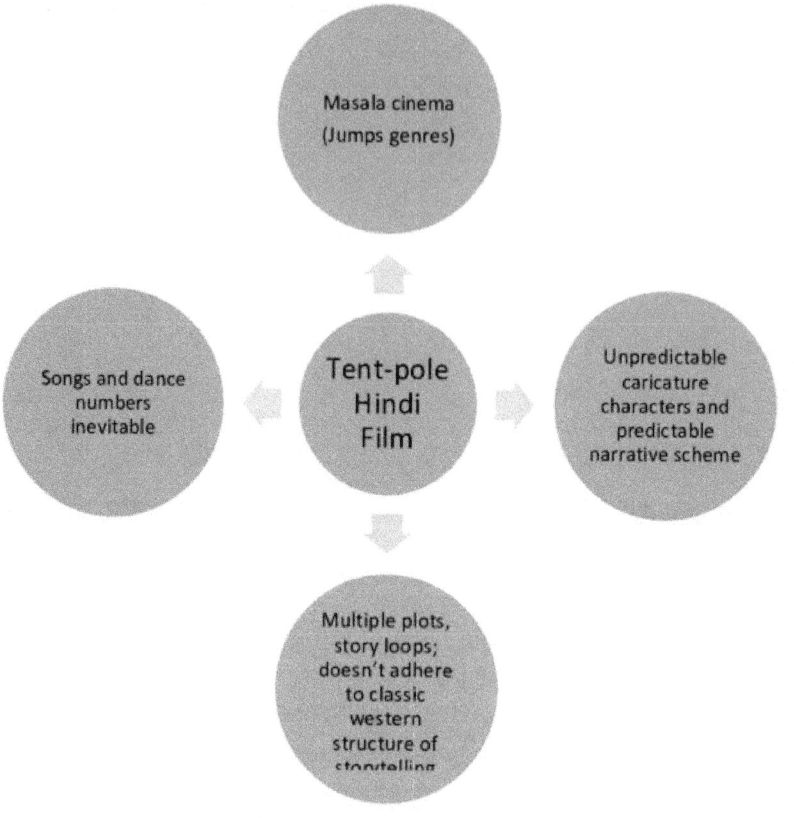


*Figure 7. Narrative elements of exclusive to OTT Indian cinema*

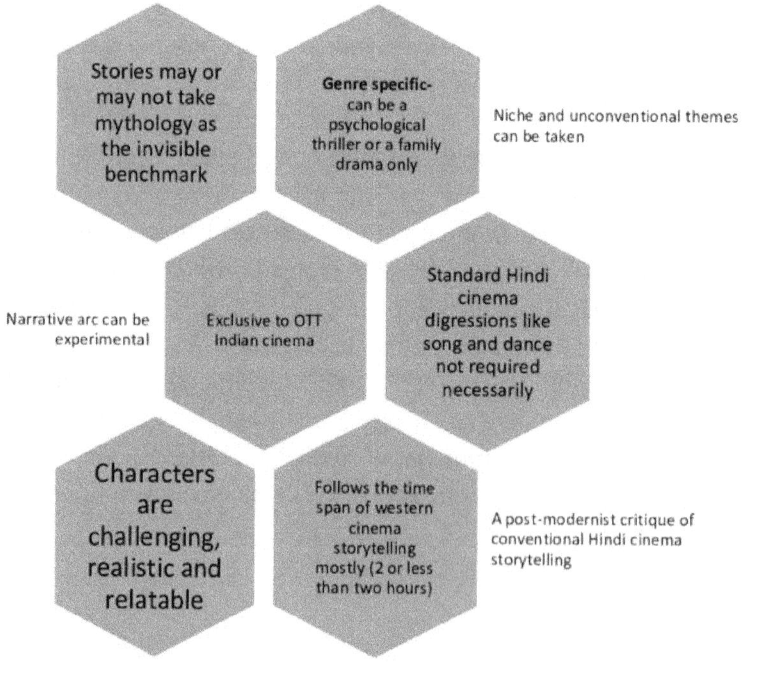

*Figure 8. Poster of Jab We Met*

## Analysing *Jab We Met* and *Meenakshi Sundareshwar*

The Indian OTT landscape is highly competitive, with platforms like Disney+ Hotstar and Amazon Prime Video contending for audience attention. Netflix stands out as a frontrunner, adopting a subscription-based model that prioritizes quality storytelling and production, distinguishing it from competitors relying on advertising revenue or bundled services.Despite facing competition from regional OTT players and digital giants stemming from television broadcasters, Netflix remains a go-to platform for both content creation and viewership. Its impact on setting storytelling benchmarks in India is acknowledged by industry influencers like Ronnie Screwvala and filmmaker Mani Ratnam, attributing Netflix with elevating production standards and inspiring creative talent.Netflix's commitment to an immersive user experience is evident through significant investments in technology. By embracing technological innovations in streaming services, offering features like HDR, Dolby Atmos, and 4K content, Netflix aims to provide not just compelling content but also a superior viewing experience.The evolving OTT industry

*Figure 9. Poster of Meenakshi Sundareshwer*

emphasizes the importance of diverse business models. Hybrid approaches, combining subscription-based (SVOD) and advertising-based (AVOD) models, prove successful, as seen in Disney+ Hotstar's case. Despite challenges, the central role of content remains constant. Compelling narratives in popular series continue to drive viewership, emphasizing the enduring influence of impactful storytelling on these platforms. (Netflix's Monika Shergill on understanding audiences, 2021; Puthiyakath and Goswami,2021; Singh,2019). Choosing Netflix as the video streaming platform and *Meenakshi Sundareshwar* as the OTT film for discussion is purposive but based on their significance, Netflix being the global giant of video streaming whose competition is only considered to be sleep; yet ironically couldn't capture the Indian markets due to reasons to do with elite and expensive subscription plans and varied audience reception and demand for quality *desi* content. *Meenakshi Sundareshwar* is a film which is a fresh portrayal of romcom genre made for OTT; actors are not stars, ambience is not expensive foreign locales; characters are not stereotypical caricatures mirroring Bollywood cliche in characterization; they are ambitious and selfish as opposed to typical Bollywood protagonists who are burdened with morals and principles. The film also received criticism for its stereotypical portrayal of Tamil people; underrated critique of long distance marriages, arranged marriages and the experience of living in joint Indian setups. The language used in songs is also fresh and a mirror of how millennials and gen Z understands poetry- free verse, simple and straight; its not Hindustani or Hindi but *hinglish* that is used in most of the songs. The *misc-en-scene* of the film is simple ; interiors of traditional households (Karaikudi shoots), cosmopolitan IT jungles have been intricately captured by the camera. Outdoors are location shoots to give the audience a feel of Madurai city (Karaikudi shoots) and Bangalore city. *Ayogya Sringara* is portrayed in foggy and dim Ooty. Protagonists' mental struggles are portrayed through closeups but also through typical Hindi film digressions (songs). It's a film which carries few elements of cliche bollywood romcom narrative but builds up on *syuzhet*, and film style with a fresh look for OTT as has been explained above; again its a purposive selection

Analyzing films requires a comprehensive understanding of various theoretical frameworks that illuminate the complexities of this performance art. Field's model and Bordwell's technique of narrative strategy, as outlined in his work on the intricate relationship of fabula, syuzhet, and film style (1985), offer valuable tools for dissecting the structural elements of a narrative. Bordwell's approach focuses on the interplay between fabula (the story's chronological order) and syuzhet (the narrative's actual presentation), emphasizing the significance of film style in conveying the intended message.Cognitive theory, as applied through Hogan's cognitive culturalist approach, provides a unique perspective for critiquing Indian films. This approach delves into the cognitive processes of perception, attention, and memory to analyze how cultural nuances influence the cinematic experience. Hogan's lens becomes particularly useful in understanding the intricate layers embedded in Indian films, allowing for a nuanced critique that considers the intersection of culture and cognition. *Rasa* theory, a fundamental concept in Indian aesthetics, adds another layer to film analysis. As Roy (2017) suggests, Rasa theory becomes inevitable in understanding how specific emotions, their portrayal, or the existence of a scene in the story create a perfect combination of *bhav* (emotion), *vibhav* (stimulus or determinant), and *anubhav* (consequent emotional response). This framework enriches the analysis by highlighting the emotional resonance intended by the filmmakers and its impact on the audience. Film analysis necessitates a multi-faceted approach that considers narrative theory, cognitive culturalist perspectives, and the rich cultural aesthetics embedded in Rasa theory. In the ever-evolving landscape of visual storytelling, the model proposed by Drennan and Baranovsky in their book "Writing for the Digital Age" introduces a crucial perspective for analyzing narrative structures in web series. Their model addresses the unique characteristics of the digital medium, providing insights that complement traditional film and television analysis.Drennan and Baranovsky's model emphasizes the dynamic nature of web series narratives within the digital age. In the context of analyzing web series, their framework likely considers factors such as nonlinear storytelling, interactive elements, and the engagement of online audiences. This is a departure from traditional linear narratives found in films or conventional television shows.Considering this model in conjunction with narrative theories like Field's, Bordwell's, and Hogan's adds another layer to the analysis. While traditional theories offer a foundation, Drennan and Baranovsky's model allows for the exploration of how web series narrative structures adapt to the unique demands of the digital platform. It prompts considerations of user engagement, audience interaction, and the impact of the digital environment on storytelling techniques.In essence, the integration of Drennan and Baranovsky's model expands the analytical framework to encompass the specificities of digital storytelling, recognizing that web series often present narratives in a more fragmented, interactive, and audience-oriented manner. This approach is particularly relevant in the contemporary media landscape, where digital platforms have reshaped not only how stories are told but also how audiences participate in and experience narratives.

*Jab We Met* is considered to be a rom-com, while as its difficult to put Hindi tent-pole cinema in genres; but for dominant discourses that's where it is compartmentalised. It is an Imtiaz Ali film, someone who turned the romantic narrative of popular Hindi cinema which at that time was ruled by bourgeois romances of Chopras and Johars (Arya, 2017; Arya and Verma, 2021) upside down. His films projected a cosmopolitan and glamorous lifestyle but also had psychological investigations into the protagonist's mind-set; the confident city goer and their internal battles were laid bare in front of audiences with films like *Jab We Met*, *Highway* (2014), *Rockstar (*2011) and *Tamasha* (2015). Travel, especially, road travel became his symbolic motif in film after film and Rahman's music and Irshad kamil's *Hindustani* lyrics mostly (*pataka gudi ho, tirkit kaal se chali kahani* or *kun faya kun*) through these journeys led to a self-revelation of the characters; of course romance and its moments intact throughout (demand of this (popular) cinema, as Mishra (2002) says Grand Syntagm that is Romantic Melodrama). His films de-glamorised the protagonists laying bare their imperfections; but also redeeming them by ending with some kind of mystic experience; romance and standard digressions keeping the rom-com façade intact. The *fabula* in his films is at crux the story of self evolution which through the *syuzhet*, the essentials of bollywood rom-com gets veiled and the texture of bollywood romance becomes dominant supported by unique filmmaking style of Imtiaz which will have non linear storytelling, song and dance digressions which have their own subplots, for e.g- in *Tamasha, Tirkit kaal se chali kahaani* is an ode to the storyteller infused and veiled in the romance of Ved (Ranbir kapoor)and Taara (Deepika Padukone)which is the dominant string of narrative but crux is evolution of Ved's storyteller self. Imtiaz's film style will also include a montage of road travels in India, sufi music and internally perplexed characters on a journey. Graffiti art in his *misc en scene* though a western art form brings strong rasa infused performance to the fore ; for e.g.- in the song *Agar tum saath ho,* the post modernist art behind the protagonist evokes strong *vipralambha* rasa.

*Jab We Met* though not as intricate as a *Tamasha* or *Highway* is a film whose narrative can be deciphered through Field's (2005) model but requires thorough understanding of popular Indian cinema storytelling, its tradition and its various elements like folklore and mythology to understand the film's narrative arc. Introduction is exposition of both protagonists and their world, we reach the first important story beat with Geet (Kareena) missing her train and the frenzy that follows with it; Aditya (Shahid) realises he's in love with the song *Ye Ishq Haaye* ; this is another important point in the story immediately after Geet and Aditya run away from her home so that she can meet and settle down with her lover ; the moment he leaves her near her destination is the first plot point, that's when middle starts and then Aditya's rise and his celebration of esoteric love with Geet and and then his realisation that Geet is not in touch with her family that's climax; second plot point arrives with Aditya meeting Geet (she is oblivious of his presence) with the song *Aoge jab tum o saajna* (*vipralambha* rasa is evoked); conclusion section is about Geet's realisation of Aditya's love and the satiric trajectory this realisation follows. Songs and dance comprise important digressions in this story but without them the true *rasic* experience of the story cannot be realised. Also, Aditya's and Geet's characters are more of caricatures; Pawan Malhotra's character as Geet's uncle looks more authentic though w.r.t. his mannerisms, costume and use of Punjabi language. Thus, a storyline infused with song and dance numbers, caricature characters, simplistic narrative which is melodramatic, released with huge theatrical reception and grand marketing strategy for its times places this film among the tent pole Hindi cinema. *Fabula's* main aim is the love story between the two characters but *syuzhet* uses travelogues, self realisation plot and hindi cinematic digressions to render the story to audiences with the unconventional film making style of Ali.

Contrastingly, *Meenakshi Sundareshwar* is a new age rom-com which tries to stay within the rom-com genre parameters; discusses a never before theme of long distance relationships and the couple's struggles in a typical joint Indian setup. Meenakshi's character is not a caricature though the mildly stereotypical portrayal of a tamilian family was severely criticised by many reviewers and critics (Meenakshi Sundareshwar Review: A Languid Film Let Down by Its Unbelievable Conflict, 2021; Meenakshi Sundareshwar Treads off the Beaten Path, 2021); Meenakshi's character (Sanya Malhotra) is bold, has independent thinking and yet as many Indians do, gets married in a traditional way after one meeting with Sundareshwar (Abhimanyu) and his family. In contrast to Geet, this character has more complexity and realistic tone; her difficulties in marriage or her incompatibility with the joint set-up doesn't break her; instead it puts her in a state of contemplation about the future. She is a thinker and someone who wants to achieve and carve out a niche for herself; she is not a typical arm-candy character of Hindi cinema. Sundareshwar's character is subtle and a simpleton his character arc grows with the experiences he undergoes with staying away from family and dealing with a long distance marriage; gradually we get to see shades of this character who is subtle and looks like a typical engineer graduate from a middle class family sans the melodrama and theatrics we encounter in Aditya's portrayal (Shahid). Sundar is no glorious hero who has extraordinary skills, he cannot fight like a typical popular cinema hero; his personality defies any chocolate-boy image characteristics that is so typical of popular cinema. Sundar cannot write poetry and sing and have traits of an eternal romantic. Sundar is very real and simple like any of us who wants to reach a balance between love, relationships and work; his aspirations are as limited as any middle class Indian's; and that is what connects the streaming audience with this character.

The film has been appreciated for taking a topic like long distance marriage but storytelling falls short of creating significant conflicts and their resolution as per critics though it has been among Netflix top ten watch-list for India for a significant time.The story can be critiqued through the model of OTT video streaming storytelling that Baranovsy and Drennan (2018) give us; though they give it for web-series narrative but it has some similarity to exclusive to OTT Indian popular cinema narrativeas well but only this model wont be sufficient for analysis cognitive culturalist lens, rasa elements and Field's model also need s to be used for a near to satisfactory analysis. Both the characters are not established at the onset but exposition shows us both the characters getting married the first milestone (story beat) of the story; later as Sundar leaves his newly wedded wife for a job in Bangalore the characters start unfolding; we see Meenakshi's happy go lucky nature getting stifled with the *ayogya sringara*[12] moments she has to go through. This is a simple rom-com which tries to deal with issues of long distance marriage, millennial in a typical Indian joint setup; friction between the old and new mind-set and a fresh perspective of how a futuristic woman looks at marriage as not the start and end of her life.

There are no melodramatic moments, romance is subtle and so is the pain of separation; from the exposition the narrative moves to small star moments which are like the mini conflicts or pivot points of story leading to the starburst moment when Sundar's lies are exposed in front of Meenakshi and that is like a death blow/ death star moment to this relationship. The ending is a bit abrupt and could have been written better; the subplots like Meenakshi's friendship with Ananthan (Varun Rao) could be explored in a much more nuanced way. The movie has subtle protagonists with realistic tone, minimal melodrama in narrative or character portrayals, it has songs which is not a necessity in this kind of cinema; besides it takes a fresh theme of long distance marriage and is extremely conscious of its female protagonists internal world.

The recurrent motifs of popular Indian cinema; a chocolate boy image man and a fair and fragile female; economic divide among the protagonists. Surreal qualities of the main leads –look good, can sing and dance and can be achievers; the man has to come and rescue the woman; as soon as the romantic relationship breaks, the woman loses her focal point in life. There will be a necessary rain dance and the umbrella scene. The start and end of the woman's life is her relationship status; the man has priorities such as economic stability and a romantic relationship and the complete man image to be managed. *Jab We Met* though an unconventional romance for its times but still follows the mainstream storytelling trajectory which *Meenkashi Sundareshwar* doesn't.

Contrastingly, *Meenakshi Sundaresheshwar* has two average looking protagonists who have been put in a very realistic setting of joint Indian households which are still prevalent in tier two and three, small towns and villages. Their first *Sambhoga sringara* moment is not in Swiss Alps but near the window of their room which is a barely private space. They have open conversations about sex and its pleasures which some years back would have been impossible in a popular cinema narrative. Marriage is a phase in life, it's not something that is the- be it and end it of the narrative. Sundar wants to be a successful engineer and Meenakshi wants to spend her time joyously in a joint setup. She has romantic fantasies but her career is also important for her. She is of an independent mind, considers this a marriage of equals and when she senses that her freedom and feelings are at stake, she decides to give this relationship a break and that also a thoughtful one. We have grey characters, thinking breathing souls in the film instead of caricatures or stereotypes. No vile mother in laws or stunted sister in laws. The *misc-en-scene* of the film is not exaggerated sets or exotic locations but small cities and traditional households and hustling and bustling gullies and lanes of exotic foods and simple humanity versus the unemotional construction jungle kind cities which look sleek and glamorous but are bereft of human touch and real emotions. The protagonists are simply dressed and shown normally instead of creating an extraordinary bubble around them like Aditya's stingingly rich persona which makes him free of any financial troubles and Geet's *ayogya sringara* or *viraha* shown in beautiful landscapes. For Meenakshi and Sundareshwar's *sringara* moments are their bedroom spaces, video calls and simple conversations and *nazar milana* moments are enough

Comparing dialogues of both the films gives a insight into the character creation as well as use of language in the films,

Sundareshwar explaining his worth as a husband to Meenakshi in a light Sringara moment:

*Engineers make the best husbands ... how ... we don't give up that easily ... engineering ho ya relationships, ek baar decide kar liye toh poora karte hai ... we're great problem solvers ... engineer ko ek baar answer bata do ... wahan tak pahunchna kaise hai, woh figure out kar lega ... we're experts in time management ... ek raat mein exam ke liye prepare karna ho ya ek date mein girlfriend ko imrpress karna ho ... no one can do that better than an engineer*

Meenakshi responds back*: I want to work at a small firm, where I can make a big difference ... rather than working at a big firm, where I make no difference*[13]

The dialogue is an insight into the characters mind, Sundar is a simple man with limited expectations from life but proud to have chosen engineering as his career ; while as Meenakshi is smart and cosmopolitan and marriage is not where her life ends, she has career plans. Dialogues are written in Hinglish, tone and intentions are bereft of any moralistic or familial expectations.

A dialogue from Jab We Met:

*Geet: Main apni favorite hoon Bachpan se hi naa ... mujhe shaadi karne ka bahut craze hai, by god...*

*Aditya: Tum ho meri problem. Bolti hi ja rahi ho bolti hi ja rahi ho. Tumhe samajh mein nahi aa raha hai ki mai tumse baat karne mein interested nahi hun. Mai nahi karna chaahta tumse baat, mai nahi jan'na chahta hun ki tum Bhatinda ja rahi ho ya Banaras, kisi hostel mein rehti thi ya brothel mein, I just don't care alright. So just please leave me alone.*[14]

Geet was a character different for its times because there is no *sharm*[15] or *lihaaz* [16]or *ghar ki izzat* [17] kind of thoughts that create this character, she is unapologetic but still not an economically independent character or someone for whom marriage is not vital; she lives and breathes to be in a marriage. She is melodramatic and exaggerated in her manner as well. Aditya is a typical Bollywood male protagonist who has a *bechara*[18] complex which he is able to come out of because of his companionship with Geet. He is rich and in love, good looking and traditional with a huge amount of moral baggage which is not the case with Sundar.

Meenakshi Sundareshwar's f*abula* is a modern romcom infused with questions of long distance relationships and habitation in joint Indian setups but its *syuzhet* explores the story through a millennial point of view, where open discussions on sex are normalised, characters are realistic and not grand; career and self centeredness is prioritised and being out of love is normalised rather than creating huge moments of *vipralambha*. Film style gives it a very middle cinema kind of a feel, where settings are small Indian cities which are realistically captured rather than creation of huge sets. Characters are not greek gods but normal looking and dressed people, especially protagonists; to show their love fancy locales are not needed just private space and some erotic conversations in close ups. Indian cities are hill stations are used to create misc en scene which is rasa infused like Ooty for Meenakshi's *ayogya sringara* . The story chooses a rarely portrayed topic of long distance relationships and romance in joint family spaces in India. It's a new age romance for the OTT medium.

## CONCLUSION

Storytelling has been an ancient practice, evolving with the development of various communication mediums. The advent of cinema provided a tangible form to imaginative narratives, captivating audiences with visual and sensory experiences. Pioneers like Melies in the West and Phalke in India introduced magical and science fiction stories, setting the stage for diverse cinematic expressions. Indian cinema, from its inception, strived to reach the masses, cultivating a vibrant market that continuously attracted aspiring talents and eager audiences. It transitioned from mythological tales to a myriad of genres, with *masala* cinema emerging as the most dominant and commercially successful. However, with the digital medium and the rise of OTT platforms, storytelling underwent a transformative journey, becoming more diverse, dense, and unconventional.

The exclusive presence of Indian popular cinema on OTT platforms is a refreshing development in the industry. Unlike traditional narratives, these films eschew popular storylines and characters, resonating with both cosmopolitan and small-town audiences. They adopt experimental storytelling techniques, breaking away from the conventional reliance on star power and fandom. Unlike their mainstream counterparts, these films may not adhere to the traditional elements of songs and dances, and they might prioritize realism over glamour. The narrative structure of OTT cinema allows for greater flexibility, unburdened by the constraints of conventional production, distribution, and release strategies. This shift prompts an exploration of narrative structures in Indian cinema. It involves understanding how traditional narratives differ from the unconventional and experimental approaches adopted by OTT cinema. The chapter delves into the contrasts between these two forms, examining the impact on storytelling, character portrayal, and audience engagement. It seeks to unravel the unique qualities that distinguish OTT cinema from its mainstream counterparts, shedding light on the evolving landscape of Indian storytelling.

In this chapter, our focus has been on unraveling the intricacies of storytelling in cinema, delving into its creation, comprehension, and critique. We traverse through diverse narrative models, exploring the interplay between *fabula, syuzhet*, and film style to elucidate how their nuanced relationship shapes cinematic stories for audiences. The significance of narrative in the realm of cinema is thoroughly examined.Our investigation extends to the demystification of the narratives inherent in typical Bollywood films and their contrasting counterparts on OTT video streaming platforms. A particular emphasis is placed on the narrative dynamics of Bollywood films on OTT platforms, with a spotlight on Netflix as an elite storytelling platform. We draw distinctions between the traditional tent pole Bollywood narrative and the more contemporary Bollywood rom-coms on OTT platforms, using films like *Jab We Met* and *Meenakshi Sundareshwar* as case studies.

While our study is intentionally limited in scope, focusing on Netflix and a purposive small sample, it serves as a critical essay that initiates exploration into the subject. We shed light on the evolving narrative strategies employed in new-age OTT films, acknowledging the potential for further, more extensive research. Notably, this study cannot be generalized, leaving ample room for in-depth investigations with larger sample sizes.Moreover, avenues for future research are proposed, suggesting a broader examination of various stories on different video streaming platforms. The rendering of *fabula* through *syuzhet* and style in OTT content can be further explored, alongside investigating audience reception through ethnographic or netnographic studies. The exploration of textual aspects in web series, documentaries, and films on OTT platforms presents a rich field for future research endeavors. Overall, this study serves as a foundational exploration, beckoning further in-depth and expansive studies into the multifaceted realm of OTT cinema narratives.

## REFERENCES

Dwyer, R. (1998). The Hindi Romantic Cinema. *Journal of South Asian Studies*.21(1). Retrieved from: https://www.tandfonline.com/loi/csas20

Dwyer, R. (2000). The erotics of the wet sari in Hindi films. *South Asia: Journal of South Asian Studies*, 23(2), 143-160, DOI: doi:10.1080/00856400008723418

Dwyer, Rachel (2012). Fire and Rain, The Tramp and The Trickster: Romance and the Family in the early films of Raj Kapoor. *The South Asianist.* 2(3), 9–32. Retrieved from: www.southasianist.ed.ac.uk

Hogan, P. (2009). Hindi Cinema as a Challenge to Film Theory and Criticism. *Projections*, 3(2), 5-9. Retrieved from: https://doi.org/ doi:10.3167/proj.2009.030201

Vasudevan, R. (2008). Shifting Codes, Dissolving Identities. *Third Text,* 10(34), 59-77. Retrieved from: http://dx.doi.org/. doi:10.1080/09528829608576595 62

An Unruly Modernity: The City in Bombay Cinema. (n.d.). The Book Review, Monthly Review of Important Books. https://www.thebookreviewindia.org/an-unruly-modernity-the-city-in-bombay-cinema/

Arya, K. K., & Verma, M. (2020). Contemporary Bollywood's Tryst with Bourgeois Romances of Hindi Cinema. Journal of Content. *Community and Communication*, *13*, 308–315.

Arya. (2017). Imtiaz Ali as an Auteur. Amity Journal of Media & Communication Studies, Vol.7(1), 50–53.

Booth, G. D. (1995). Traditional Content and Narrative Structure in the Hindi Commercial Cinema. *Asian Folklore Studies*, *54*(2), 169–190. https://www.jstor.org/stable/1178940. doi:10.2307/1178940

Bordwell, D. (1985). *Narration in the Fiction Film*. University of Wisconsin Press.

Bordwell, D. (2008). *Poetics of Cinema*. Routledge.

Bordwell, D., & Thompson, K. (2010). *Film Art: An Introduction*. McGraw-Hill Companies,Incorporated.

Cutting, J. E. (2016). Narrative theory and the dynamics of popular movies. *Psychonomic Bulletin & Review, 23*(6), 1713–1743. doi:10.3758/s13423-016-1051-4 PMID:27142769

Das Gupta, C. (2008). *Seeing is Believing: Selected Writings on Cinema*. Penguin Viking.

Drennan, M., Baranovsky, Y., & Baranovsky, V. (2018). *Scriptwriting for web series: Writing for the Digital age*. Routledge. doi:10.4324/9781351237857

Dwyer, R. (2010). Bollywood's India: Hindi Cinema as a Guide to Modern India. *Asian Affairs, XLI*(III), 381–398. https://www.tandfonline.com/loi/raaf20. doi:10.1080/03068374.2010.508231

Dwyer, R. (2014). *Picture Abhi Baaki Hai: Bollywood as a Guide to Modern India*. Hachette India.

FICCI FRAMES 2023 - Industry Reports. (n.d.). FICCI FRAMES 2023 - Industry Reports. https://www.ficci-frames.com/reports.html

Ganti, T. (2013). *Bollywood: A Guidebook to Popular Hindi Cinema*. Routledge. doi:10.4324/9780203834411

Hogan, P. C. (2009). *Understanding Indian Movies: Culture, Cognition, and Cinematic Imagination*. University of Texas Press.

Hugo Munsterberg's The Photoplay. (2021, February 4). YouTube. https://www.youtube.com/watch?v=vd32nkyUjgM

Kakar, S. (1981). The Ties that Bind: Family Relationships in the Mythology of Hindi Cinema. *India International Centre Quarterly*. 8:1, 11-21. Retrieved from- https://www.jstor.org/stable/23001932

Kaur, S. P., & Sharma, V. "A transition from Male Mas- culinity to Female Androgyny in Hindi Cinema." Media Watch,vol. 8, no. 2, 2017, pp. 287-94. https://mediawatchjournal.in/325mwj/may17-12.pdf

Kohli-Khandekar, V. (2022). *The Indian Media Business*. SAGE Publications.

Kohli-Khandekar, V. (2023, July 25). *Farzi, Citadel top OTT charts in first half of 2023: Ormax Media report*. www.business-standard.com. https://www.business-standard.com/entertainment/farzi-citadel-top-ott-charts-in-first-half-of-2023-ormax-media-report-123072500905_1.html

Laxmii to Dil Bechara, big Bollywood films released on OTT in 2020. (2020, December 22). India Today. https://www.indiatoday.in/movies/bollywood/story/laxmii-to-dil-bechara-big-bollywood-films-released-on-ott-in-2020-1751978-2020-12-22

Lutgendorf, P. (2005). Sex in the Snow: The Himalayas as Erotic Topos in Popular Hindi Cinema. *Himalaya: Journal of the Association for Nepal and Himalayan Studies, 25*(1-2), 29–37.

Lutgendorf, P. "Is There an Indian Way of Filmmaking?" *International Journal of Hindu Studies*, vol. 10, no. 3, 2006, pp. 227–56. *JSTOR*, https://www.jstor.org/stable/20106974. Accessed 24 Aug. 2023.

Malhotra, S., & Alagh, T. (2004). Dreaming the Nation: Domestic Dramas in Hindi Films Post 1990. *South Asian Popular Culture, 2*(1), 19–37. doi:10.1080/1474668042000210492

MAMBROL. N. (2018, January 24). Key Theories of Mikhail Bakhtin. Key Theories of Mikhail Bakhtin – Literary Theory and Criticism. https://literariness.org/2018/01/24/key-theories-of-mikhail-bakhtin/

Mazumdar, R. (2007). *Bombay Cinema: An Archive of the City*. University of Minnesota Press.

McQuail, D., & Deuze, M. (2020). *McQuail's Media and Mass Communication Theory*. SAGE Publications.

Meenakshi Sundareshwar Review. Meenakshi Sundareshwar treads off the beaten path. (2021, November 5). Meenakshi Sundareshwar Review: Meenakshi Sundareshwar Treads off the Beaten Path. https://timesofindia.indiatimes.com/web-series/reviews/hindi/meenakshi-sundareshwar/ottmoviereview/87282604.cms

Meenakshi Sundareshwar review: A languid film let down by its unbelievable conflict. (2021, November 5). The Indian Express. https://indianexpress.com/article/entertainment/movie-review/meenakshi-sundareshwar-movie-review-a-languid-film-let-down-by-its-unbelievable-conflict-7609080/

Mishra, V. (2002). *Bollywood Cinema: Temples of Desire*. Routledge.

Nandy. Ashis. (1981). The Popular Hindi Film: Ideology and First Principles.*India International Centre Quarterly*. Vol. 8(1), 89-96. Retrieved from:https://www.jstor.org/stable/23001938

*Netflix's Monika Shergill on understanding audiences, stories and the storytelling journey*. (2021, May 6). Afaqs! https://www.afaqs.com/news/media/netflixs-monika-shergill-on-understanding-audiences-stories-and-the-storytelling-journey

Pattanaik, D. (2010). *Jaya*. Penguin Books.

Puthiyakath, H. H., & Goswami, M. P. (2021). Is Over the Top Video Platform the Game Changer over Traditional TV Channels in India? A Niche Analysis. *Asia Pacific Media Educator*, *31*(1), 133–150. doi:10.1177/1326365X211009639

'Qala' movie review: Anvitaa Dutt's mother-daughter tale is poignant and admirable. (2022, December 2). The Hindu. https://www.thehindu.com/entertainment/movies/qala-movie-review-anvitaa-dutts-mother-daughter-tale-is-poignant-and-admirable/article66215167.ece

Rajadhakshya. (2016. Indian Cinema: A Very Short Introduction (1st ed.). Oxford.

Rajdhakshya, A. (2003). The 'Bollywoodization' of the Indian Cinema: Cultural Nationalism in a Global Arena. *Inter-Asia Cultural Studies*, *4*(1), 25–39. doi:10.1080/1464937032000060195

Ray, S. (1976). *Our films, their films*. Orient Longman.

Review, F. Kartik Aaryan is a revelation in this dark psycho-thriller. (2022, December 2). Freddy Review: Kartik Aaryan Is a Revelation in This Dark Psycho-thriller. https://timesofindia.indiatimes.com/entertainment/hindi/movie-reviews/freddy/ottmoviereview/95835669.cms

Roy, P. (2016). The Aesthetics of Emotional Acting: An Argument for a Rasa-based Criticism of Indian Cinema and Television. (Doctoral Thesis, University of Edinburgh). Retrieved: https://era.ed.ac.uk/handle/1842/22910

Rudolf Arnheim's Formalist Film Theory. (2021, February 8). YouTube. https://www.youtube.com/watch?v=_qvrzA1b-p0

Sarrazin, N. (2008). Celluloid love songs: Musical *modus operandi* and the dramatic aesthetics of romantic Hindi film. *Popular Music*, *27*(3), 393–411. https://www.researchgate.net/publication/291909806_Celluloid_love_songs_Musical_modus_operandi_and_the_dramatic_aesthetics_of_romantic_Hindi_film_Popular_Music_2008_27_3_393_DOI101017_S0261143008102197.doi:10.1017/S0261143008102197

Schechner. (2001). *Rasaesthetics*. California State University, Northridge.

Schneider, A. (2009, January 1). *Hum Aapke Hain Koun. . .!: An Example of the Coding of Emotions in Contemporary Hindi Mainstream Film*. Projections; Berghahn Books. https://doi.org/ doi:10.3167/proj.2009.030205

Sharma, S. (2017). Transnational Publics, Nationalist Ideology, and the Women Question in Hindi Cinema. *South Asia*, *41*(1), 106–120. doi:10.1080/00856401.2017.1357094

Sharma, V. (2022, July 19). How OTT has changed the story narrative gameplan of content creators. How OTT Has Changed the Story Narrative Gameplan of Content

Singh, P. (2019). New Media as a Change Agent of Indian Television and Cinema: A Study of Over the Top Platforms. *Journal of Content, Community & Communication*, *9*(5), 131.

Standard, B. (2022, August 17). Partho Dasgupta: Change in the strategy of OTT platforms needed to succeed in India. https://www.business-standard.com/content/specials/partho-dasgupta-change-in-the-strategy-of-ott-platforms-needed-to-succeed-in-india-122081700945_1.html

T, S. (n.d.). *An Overview of India's Booming OTT Market | marketfeed*. https://www.marketfeed.com/read/en/an-overview-of-indias-booming-ott-market

Today, B. (n.d.). *Netflix Next*. Netflix Next. https://www.businesstoday.in/interactive/immersive/netflix-india-popular-ott-platform-in-india/

Vemireddy, V. (2011). Ram Gopal Varma as an Auteur. *Journal of Creative Communications*, *6*(1-2), 195–202. doi:10.1177/0973258613499218

Virdi, J. (1999). Reverence, rape- and then revenge: Popular Hindi cinema's 'women's film'. *Screen*, *40*(1), 17–37. http://screen.oxfordjournals.org. doi:10.1093/screen/40.1.17

Wright, N. S. (2015). *Bollywood and Postmodernism: Popular Indian Cinema in the 21st Century*. Edinburgh University Press. doi:10.3366/edinburgh/9780748696345.001.0001

Zankar, A. (2013). *Mughal-E-Azam*. HarperCollins Publishers India.

## ENDNOTES

1 According to the Rasa theory when the actor performs and their *bhava, vibhava* and *anubhava* create a sense of *rasa* in the audience the receptor becomes a *rasic*.

2 Refers to the phenomena of change in this cinema w.r.t its narrative, aesthetics, technique or marketing of it through this term

3 Masala is a balanced combination of spices which when used in proper quantity and used in cooking following age old culinary rules can make a food item aromatic, healthy and tasty, this connotation

was used by Prof Zankar in a conversation with me to describe how important it is for Hindi film to break genre rules and concoct a story which satisfies the age old storytelling traditions of our country and includes some techniques of world cinema as well to create something that has been alive and running India cinema storytelling for over 100 years now.

4 Technology that delivers streamed content over internet

5 Cutting. (2016). Four representations of narrative theories of literature and films. https://link.springer.com/article/10.3758/s13423-016-1051-4#citeas.

6 *Roshomon* (Kurosawa,1950) is a Japanese film which experiments with the narrative in a typical oriental folklorish streak (Pattanaik in conversation with Diwedi about oriental story motifs (Lallantop Newsroom,Dec,2022))

7 The *Natyashastra* is one of the oldest surviving ancient Indian works on performance arts.

8 New new wave cinema of filmmakers like Kashyap and Bhardwaj

9 Kumar, A. (2022, December 2). *'Qala' movie review: Anvitaa Dutt's mother-daughter tale is poignant and admirable*. The Hindu. https://www.thehindu.com/entertainment/movies/qala-movie-review-anvitaa-dutts-mother-daughter-tale-is-poignant-and-admirable/article66215167.ece

10 Drennan and Baranovsky. *Scriptwriting for Webseries*. Routledge.

11 Drennan and Baranovsky. *Narrative structure of epic inspired stories on digital medium, Scriptwriting for Webseries*. Routledge.

12 Separation in love and its depiction in a rasa performance

13 Sundar and Meenakshi talking about themselves whereas Sundars explains he is a simple man and a proud engineer, Meenakshi replies that though she wants to get married in a joint setup she aspires to work as well.

14 Geet says that she loves herself a lot and Aditya, exasperated with her verbal exaggeration angrily tells her that he is not interested in her

15 Shy

16 Respect for elders and family

17 Women's chastity equals a family's respect in many orthodox Indian setups

18 Someone who is unable to overcome a breakup; this term has been used by Mishra (2000) to explain traits of a typical Bollywood male protagonist

# Chapter 10
# The Impact of OTT Media on Comedy in India: Changing Trends and New Opportunities

**Sonali Jha**
*Ohio University, USA*

**Tamanna M. Shah**
https://orcid.org/0000-0001-9609-0191
*Ohio University, USA*

## ABSTRACT

*The chapter emphasizes the impact of OTT media on the comedy industry. A comprehensive exploration of various OTT platforms such as Netflix and Amazon Prime Video along with social media platforms, we highlight how OTT has transformed distribution and consumption habits, enabling comedians to reach a global audience. Employing content analysis of four case studies of comedians from India, this chapter provides an analysis of the trends and opportunities of digital platforms for comedians to experiment with content creation, reach varied audiences, and build personal brands. Lastly, it examines the comedian's comedy style in adapting to a changing landscape and audience preferences; providing insights into the shifting comedy ecosystem since the emergence of OTT media through an analysis of comedians who have evolved with the OTT platform.*

## INTRODUCTION

The advent of Over-The-Top (OTT) media has revolutionized the entertainment industry, including the comedy genre. Over the past decade, platforms such as Netflix, Amazon Prime Video, and others have disrupted traditional distribution models and provided a direct opportunity for comedians to showcase their talent to the global audience. There has also occurred a change in the consumption of comedy content, especially with the advent of the internet age, social media, and subscription-based models. With the growth of the internet in both rural and urban areas, people are now able to binge-watch entire seasons of sitcoms and stand-up specials on streaming services like Netflix and Amazon Prime Video.

DOI: 10.4018/979-8-3693-3526-0.ch010

Consequently, a new subgroup of comedy fans has emerged like never before, one that seeks to enjoy their entertainment free from the restrictions of conventional broadcast scheduling. Like social media, these streaming services have created employment opportunities for new comedians to showcase their talents and imparted greater creative freedom, access to a global audience, and the ability to build personal brands domestically and at international platforms such as the Cannes Festival. These new budding comedians are now able to reach more audiences and experiment with different formats of delivery, while also earning a decent living. It is this transformative influence of Over-The-Top (OTT) media on both production and consumption of the comedy industry that this chapter explores. With the advent of OTT platforms, comedians have unprecedented opportunities to reach the world's audience and experiment with new forms of content creation and distribution. However, this digital revolution also presents unique challenges for comedians who must navigate changing landscapes and adapt to audience preferences. By analyzing the emergence of comedy specials on OTT platforms and examining the effects on video production and consumption happening since a decade, this chapter aims to provide valuable insights into the evolving comedy ecosystem.

The first section of the chapter provides an overview of significance of comedians in the past, its arrival, and its importance. The second section provides a glimpse of the emergence of OTT in India, opportunities created through OTT for its main players (comedians), the creation of market trends, and demographics of the audience. Additionally, it discusses how OTT platforms have democratized comedy by providing a platform for emerging comedians with no discrimination on language or gender and diversifying content available to the world's audience. The chapter concludes with a recap of the chapter along with a case study to provide an example of the changing landscape of comedy with the demand of different genres and maintaining loyal viewers.

## History of Comedy

*"Comedy delights in itself as it delights in the shows and vanities of the world, as it delights in concupiscence. Comedy beckons us over the brink of laughter as over the brink of temptation. It laughs at the Wise Man, at the Christ or the Devil. It laughs for the sake of laughter and for the sake of freedom, of liberation from the constraints of righteousness, wisdom, and holiness. It laughs at freedom too and finally, at itself. That is its great accomplishment, it takes nothing seriously. Comedy offers laughter as its alms for oblivion, a shriek of mirth to echo through the infinite silence of the Budhha"*

(Siegel, 1989, p. 5)

Throughout history, humor has drawn the interest of intellectual thinkers and played a crucial role in human interaction. Humor has long been a topic of intense reflection, from the philosophical musings of Plato and Aristotle, who pondered the nature of comedy while influencing Western philosophy, to Charles Darwin's investigation of the origins of laughter through the delighted sounds of tickled chimpanzees, and Sigmund Freud's quest to unearth the hidden motivations behind jokes within the depths of our unconscious. Comedy has been an integral part of our everyday life.

In more recent times, we find a plethora of multi-million budget movies, web series, and documentaries that have swarmed through Amazon Prime, Netflix, Hulu, and many other OTT audiovisual media platforms. Most content is comprised of plotlines, and exotic geographical locations, accessible in multiple languages, music, melodrama, suspense, and fashion. However, the one genre that requires nothing but a microphone and a spotlight and the person talking on a stage that is not even visible is the *Comedians*, who deliver nothing but the power of their humor - *relief.*

In Europe, China, India, Japan, Russia, and Africa comedy emerged in literature and shows like the Minstrel - Court Jesters of the Medieval and Renaissance periods. They were officially sanctioned and paid comedians whose sole responsibility was to perform jokes for the King (Westwood, 2004). The open forum of just making people laugh allowed the comedians to work and not worry about the boundaries of any differentiation. In the court, jesters serve as adored figures rather than objects of tolerant behavior. Despite their sharp attacks, their comments were understood in the spirit of compassion and goodwill. Jesters prevented the monarch from acting unjustly by rescuing him from his moral failings as well as shielding him from his own rage (Otto, 2007).

The popularity of court jesters in India, as depicted in novels and comic books, equals to that of Europe. In southern India, a king was barely considered whole without his jester. Jester's objective viewpoint permitted them to support the downtrodden to rein in the system's excesses without trying to topple it. Instead of replacing one system with another, their goal was to free people from all restrictions. Two of India's famous jesters are Birbal and Tenali Raman. Tenali Raman is rumored to have gained his position by making King Krsnadevaraya laugh. The jester's legacy is still influential with the wandering theater troupes (Singh & Sharma, 2023).

Comedic elements can be found in classic literature and stand-up performances. Raghavan (2020) provides evidence of the humor component in Sanskrit literature - Bharata's Natyasatra, the earliest treatise on drama to Nyayas by Akṣapāda Gautama, Naishadhiya Charita, and Kalidasa's characterization of Vidusakas in Mrcchakatik. Modern theatre emergence began with the plays of Girish Karnad and Badal Sarkar incorporating satire into comedy for social issues. By the 1920s, radio gained popularity to become a medium of communication for comedians at the local level. Gradually, social, and political satires became the trend.

## Democratization of Comedy and Indian OTT Platforms

Humans are so adaptive to nature that it seems like we have been consuming content on our personal screens for as long as our memory goes. We have witnessed the rise of smartphones, OTT, and a dynamic shift in content consumption in the last two decades. OTT emerged as a method of delivering television and movie content via the Internet (Puthiyakath & Goswami, 2021) to cater to the needs of consumers. The word "over-the-top" itself denotes that a content provider is going above and beyond what is already offered on the internet. With swipes, taps, clicks, gestures, likes, and dislikes, comedy is now ingested, experienced, and regurgitated in non-linear flows across a variety of social media platforms. A story, whether personal, cultural, or social, is at the heart of locally produced content, and this has been reimagined not only for a new media world but also for a post-colonial, millennial audience that does not identify as a homogenized group but rather as a diverse audience of global users who come to hear, see, and make their voices heard (Kay, 2018).

First, we explore the contribution of the OTT streaming revolution regardless of how advanced it seems. In the 1940s television became an integral part of everyone's lifestyle either to consume news, education, or entertainment. However, what was lacking is the flexibility of the content that could cater to the different needs of the people. Moreover, the availability of new content was never at the convenience of its consumers. It was only in 1997 that Netflix disrupted the marketplace with its idea of renting DVDs and tapes. By 1999, Netflix released its first subscription model for $15.95, allowing customers to rent up to four movies without any return policy (Shah, 2018).

In the 2000s, Netflix proposed a monthly subscription plan at $19.95, forsaking late fees and return-by dates. Other companies followed suit making similar offers. Blockbuster, as a leading player in the market, lost most of its consumers. In 2007, Apple launched *Apple TV* for movies, podcasts, music, photos, and videos. A year later Samsung launched internet TV as *Smart TV*, though it was Amazon that emerged as a competitor for Netflix. In 2006, Amazon launched *Amazon Unbox*, which later evolved into *Amazon on Demand*. Subsequently, it transformed into *Amazon Instant Video*, and by 2018, the brand consolidated as *Amazon Prime Video* (Shah, 2018). Amazon meticulously refined its offerings and successfully established its brand presence over the course of a decade. This progression vividly demonstrates how OTT has become an integral part of daily life, with widespread adoption and long-term implications.

In 2020, the United States OTT platform generated a revenue of $75.84 billion and is projected to reach $126.50 billion in 2023 (Statista, 2023). Moreover, it is anticipated that by 2027, this figure would reach $175.40 billion, with the revenue forecast to expand at an average yearly rate of 8.51% (CAGR 2023–2027). OTT Video Advertising will be the largest market in 2023, with a market value of $76.97 billion. Clearly, these projections reflect a very high number of OTT platforms being a part of the lifestyle worldwide (Stoll, 2023).

Currently, there are more than forty OTT providers in India. The first OTT platform, BigFlix, launched in 2008 and was funded by Reliance Industries. A few years later *Ditto TV, SonyLIV, Disney+, Hotstar, Netflix*, and *Amazon Prime Video* entered the market from 2013 to 2017. In terms of market shares of OTT platforms in India Netflix holds 20%, Amazon Prime Video with 20%, and Disney Hotstar with 17% of the market share (Buchholz, 2020). Other platforms such as *Zee5, JioCinema, Sony Liv*, and *ALT Balaji* have market shares between 4% to 9% (Research and Ranking, 2021). The convenience of instant gratification and the absence of advertisements make OTT platforms attractive to viewers. Furthermore, OTT platforms provide a wide range of content of higher quality.

Many factors have contributed to the popularity of OTT among audiences. These include low data prices, various pricing structures to accommodate diverse user budgets, premium quality material on OTTs, and, most recently, the global pandemic. OTT platforms have emerged as champions in the COVID-19 era, especially with multiplexes operating at no capacity because of the lockdown and later at limited capacity. Internet dissemination surpasses most other sectors such as banking in urban and rural areas with a growing preference for handheld devices and high-speed viewership. Independent content producers can experiment with new formats and adjust to consumer demands, which is both entertaining and educational. The Indian OTT Platforms Report 2020[1] suggests that the comedy sector has benefited the most from the pandemic. About 75% of the survey participants claimed to watch comedy content (CC) using OTT platforms, whether it be in the form of TV shows, web series, movies, or stand-up performances (Rahman, & Arif, 2021; Mpofu, 2021; Shastri, 2021). Today, the comedy environment has changed and so has the platform and the way it is consumed. OTT and various social media platforms serve as a platform for stand-up comedians to thrive and stay relevant with their audience.

Indian comedy was dominated by the upper class amusing themselves with *Russell Peters'* jokes, a small group taking pleasure in *Kavi Samelans* (poetry gatherings), and the public finding happiness in the comedic performances of actors like *Johnny Lever and Laxmikant Berde* on the big screen, along with a growing number of comedies shows on television. A new spirit and chances for emerging comedians were introduced with the opening of *'The Comedy Store'* in Mumbai in 2010 and the rise of *Papa CJ* in Delhi-NCR[1]. The growth of OTT platforms, the spread of the internet, and the low cost of data have all contributed significantly to the democratization of comedy by giving rising comedians a stage to display their skills, engage with audiences, and reshape the business (Seigel, 1989; Paul, 2017; Sahoo & Das, 2018). There are multiple shows that accounted for providing comedians with the opportunity to write and direct their shows. For instance, *'Comicstaan', 'Laugh Out Loud Network',* and *'The Improvisers: Something from Nothing'* among others. Some comedians such as *Bhuvam Bam and Kenny Sebastian* capitalized on the benefits of OTT by launching their web series (*Bhavam Bam*) and song albums (*Bhuvam Bam* and *Kenny Sebastian*).

These transformations have produced a generation of Indian comedians who continue to reshape the comedy industry. There has been a shift in how jokes are perceived today – the transition from consumption for entertainment to a source of conflict and disagreement, with audiences finding humor, and sometimes discord, in the jokes that originate from their culture and ethos. It is because these OTT platforms opened doors for a wide range of styles, languages, visibility on a global scale, and perspective in comedy since they are not required to pass through a censorship board. However, not all comedians pave the same road as they start but some go on to build a legacy bringing freshness in the content and way to curate comedy specials with their own style.

## Research Methodology

This chapter uses a case study approach to explore stand-up comedy and the growth of the Indian comedy industry through various media, particularly Over-The-Top (OTT) platforms. The study focuses on comedians and artists who have created a space within the comedy industry over the last decade by utilizing the benefits of subscription and non-subscription-based platforms. Yin (2009) argues that case studies are best used to "explain, describe or explore events or phenomena in the everyday contexts in which they occur."

Data was collected from podcasts (both audio and video), comedian's YouTube channels, OTT platform comedy specials, discussion shows of various brands on multimedia platforms of stand-up comedians, including OTT and television shows, interview appearances of reality television shows, individual comedian's social media accounts, and conference appearances. Through content analysis, we analysed the case studies on the basis of the thematic elements, character portrayals, and dialogues and content delivery pattern and medium to recognize the messages being delivered on the artist's content available. Secondly, the adaptability and challenges faced by comedians to the fastest changing landscapes were given special attention to buil a comprehensive understanding of their career progression and chnage in comedy style.

This method was useful in understanding their strategies for staying relevant in the comedy industry. The comedians were selected for analysis are - Kusha Kapila, Prajakta Koli, Vir Das, and Kenny Sebastian - who have carved out their own niche in the Indian comedy industry through OTT platforms: showcasing the opportunities available to comedians in the digital age with a purpose of change. Additionally, using their fame to deliver knowledge and exposure, they have effectively adapted their humor to a variety of formats, including stand-up specials, web series, and podcasts.

Examining their careers and content can provide useful insights into the changing landscape of comedy consumption, the impact of digital platforms on humor preferences, the representation of diverse voices in comedy, and the role of content creators in shaping public perception and entertainment trends.

## Case 1: Kusha Kapila

Kusha Kapila is a digital Indian comedian, content creator, fashion editor, actress, and YouTuber. She shot to fame in 2017 with her second video where she played a fictional character called "*Billi Maasi*" for an Indian media platform called iDiva, a women's lifestyle portal featuring content about lifestyle, Bollywood, fashion, beauty, health, and relationships.

She is a graduate of the National Institute of Fashion and Technology (NIFT). She began her career as a trend analyst and content developer working for many firms before she joined to work for iDiva. Kusha in an interview with Vogue magazine revealed that the rise of her as a content creator and comedian began all by accident. She joined as a stylist and soon transitioned to producing creative video content for the firm which aired on Facebook and YouTube (Vaidya, 2019). During the rise of her career, Kusha and other comedians toyed with humorous short skits about South Delhi girl's accent and their unawareness of lower middle-class struggle. The videos captivated the audience with her unique brand of observational comedy. Most of her initial content with iDiva falls under the category of, the now renowned, *"Types"* video series that covers a wide range of relatable scenarios, from *'Types of South Delhi Girls: Part 1- Every South Delhi Girl In The World'* to *'South Delhi Aunties Try Living The Organic Life.'* In the former, Kusha fakes her accent and shows her inability to speak the Hindi language, a popular language spoken in Delhi. The conversation in the video goes like this:

*"Kaha se aati ho, Pritampura se"*

*"Koi chaat ka naam hai, Pritampura, I mean"*

*"Mujhe toh pata hai, jaise panchsil, GK, golf link pata hai...aur kuch hota hai Delhi mein?'*

*[Where do you come from? From Pitampura?*

*Is Pritampura a snack name*

*I know places like Panchshil, GK, and Golf Link, is there anything else in Delhi?]*

The tone of the content was faking the English American accent because you are from the rich side of the city. Her videos stood out with her savage and bold approach, addressing issues such as women being labeled as crazy and facing eve-teasing on the streets, and how South Delhi elite class girls talk (Vaidya, 2019). She injects humor into her content to deliver socially relevant messages such as classicism and narcissism. Her content initially circulated on the Facebook pages of iDiva and with the release of two videos, after which she became a household name.

Her characters were inspired by people she met in her daily life. Her vlogs are representations of her life as an observational comedy, which helps her audience to feel a part of her journey. In terms of her approach to content creation, she understands the necessity of being authentic and real. For instance, her famous phrases like *"Gorom lagche"* (*It feels hot*) were inspired by a Bengali editor Santu, and her character "*Naina*" is a characterization of South Delhi girls. Other videos produced by Kusha include *DJ Arti, Zaalim aunty, Guru maa, Coco Bhaiya, Maa Wokeananda, and Sunita* are evidence of her focus around her but showing elitism and high standards of people living in posh areas.

Given her success at iDiva, she was able to optimize her Instagram reels, reaching millions of subscribers in a few months, and subsequently created content for the YouTube audience. OTT platform gave her the opportunity to develop content related to her area which was to empower women. She also makes comic content that talks about *"How massage is better than sex?", "Why does poop have the worst timing", "Indian parents and their love for documents", and "You wear too much makeup."* Her sarcasm is not hard to understand, for instance, in one of the videos she becomes a beautician and talks about what should one do to 'become' beautiful.

**Video 1**[2]: This homemade drink makes your skin glow!

**Dialogues***: Hello Everyone! This is beauty secrets with Gehnaaz. Agar aap apne skin ko leke conscious hai and aapko glow chahiye. Toh roz subah uth k garam paani mein nimbu, adha chammach ajwain, haldi powder, garam masala, laal mirch powder, apne ex- k baal, apna shampoo aur apne chappal pe lagi miti m daal k fek de. App jaisi hai bahot khoobsurat hai!*

**Translation:** *Hello Everyone! This is beauty secrets with Gehnaaz. If you are conscious about your skin and you want to glow in your skin. You need to drink warm water in the morning mixing, half lemon, half tablespoon of carom seed, turmeric powder, Mixed spice powder, red chili powder, your ex's hair, cow's dung, and the sand on your slippers, mix it and throw it. The way you are, you are very pretty.*

**Meaning and inference:** *There are a lot of beauticians and industrial products that people buy and are dupped into becoming beautiful without understanding their self-worth. They do not feel confident because of a certain complexion they are born with. As a comedian, Kusha mimics them, to convey the opposite – women need to accept who they are, how they are, and their self-worth and self-acceptance are key to their success in life.*

In another video of hers, she is seen talking about the current generation's activism and how it is not implemented in real life. She mimics the Indian mothers who are not given much space because of the generational gap between their children, also bashing the misunderstood feminist approach/activism by the youth using social media.

**Video 2**[3]**:** *Help me get you, if moms were truly honest!*

**Dialogues:** *kids be like mummy, you don't get it. Help me get no, help me get it. Kyun ek paani ka glass uthane m tumhe maut aati hai, kyun tum dhai dhai ghanta lagate ho potty, aisa kya business chal rah hai bathroom k andar? Kyun tum internet pe itne activist bante ho, par mummy ki do rupye ki help nai krte. Kyun tum subah char chaar baje tak aashiqi krte ho? Par meri jamuna paar wali bua ji ko hello bolne m, JCB tumhari jubaan lekar chali jaati hai. Help me get it, no, help me get you. Patrirachy hatayenge mummy ko chai bana do bol kar, chup kar!*

**Translation:** Kids behave like mothers you do not understand. Help me understand you. Why cannot you take glass of water by yourself? Why you take two and half an hour in that bathroom? What kind of business you run there? Why you act like an activist on the internet but do not even help a penny to your mother? Why you cannot you say hi to me far of relative as if some truck took you tongue out? Help me understand you. They are going to remove patriarchy asking their mother to make tea for them. Shut up!

**Meaning and inference:** In this video, she dresses up like a mother in an Indian household and questions the child on why he/she/they think she will not understand them. She further questions why taking a glass of water for yourself is a burden. The other thing questioned by the mother is the lack of greetings for her relatives when children can spend more than two hours in the bathroom. The video also shows how the child is so active and busy on the internet that they ask their mother for a cup of tea. Is that how they are going to bring equality and do away with patriarchy, and is that what their activism is all about? Her sarcasm about empowerment and breaking stereotypes in online activism comes at the end of the video. She adds a song in the background, music that builds up suspense. In her videos, most of the dialogues takes on a sarcastic and ranting tone as she addresses social issues. The manner of communication often involves framing questions filled with clever puns, intended to provoke the listener in a relatable manner. Her videos showcase different people in their everyday lives, sometimes, a mother, a make-up artist, or a character from a film, or highlight the various traits of a person from being lazy, to sad, or to feeling elated. She likes to question and uses sarcasm to make her audience laugh and ponder over important societal issues and injustices. Most of her videos, even those in collaboration with celebrities, follow the same script. For instance, in a promotional video for the movie *Double XL* (2022), she used the theme of the movie on weight and body size to stress issues of self-worth, career, and questions of marriage.

**Video 3[4]:** My body is my body, none of your body

**Dialogues:** *Ye sab jo milte hi body pe comment krte hai, ye sab jo ye bolte hai uske paise, uske paise toh humko charge krne chahiye. Kyunki din toh hamara khraab hota hai. Tumhe small talk hi krni hai toh Mausam ke baare me kar lo, world economy k baare m kar lo, meri anatomy se tumhara kya lena dena hai? Mere love handle aapse kyun nai handle ho aa rahe hai. Meri ass se ashes ho rahe hai, what is expectation dude? Hume body shame kar k ye log apna time vyarth kar rahe hai, kyunki meri worth unaffected hai. Aur mujhe dekh k jo tum full BMI aur full medical report nikalte ho, meri jindagi ka gall bladder stone na aap hi ho. Aap ho meri pathri…baba, we know what our body looks like, we live in it every day but agar aapko jada problem hori hai toh aap possess kr lo…!*

**Translation:** All of you who right after the meeting start commenting on my body, do you charge money for it, but in fact, I should be paid because it is my me who gets affected the whole day. If you want to have small conversation talk about the weather or world economy, what do you have to with my body? Why is it difficult for you to handle my love handles. Why looking at my body makes you into ashes, what are your expectation game? Body shaming me does not affect me, it is you whose time gets wasted because I am unaffected. Stop printing or providing me with my medical reports based on your assumptions, you are the gallbladder stone in my life. We live in this body, and we know how we look like but if you have problem with it possess my body, do it.

**Meaning and inference:** Body shaming women or men is a common thing in India (Kaul, 2019). Many women deal with comments about not being of a certain color, having a good height, not being pretty, or having a certain figure. Many people express having faced a lot of taunts, statements, advice, and suggestions to lose or gain weight. It starts at an early age for girls. This problem is exacerbated by the corporate world. Advertisements often portray women and girls as *desiring* a different self and body shape. Advertisements for green tea or oats start by showing women struggling to lose weight. Kusha, among other comedians, questions such a portrayal of women and girls in media. In many videos, she has spoken about her issues with Polycystic Ovarian Syndrome (PCOS), to build a rapport with her audience. Through her videos, Kusha has attempted to break stereotypes around body image and question social norms. Similarly, her content created since lockdown demonstrates her efforts to break stereotypes about women, parenting, and independent women in India. The opportunity provided by OTT materialized into a debut in the Netflix anthology film *Ghost Stories* (2020), as well as on Netflix India's YouTube show *Behensplaining*.

She appeared on Amazon Prime in the comic reality show "*LOL: Hasse Toh Phasse*." The following year, she participated on Amazon MiniTV in the reality sketch comedy show "*Case Toh Banta Hai*" and in Season 2 of Netflix's TV series "*Masaba Masaba*". She co-hosted *Comicstaan* Season 3 and appeared on "*Plan A Plan B*" and as a jury member on Season 7 of *Koffee with Karan* and later represented women comedians on *The Kapil Sharma Shows*.

In 2023, she made an appearance in the Cannes film festival and contributed to shows titled "*Minus One: New Chapter*" and "*Selfiee.*" Lastly, she has been the host for the show Tinder India called *Swipe Ride ft.* In a realistic setting, the show begins with conversations between Kusha driving, a girl for their Tinder date with actresses and influencers like Sanam Bjawa, Pooja Bhatt, Urfi Javed. Kusha probes women to talk about the person, why they chose them, to what they expect from the dates. For instance, Kusha asks in one video:

*Kusha: "So if this is your first date, why did you choose this Sid?"*

*Kusha: "We appreciate a good welcome joke."*

And in further conversation about the venue of the date, a cat café, instead of trekking, Kusha is appreciative of the participant's self-worth: "*I appreciate that you prioritize yourself, you have to prioritize yourself.*"

Kusha repeats these questions in every episode. She is appreciative of the women whom she speaks to who are unafraid of breaking stereotypes and taboos to prioritize themselves, voice their concerns, and be independent. Kusha with her humor and skills has honed the opportunities coming her way from social media to OTT platform, to becoming a household name with her acting skills.

An analysis of her work is evidence of the challenges for comedians who must navigate changing landscapes and adapt to audience preferences. It is also evidence of how OTT empowers the consumer while revolutionizing production. While Kusha welcomes male viewers and criticism from her male counterparts, which serves as an encouragement and is proof that her content is socially relevant that raises awareness (Vaidya, 2019). There is also the challenge of dealing with trolls and negative comments, which Kusha understands is part of the industry (Anjum, 2021). As a comedian, she approaches topics such as #MeToo and #Covid-19 in a humorous way rather than being sentimental about it, which instinctively awakens people to ask questions (Vaidya, 2019).

## Case 2: Prajakta Kohli

Prajakta Koli is an Indian YouTuber and comedian renowned for her channel "*MostlySane.*" YouTube was founded in February 2005 as a video-sharing platform, which allowed users to upload and share videos with a global audience. In April 2022, there were around 451 million unique visits to YouTube's video over-the-top (OTT) platform in India. YouTube was largely an advertisement-driven OTT video platform at the time, with the biggest number of unique visitors among other streaming sites (Basuroy, 2023).

She was a radio Jockey before entering the space of multimedia in 2015. She started her channel with sketches *like 5 Types of Singles on Valentine's Day, 10 things you see on the Indian street, Indian Moms Preparing for Festivals! and #SawaalSaturday*, which is a day dedicated to talk to her subscribers. Similar to Kusha, she drew inspiration from her daily life and the people around her. However, her approach and use of comedy are different from Kusha. She uses comedy as a source of activism, which allows her to create content to address women's issues in India. For instance, she created a song in 2017 called *shameless* describing it as *'be shameless, be happy*[5]*'* to empower women. The lyrics address a lot of issues:

"Have you seen yourself in the mirror, what do you see?
Too Fat, Too Thin
Your dressing sense is bad
OMG she's a dork, do a little makeup.
Some M.A.C., some Lauder, some Dolce Gabanna
How will you get married? You are dark-skinned.
Nobody will take a daughter-in-law who has a dark complexion.
All your degrees and fancy English are of no use
Use some gram flour and some bleach along with turmeric and fairness cream
How much do you weigh? Did you eat a whole elephant?
How are you this thin? You only have bones
You should eat less. Start eating a little more
So, fat! So thin! Girls aren't like this
Love handles on your waist.
12 pimples on your face
Dark-skinned and stout
Don't care be shameless.
I'm Shameless
Uncle, Aunty, Sister, come listen to what I have to say
I'll tell you this with love and without getting physical
I am me, you are you.
Different people different minds. Don't judge me straight
I am one of a kind"

The few lines of the lyrics reflect the lens employed by society to determine the worth of a woman's body. The standards of what a good woman and wife is decides how well she fits in not just society, but also the family. The rest of the song is about the acceptance and hypocrisy of society. In an interview with *Humans of Bombay,* Prajakta talks about the impact her song *Shameless* had on a girl fan of hers:

*"I'll never forget this one girl who had come to a meet and greet I did in Delhi and she was wearing a skirt and she came up to me and she's like…look, look at my legs they're hairy. And I was like okay, she was like I hate waxing or shaving because it hurts me but I love wearing skirts and because I watched your video Shameless, which was a song I did against body shaming I have worn this and I have taken a Metro and I've come all the way from my house and I'm not ashamed of it" (Mehta & Koli, 2023).*

Her short film *Khayali Pulao*, which premiered in July 2020, focused on female empowerment in rural Haryana. The short film is set on a schoolgirl who dreams of joining the school handball team. The dialogues highlight the myths that little young girls inculcate about purity related to hymen breakage, her freedom of choices to wear clothes and being eligible and independent to take the lead of her life, which is carried as a civilizational memory and ethics overlooking the inner desire of a girl, the freedom to choose what to wear, but subsequently highlights myths that girls grow up believing in.

In the first scene, Asha (Prajakta Koli) asks her friend (Jyoti) to participate in a Handball trial[6]. Together they expose stereotypical thinking and the taboo of sports and virginity, affecting significant life decisions such as marriage.

**Asha:** *Just come with me, be a good friend, and do this for me.*

**Jyoti:** *Have you gone mad?! Why do you want to run with a football in your hand?*

**Asha:** *Handball... It's a different game.*

**Jyoti:** *Whatever, you have to use a lot of strength in games. Mom says that girls who play sports, their hymen breaks early in life. No one will marry you. Listen to me and let's go home, Asha you've gone mad!*

In another scene we see her parent's conversation of girls playing sports:

**Asha's Father:** *Balbir's daughter, Preeti, she's made such a mess. She was walking around wearing a pair of jeans in the city.*

**Asha's Mother:** *Oh God! Did anyone see her like that?*

**Asha's Father:** *The whole village saw her. She posted a picture on Facebook! You (Asha) there..I hope you are not on Facebook.*

**Asha's Mother:** *She had gone to play sports after school today.*

**Asha's Father:** *Playing sports is a good thing, she'll get fit.*

**Asha's Mother:** *But who will marry Preeti now?*

**Asha's Father:** *And also, there is some shameless girl called Queen Deepika...She has commented on that picture saying, "Very Nice". Good that our daughter will go play in the evening.*

The whole movie reflects upon the difficulties of girls from a very young age. For instance, we see how Asha takes pride in being an extra player and wearing shorts and T-shirts. The movie is about freedom to take decision for the basic rights decide, and ability to defy societal norms. For Asha, having to choose her clothes was her step toward freedom (see Figures 1 to 3).

Figure 1. is a snapshot from *Khayali Pulao*, a short film by Prajakta Kohli. It showcases the meaning of freedom for her to wear her clothes without permission and judgment.

*Figure 1. Video image from Khayali Pulao (2020)*

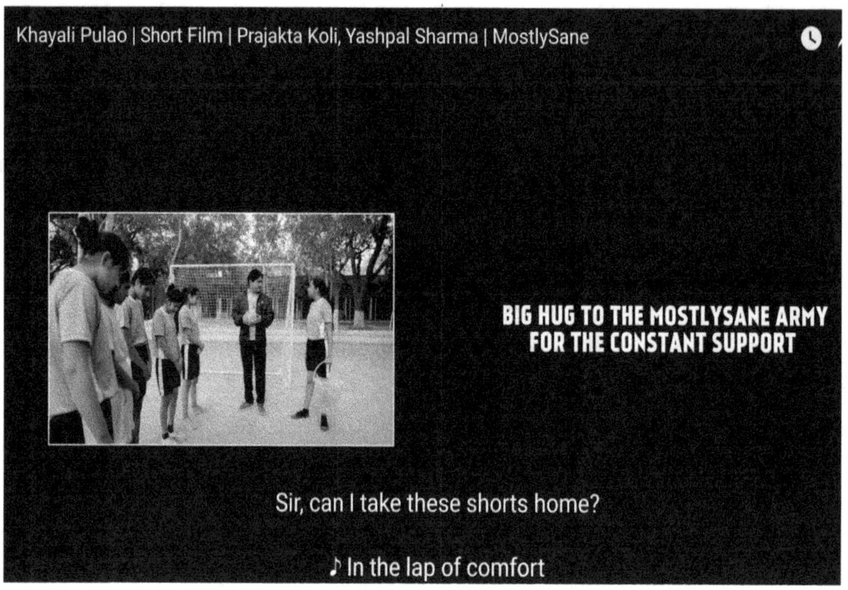


Figure 2. depicts the character wearing shorts as a sign of freedom. This opportunity was provided by virtue of the sports she loved, where she tries hard to be a part of the sport and to feel the freedom of wearing clothes which is not permitted otherwise.

*Figure 2. Video image from Khayali Pulao (2020)*

Figure 3. It is the last scene of the film, which reveals how much Asha enjoys the sport and how important it was for her to wear shorts which felt like freedom to her.

*Figure 3. Video image from Khayali Pulao (2020)*

Prajakta has gained popularity with her activism and knowledge-based content besides comedy. She is the first female comic YouTuber who turned into an influencer, writer, singer, actor, and content creator. She was appointed as a global ambassador for YouTube's Creators for Change initiative at the United Nations in 2018. Her work was recognized in 2019 on Forbes' 30 Under 30, Outlook Business Magazine's Women of Worth, and Entrepreneur India's 35 Under 35. She continued to advocate for female education by taking part in the *#GirlsCount* campaign.

Prajakta was named Cosmopolitan's *YouTuber of the Year*, subsequently making her acting debut in Netflix's *Mismatched (2020)*, establishing a powerful presence in entertainment and social campaigning.

In an interview, '*Celebrating Women's Day with the Stars of Netflix!*', Prajakta reflects on her role in the Netflix movie - *Mismatched* (2020). She reflects on her authenticity and appreciation s for the writer and how important it is for the viewers to know of various issues related to women: "Because *I was not used to playing characters I have not written before playing Dimple. she was served to me on the platter she was I knew exactly what she likes what she doesn't like. I remember one of the comments and I'll never forget it said that thank you so much for finally getting a character that looks like me yeah and I was just like yes and that was also what I love the most about her job.'* She further adds, *'I fell in love with her because I was so happy to see that the sketch that came to me had nothing to do with the way dimple looks right called bubbly oh I hate bubbly I'm not soda compliment and then when I read that it said she is a gamer and she wants to be a coder she's stubborn and she not very likable flawed or very focused very like future-oriented and I loved that this is a great character I know her I want to be her yeah so I think that was my favorite part about being dimple."*

Prajakta's YouTube channel now has about 7.04 million followers. She is invested in her podcast #RealTalkTuesday where she interviews guests such as Bill Gates, Priyanka Chopra, and Anoushka Shankar. Her path showcases the authenticity of owning the content as a completely different genre, and role to connect with the audience which helps her to grow as a creator as well (Duggal, 2023).

OTT platforms like YouTube are potent tools for comedians to display their skills and connect with their audience. It allows for the monetization of the content with advertisements. Many of them support their artistic activities through merchandise sales, crowdfunding campaigns, and brand collaborations becoming a grand value. This strategic use of social media enables comedians to expand their digital presence and connect with a broader demographic, selling tickets, and ultimately increasing their influence in the comedy industry to help them stay relevant.

## Case 3: Kenny Sebastian

Kenny is an Indian YouTuber and comedian. He gained recognition after winning a stand-up comedy competition at the Indian Institute of Technology (IIT) fest in Mumbai. In 2008, he began his journey on YouTube as "Kenny Sebastian." He was also a member of Clubs like *Canvas Laugh Club* and *The Comedy Store* that allowed him to perform live and later he used it on his YouTube channel like many other stand-up comedians to reach more audience (Xalxo, 2020). Later in 2014, he wrote and directed a season of "*The Living Room*" (2014-2015), India's sketch comedy television show with other stand-up comedians such as Abish Mathew and Kanan Gill. In the same year he released his first stand-up comedy *"Journey to the center of my brain,"* a one-hour stand-up comedy special (now private) on his YouTube channel. In this special, Kenny combines conventional stand-up with music humor, covering topics such as being a 23-year-old Malayali (South Indian language) boy, dealing with break-ups, the magic of the guitar, and a love song suitable for a horror film (YouTube, 2014).

One of Kenny's early successes came from creating songs based on tweets, initiating the #KennySing4Me campaign on Twitter. He further expanded his presence in the OTT platform with shows like *"Comedy Central", "Humorously Yours," "Better Life Foundation," "Son of Abish," and "Pushpavalli."* He co-wrote and acted in the sci-fi web series *"Star Boyz,"* and owns original series like *"Die Trying"* and a sketch comedy show titled *"Sketchy Behavior"* on OTT like Amazon Prime Video (Bhatt, 2023). He discusses relationships, his childhood, cultural differences, middle-class Indian culture, and encounters with technology. His humor is clean and family-friendly, and it appeals to a wide audience. For instance, his comedy special called *Don't Be That Guy* on Amazon Prime Video in 2017 has dialogues that are everyday conversations with himself, his mom, and the maid stealing the television.

**Show name:** *Don't Be That Guy*

**Segment name**[7]**:** Me, My Mom, and My Maid!

Dialogues:

And then it's crazy like we're at home and I'm chilling at home.
I am chilling at home.
My mom catches me.
"Kenny, what are you doing?"
"Mom, I live here?"
"What are you doing in this room?"
"Existing?"

"Don't act smart with me Kenneth.
Where is the maid?"
"She is in the top floor."
- "So why are you not with her?"
I don't know, I just want to be single for a while.
"Why not with her in the room?"
"Why would I be with her in the room?'
"What if she steals the TV?!"
"She's been with us for 28 years, mom.
Why would she steal the TV?"
"I've heard stories.
Go watch her clean!"
Now I have to do this weird thing.
Where I enter the room.
To evaluate my maid's work.
I have no purpose of being there.
So, I'm just walking like that.

**Meaning and Inference:** Kenny's introductory conversation, "And then it's crazy like we're at home and I'm chilling at home," is casual and relatable. "I'm chilling at home," provides a relaxed and familiar tone, immersing the spectator in an everyday, ordinary situation. Kenny's interaction with his mother is full of lighthearted banter and sarcasm. When his mother confronts him about his whereabouts, Kenny's hilarious retort of "Existing?" lightens the mood.

Additionally, each dialogue is delivered in a calm tone with a pause of a few seconds to add irony and humor. Kenny's mother expresses concern about the maid, questioning why he is not in the room with her. He responds: "*I don't know, I just want to be single for a while.*" Here Kenny addresses the issue of dependence on domestic help and also the social norms around how to treat domestic help in a humorous yet effective way. Kenny's amusing answer that the maid has been with them for 28 years and would never steal the TV calls these assumptions into question. Lastly, Kenny explains the humorous situation in which he is forced to enter the room to "evaluate" the maid's work, emphasizing the inconvenience of this superfluous task and adding to the comedy. He pauses for a moment to express the awkwardness of the situation, to which the audience relates, laughs, and applauds. The conversations in "Me, My Mom, and My Maid" by Kenny Sebastian demonstrate his ability to find humor in daily circumstances as well as his ability to provide realistic information that resonates with his audience.

He went on to become a judge on comedy shows such as *Comicstaan* in 2017, became a writer for many reality television shows and directed episodes of other comedians. Kenny owns his production studio called *"SuperHuman Studioz,"* where he produces content for his YouTube Channel. Additionally, he hosts a monthly live video blog named *"Chai Time Comedy with Kenny Sebastian,"* where he chats with people over a cup of tea. With over millions of followers online and popular live vlog shows on OTT platform, he has carved a niche for himself in the comedy space. After performing six stand-up specials, Kenny's special *"The Most Interesting Person in the Room"* became available on Netflix in 2020 (Xalxo, 2020).

Apart from his comedic ventures, Kenny is also the creator and host of "*Simple Ken*," a podcast tailored for millennials. His diverse talents and unique approach to comedy have solidified his position as a prominent figure in the media and entertainment industry.

## Case 4: Vir Das

Vir Das is an Indian stand-up comedian, actor, scriptwriter, and musician born in India. Vir Das' comedic career began at Knox College, where he graduated with major in acting and economics also getting recognition after he wrote '*Brown Men Can't Hump*,' a play inspired by the 1992 film "*White Men Can't Jump*." He has had a lot of success since then, releasing four Netflix comedy specials, including "*Landing*," which was broadcasted in 2022. Das has created a reputation for himself in both, the Indian and American cinema sectors, in addition to his comedic abilities. He has been part of American television shows such as "*Whiskey Cavalier*" (2019) and in films such as Judd Apatow's "*The Bubble*" (2022). In addition to his humorous endeavors, Das is the vocalist for the comedy-rock band Alien Chutney (Wing, 2023). Das has been in several funny television shows - Cricket Firsts, a sports humor show on Zee Sports (now TEN Action+), and SAB TV's Lo Kar Lo Baat, he performed improvised comedy. He joined the cast of The Great Indian Comedy Show on STAR One. He has been in 18 films at the national and international levels. However, he is most recognized for his stand-up comedy specials on the OTT platforms such as *Canvas Laugh Factory on YouTube (2014), Vir Das: Abroad Understanding (2017), Vir Das: Losing It (2018), Vir Das: For India (2020), Vir Das: Landing (2022).*

Das is best remembered for a six-minute monologue performed at the close of a 2021 performance at Washington's John F. Kennedy Center for the Performing Arts. Das recited the poem *"I Come from Two Indias."* The performance emphasized the harsh realities and appreciation of a society in which values and deeds are not always in sync, containing equal parts pride and criticism.

**Scene 1:** *I come from Two Indias*

Poem lyrics[8]:

*I come... from an India where children in masks hold hands with each other... and yet, I come from an India... where leaders hug each other without masks.*

*I come from an India where the AQI is 9000, but we still sleep on the roof and look up at the stars.*

*I come from an India where we worship women during the day and gangrape them at night.*

*....I come from an India where we claim to fully support the troups until it comes to their pension plan.*

*I come from an India where we can never be on time no matter where we go but yet we are always early on the Cowin website for some reason.*

*I come from an India where we have maids and drivers and yet want to come to America to do their job.*

*I come from an India that self-serves, and I come from an India that self-preserves.*

*I come from an India that will not shut up and I come from an India that will not speak up.*

The performance went viral, with million views on YouTube, though it received mixed reviews. While some praised him for his courageous criticism and appreciation of these issues, others blamed him for tarnishing the image of India, labeling him a terrorist, and threatening legal action against him. Through his comedy, Das has spoken about the duality of thoughts, secularism, and attempts to bring about change by changing perspectives. In his comedy special, *Losing It,* he tries to explain the doctrine of religion and gods.

*"What is religion? It's a really old comic book. It's a really old superhero story. Muslims, Allah is your Batman. Christians, Jesus is your Superman.... But Hindus, we created the Avengers... There're too many guys. Nobody knows what the story is. And don't eat beef. No matter what we say, we don't understand any of it. We just end it with "don't eat beef." "Don't eat beef" is our "Despacito*[9]*."*

Within the same comedy special, he reflects on feminism and judgments about women's clothing in a sarcastic, and witty way:

*"I am proposing where all children cross dress, all boys were girls clothing every single item all girls were boys, clothing every single item, so that the right thoughts connect in your mind as an adult. The next time you Indian boys you go out to a club and you see a beautiful girl walking towards you just miniskirt, cleavage, and heels your first thought is an Indian male is her feet must be tired man 5 inch. (Audience applauses and laughter) Her underboob must be so sore with the wire or tight tight jammed in there, see most guys aren't laughing because you don't know there's a wire in the undertow you just thought there was 300 bucks in there for a rainy day right every time a woman in India wears something revealing like many of you are doing tonight. Indian men say things like oh she's asking for it. Am I wrong ladies? Have you heard that in your lives? Yeah (Audience laughter and girls applause heard) The infinite beauty and fun of being a woman is, if she's asking for it, she can just ask for it. Because it is available, I promise you she can just think of it in 28 men will line up outside her door for the woman. The world is Amazon, if you can ask for it. Compare it, size it, delivery date it, prime it, non-prime it, cash on delivery whatever you like that's the beauty of being a woman it's so specific you get to ask for it. (audience applause) Man, we just want to get some. (audience applause) but nobody's looking for hidden messages in men's clothing are they? (audience applause) I could wear a T-shirt that said **** me in the *** (audience applause) I give you full permission to **** me in the *** and nobody would do anything about that they just be like well that's the cool souvenir. Yet men have the freedom to wear whatever we want within limits you know you can't just drape 2 bed sheets around yourself and run an investment bank but you can run Uttar Pradesh..."*

He was the first stand-up comedian to have his special on the OTT platform, Netflix, and now has eight comedy specials, including World Tours, currently for his new comedy special called *MindFool.* He is known for his enthusiastic interaction with his audience and empowering comedy sketches using the OTT platform. He has a 10-episode show titled *10 on 10* on his YouTube channel, delving into issues surrounding religion, freedom of speech, how and if the West is privileged, tribalism and cancel culture, and brother and others where others is defined as the one, we fight with to own a piece of land or be superior to.

His idea through comedy is to make people aware of where the perspective is flawed. For instance, he talks about freedom of speech in one of the episodes of his series #TenOnTen titled Who has Freedom of Speech?

*"oh you feels like they have enough freedom of speech never got to get into trouble making noise they have freedom of speech (audience clapping and shouting) Clap who think they don't have freedom of speech. (silence) You just have uninteresting thought, nobody's coming after you because all you tweet about is talking football and intermittent fasting. And that's the way for those of you that are watching who don't know what intermittent fasting is it's where you know some people will like not eat for 16 hours a day it's like a Western thing. Except in the West intermittent fasting is a choice in India is this life[10]."*

Most of his comedy intends to educate the youth with references from history, law, current affairs, and trends.

## DISCUSSION AND CONCLUSION

Comedy has existed since 400 BC depicting society, and its behaviors, filled with parodies, songs, sarcasm, mimicry, sexual jokes, and surreal fantasies. It has not just provided relief but has diverted society on the right path with humor. Aristophanes' plays like '*Peace*' and Sanskrit literature, such as, "*Abhijnanasakuntalam*" (The Recognition of Shakuntala) and "*Malavikagnimitram*" (Malavika and Agnimitra). The plays were full of wit, comedy, and satire, with the goal of entertaining and educating the audience (Seigel, 1989).

Comedians today employ a slightly different approach from the past. They have various mediums to interact with, and to present themselves. In the current era, there are no minstrel shows or jesters with government officials like before, but comedy has reached the mass in the most accessible way. The medium of comedy performance has changed since the 1800s when comedians adopted various streaming services. For instance, comedians went from public parks, backyards, courtyards, parking lots, and comedy clubs to Zoom, YouTube, Amazon Prime, Instagram, and Netflix to name a few.

OTT platforms revolutionized entertainment consumption in the 2000s with the fourth screen (mobile phone) introduction, offering convenience and varied content. In India, numerous platforms, including Netflix, Amazon Prime Video, and Disney+ Hotstar, cater to diverse audiences. The pandemic further boosted OTT popularity, particularly in the comedy genre, with 75% of viewers consuming comedy content. Stand-up comedians now thrive on OTT and social media, connecting with audiences and evolving the comedy landscape.

The OTT platform specials are easy to watch and often last 20 to 30 minutes compared to dramas, which may last up to hours per episode yet remain unfinished. For viewers, content on streaming services, comedy, and lighthearted programs are popular given the quantity that each platform owns in terms of the comedy genre. Additionally, due to the variety of enjoyable content, accessible anytime, OTT provides appeals to various audiences. OTT platforms have become popular and the subject of frequent conversations. A lot of genres can be seen prospering let alone comedy specials. OTT now features comedy talent hunt shows, comedy sketches, and a series of comedy shows hosting and promoting movies, book publications, and song albums.

The acceleration of online streaming dominant with comedy seems to have just begun. However, what has not changed is that to understand the nuances of it, to be a part of it, to react, respond, and enjoy it, no one really requires training or a rule book. Over the years, stand-up comedians have served different purposes, from providing simple laughs as filers in multiple shows to addressing important social issues on their shows and channels. The comedians and comedy alike, have traveled and created multiple opportunities for them to survive. For instance, with the advancement in technology, comedians such as Kusha Kapila and Prajakta Kohli did take advantage of connecting with the audience. Eventually, with talk shows on the radio to comedy specials on OTT platforms, there has been a revolutionization in terms of content creation and consumption. Platforms such as YouTube, Facebook, Hotstar, Jio TV, Netflix, and Amazon Prime have transformed traditional media distribution and consumption while expanding the global knowledge of comedy for both the comedians and the audience.

One of the reasons comedians become significantly prominent faces is because of their particular style and approach, especially in medium like YouTube and Instagram. These platforms allow comedians to share their life updates and even receive direct messages and comments from the audience. Comedians have carved out a distinct and thriving position in the comedy industry. They interact with audiences through shared experiences of Indian rituals and social conventions, which are infused with cultural references. Many working on specials connects with their audience either in the same language as Kenny Sebastian and Prajakta Kohli. Their multilingual performances extend their reach, allowing them to connect with audiences across the globe. Another reason is understanding the audience, comedians as analysed in this chapter - Kusha Kapila, and Kenny Sebastian - use observational humor to find humor in daily circumstances that are widely relatable.

Furthermore, Indian comedians embrace political commentary without reservation, employing humor as a platform for social commentary on current events and important concerns in the country. Their performances are energetic and theatrical, with physical comedy and imitation typical of Bollywood entertainment. With such popularity with OTT platforms and social media content, these comedians have become a significant component of India's cultural landscape. However, what separates the comedians today is that these comedians not only entertain but also provide a new viewpoint to the audience through their humor. For instance, Vir Das and, Prajakta Kohli, conduct their work with activism and enlighten the audience. Their comedy is viewed as an epistemological challenge, calling into question established called "regimes of truth." Comedians that use this medium question their social surroundings and contribute to greater cultural shifts. Internet humor opposes traditional media power dynamics by leveraging YouTube and OTT as a public sphere, enabling crowd engagement in the conversation. Within this setting, the opportunities for comedy appear to be endless, providing an exciting and transformational space in the digital age (Kay, 2018).

Comedians are taking advantage of changes in media production to reach wider audiences with their messages, and while some of those messages may consist primarily of jokes or the reproduction of misogyny and hegemony through ridicule and other aggressive humor uses, some of those messages can be filed squarely under the category of *'semiotic Robin Hoodism'* (Jenkins, 2017). A concept or practice in which individuals or groups employ dominant or privileged symbols, signs, or cultural aspects and reinterpret them in a subversive or alternative fashion to challenge power systems and distribute meaning to marginalized or oppressed communities. For instance, the documentary of Prajakta Koli, titled *Khayali Pulao* and Vir Das works such as *Who Has the Freedom of Speech?*

The COVID-19 outbreak, and the emergence of OTT platforms unfolded a cost-effective way of accessing entertainment options for comedians and viewers. Many comedians created jokes about COVID-19, for instance, types of people wearing masks, government actions, and the conversation among family being at home. Additionally, a few comedians including Gaurav Kapoor and Rahul Dua went on making vlogs of their daily routine, mocking the trend in the COVID-19 of cooking everything at home and so on. They leveraged the time and optimized the media to become more relatable and hence, post-pandemic has comedy specials, world tours, and Television shows on their way. Stand-up comedians including Vir Das, Prajakta Koli, online or offline, are delving into economic models and adapting to new trends in information sharing, as indicated by their participation in transmedia communication.

Their comic element has extended beyond simple laughter, as they expertly include thought-provoking themes in their shows. They engage the audience's intellect and emotions by questioning traditional thinking and presenting unexpected ideas. Their puns serve as a board for thought and reflection, leaving the audience both entertained and enlightened by the end of their presentation.

Additionally, consumers' interaction with and assessment material has also changed because of OTT and social media platforms. Viewers now voice their ideas and feedback on a variety of venues with greater critical understanding. For instance, posts on social media, likes, number of views, reactions on posts with emojis, direct messages, and share options. Given OTT, the makers now extract information via a number of shares, likes, and algorithm metrics to learn about which comedians have a larger chance of being most watched in bringing revenue on a particular OTT platform. Additionally, assessing the audience preference on different platforms before the release of shows. It helps makers better grasp their strengths and flaws thanks to this criticism and analysis, whether it's frank or constructive, which motivates them to improve their work. It also, helps them learn about the brand value of a particular comedian.

The Indian stand-up comedy scene has grown dramatically in the last decade, with comedians gaining celebrity status mostly through OTT digital channels. Recognizing stand-up comedians' attraction to a younger audience, marketers have eagerly cooperated with them to include humor into their branded content and communications on platforms such as Instagram. For example, the comedy genre of Kusha Kapila is more feminine which allows her to present Tinder India where she talks about women's priorities and breaking stereotypes on relationships, independence, and self-care.

The future holds enormous possibilities for Indian comedians to keep enthralling audiences with their distinctive style of humor steeped in Indian culture and experiences because comedy is now recognized as a booming industry. For instance, Amazon Prime Video within the first year of its arrival in India signed 14 comedians for their specials because they recognized the scope to be huge. Many comedians used platforms that allowed them to share their content for free and connect to the audience.

OTT is here to stay, coexisting with other forms of entertainment. The popularity of OTT platforms is likely to continue, and the battle now lies in converting subscribers into loyal customers through continuous investment in content. Indeed, the OTT platform has provided an opportunity for everyone to follow their passion for making people laugh, but with has also accelerated the competition within them. Parallelly, providing an eminent place for comedians in the entertainment industry to showcase their talent in a wide arena.

Digital platforms and social media have fueled a dramatic transformation in the comedy environment in India, with a new generation of comedians making their impact on Indian popular culture. While internet comedy is still in its infancy in India, traditional forms of comedy, such as mimicry and cinema, have a long history in the country. However, what is to come with these streaming services and opportunities at the comedian's door is nothing to be assumed. But it would be interesting to see what other ways these comedians incorporate to keep the audience laughing and creating entertainment.

## REFERENCES

Anjum, A. (2021, November 16). Kusha Kapila: Anything that exists will be trolled. *Outlook India.* https://www.outlookindia.com/. https://www.outlookindia.com/website/story/entertainment-news-kusha-kapila-anything-that-exists-will-be-trolled/401191

Basuroy, T. (2023, March 20). *Number of unique visitors of video over-the-top (OTT) platform in India in April 2022, by platform.* Statista. https://www.statista.com/statistics/1373116/india-video-ott-unique-visitors-by-platform/

Bhatt, A. (2023, April 11). *"Professor of Tomfoolery" is Kenny Sebastian's most personal show yet.* 5X Fest. https://www.5xfest.com/5xpress/professor-of-tomfoolery-is-kenny-sebastians-most-personal-show-yet

Buchholz, K. (2020b, September 23). *Infographic: International Brands Rise to the top of Indian streaming market.* Statista Daily Data. https://www.statista.com/chart/23009/market-share-streaming--ott--platforms-in-india/

Duggal, S. (2023, April 13). *The insane journey of creator Prajakta Koli.* Entrepreneur. https://www.entrepreneur.com/en-in/entrepreneurs/the-insane-journey-of-creator-prajakta-koli/449636

Jenkins, H. (2017). *Convergence culture: where old and new media collide. Is the Universe a Hologram?* MIT Press. doi:10.7551/mitpress/9780262036016.003.0012

Kay, K. (2018). 'New' India and the NUTTAHS. *New Indian Nuttahs*, (Sept), 1–20. doi:10.1007/978-3-319-97867-3_1

Mpofu, S. (2021). Social Media and COVID-19: Taking Humour During Pandemics Seriously. *Digital Humour in the Covid-19 Pandemic: Perspectives from the Global South*, 1-16.

Otto, B. K. (2007). *Fools Are Everywhere: The Court Jester around the World.* University of Chicago Press.

Paul, S. (2017). A new public sphere? English-language stand-up comedy in India. *Contemporary South Asia, 25*(2), 121–135. doi:10.1080/09584935.2017.1321618

Puthiyakath, H. H., & Goswami, M. P. (2021). Is over the top video platform the game changer over traditional TV channels in India? A niche analysis. *Asia Pacific Media Educator, 31*(1), 133–150. doi:10.1177/1326365X211009639

Raghavan, V. (2020). In The comic element in Sanskrit literature. Dr. V. Raghavan Centre for Performing Arts.

Rahman, K. T., & Arif, M. Z. U. (2021). Impacts of Binge-Watching on Netflix during the COVID-19 pandemic. *South Asian Journal of Marketing, 2*(1), 97–112. doi:10.1108/SAJM-05-2021-0070

Research and Ranking. (2021, April 3) *Rise of OTT Platforms in India* [Video]. YouTube. https://www.youtube.com/watch?v=Pn3paQBgRIg

Sahoo, S., & Das, R. (2018). Is English stand-up comedy in India intersectional. *Language in India, 18*(9), 240–249.

Shah, H. (2018, April 30). *How Netflix became a $100 billion company in 20 Years*. Product Habits. https://producthabits.com/how-netflix-became-a-100-billion-company-in-20-years/

Shastri, P. (2021, March 11). comedy was Ott Remedy for covid blues. *The Times of India.* https://timesofindia.indiatimes.com/city/ahmedabad/comedy-remedy-for-covid-blues-ott-platform-survey-reveals-fav-genre-during-lockdown/articleshow/81440605.cms

Siegel, L. (1989). *Laughing matters: Comic tradition in India*. Motilal Banarsidass Publisher.

Singh, P., & Sharma, P. (2023). Entertaining the Royalty. A forgotten tradition of the Vidushaka in Ancient Indian History. *Historia i Świat, 12*, 237–254. doi:10.34739/his.2023.12.14

Stoll, J. "Ott Video Revenue Worldwide 2028." *Statista*, 17 Apr. 2023, www.statista.com/statistics/260179/over-the-top-revenue-worldwide/

Vaidya, R. (2019, March 8). *What Kusha Kapila learnt from being famous on the internet*. Vogue India. https://www.vogue.in/content/kusha-kapila-interview-indian-feminism-instagram-influencer

Westwood, R. (2004). Comic relief: Subversion and catharsis in organizational comedic theatre. *Organization Studies, 25*(5), 775–795. doi:10.1177/0170840604042414

Xalxo, J. (2020, June 4). *Kenny Sebastian: "everyone warned me that comedy is a bubble."* Rolling Stone India. https://rollingstoneindia.com/kenny-sebastian-everyone-warned-me-that-comedy-is-a-bubble/

Yin, R. K. (2009). *Case Study Research, Design and Method.* Sage Publications Ltd.

YouTube. (2014b). *STAND UP COMEDY SPECIAL PROMO - Kenny Sebastian : Journey to the Center of My Brain.* [Video]. YouTube. https://www.youtube.com/watch?v=Mh-Nh9xs4pw

YouTube. (2023). *The Making of MostlySane | Prajakta Koli on Relationships, Money & Movies.* [Video]. YouTube. https://www.youtube.com/watch?v=J8iFOoUKLHo&t=2291s

## ENDNOTES

[1] See Shastri, P. (2021, March 11). *"Comedy was Ott Remedy for covid blues." Ahmedabad News - Times of India.* The Times of India. https://timesofindia.indiatimes.com/city/ahmedabad/comedy-remedy-for-covid-blues-ott-platform-survey-reveals-fav-genre-during-lockdown/articleshow/81440605.cms

1 Jha, A. M. (2021, May 25). *A brief history of Indian standup: The Early Years supercut*. DeadAnt. https://deadant.co/a-brief-history-of-indian-standup-the-early-years-supercut/

2 Kapila, K. (n.d.). *YouTube shorts*. YouTube. https://www.youtube.com/c/YouTubeShortsSHIHAB

3 Kapila, K. (2021, May 24). *Psych! moms are always honest kushakapila new reels video insta ReelsAll famous Tiktok Star* |. YouTube. https://www.youtube.com/shorts/HthoCVcEuxA

4 Kapila, K. (2022, November 3). *My body is my body, none of your body.* YouTube. https://www.youtube.com/shorts/raBSXud8ZFM

5 Kapila, K. (2017, May 25). *Shameless (शेमलेस) by Prajakta Koli Ft. Raftaar | mostlysane | #ipledgetobeme*. YouTube. https://www.youtube.com/watch?v=H3vA0EJQ2mo

6 Koli, P. (2020, July 9). *Khayali Pulao | short film | Prajakta Koli, Yashpal Sharma | Mostlysane*. YouTube. https://www.youtube.com/watch?v=8cMDYZUfW-A&t=166s

7 Sebastian, K. (2017, May 5). *Me, my mother & our maid - Kenny Sebastian | don't be that guy*. YouTube. https://www.youtube.com/watch?v=WBUYy6NnWTo

8 Das, V. (2021, November 15). *Vir Das | I come from Two indias*. YouTube. https://www.youtube.com/watch?v=5A-F9qu6c_4&t=1s

9 Das, V. (2021a, February 2). *Vir Das: Religion is really just old comic books*. YouTube. https://www.youtube.com/watch?v=og7UroJJyoU

10 Das, V. (2021b, February 10). *Who has freedom of speech? | #TenOnTen | Vir Das | EP 2*. YouTube. https://www.youtube.com/watch?v=CQbm8XKi_xw&t=100s

# Chapter 11
# A Review of OTT Platforms in India:
## Challenges and Current Status

**Nitish Kumar Ojha**
https://orcid.org/0000-0002-2236-0766
*Amity University, Noida, India*

**Abhishek Vaish**
*Indian Institute of Information Technology, Allahabad, India*

## ABSTRACT

*Watching movies and serials has been the best form of entertainment in India, and its policies, delivery of the content, production, and related features were regulated by the government however with the growth of technology, it sub-branched into many segments later on. In the last decade, the availability of internet technology with mobile phones and satellite televisions allowed a new form of stream-based content delivery model, termed OTT (over the top). This content delivery has enabled 'leisure watching' in Indian society opposite to traditional binge-watching so it raised many issues at many levels, notably content delivery, content quality, accessibility, and most significantly controlling the freedom of expression under the boundary of creativity. This chapter reviews previously passed regulations, their current status matching the current situation, and the need for a new mechanism if any. This chapter also discusses the pros and cons of censorship hinges, its challenges, and the future scope for new legislation matching the ethics in the shadow of globalization.*

## INTRODUCTION

After Covid19, one word has become very common in our life - OTT Platform, nowadays it is very popular in the world of films. In today's time, if someone is getting bored, then he can pick up the phone and watch any video of his choice whenever he wants and can watch series, shows, and movies on the OTT platform. OTT has made entertainment quite easy. OTT platform means – an over-the-top

DOI: 10.4018/979-8-3693-3526-0.ch011

platform, which provides video or other digital media content through the Internet, this OTT platform is a type of app that Google Available on the Play Store, all companies have different OTT platforms, to use these OTT platforms, and you have to subscribe to these OTT platforms (Sujata et al., 2015). OTT platform was started in America, and it gradually came to India, but now its demand is increasing very fast. It is mainly used for video-on-demand platforms, audio streaming, OTT devices, VoIP calls, and communication channels. Used for messaging etc. (Novak et al., 2000). The rapid growth of OTT platforms in India, with projections of 50 million viewers by 2023, has led to concerns about regulation and the introduction of the Information Technology (Intermediary Guidelines and Digital Media Ethics Code) Rules, 2021. These rules have faced backlash and legal challenges due to concerns about their impact on media freedom, and the draft amendments in 2022 have also sparked concerns about vague provisions and a lack of transparency [Badlani et al., 2022].

Regulating OTT content poses challenges due to its on-demand nature, the volume of content, and its global accessibility. There are also concerns about the enforcement of regulations on foreign content and the potential rise of piracy. Balancing the regulation of content with the principles of free speech and expression is crucial in a democratic nation like India [Naik, 2023].

The Indian government's guidelines for digital media content, including OTT platforms, emphasize the need for age-related classifications for content, implementation of parental locks and age verification mechanisms, and caution when depicting racial and religious groups [Mansouri, 2023].

**History:** India's first independent OTT platform was BIGFlix, launched by Reliance Entertainment in 2008 [2]. It paved the way for the growth of Over-The-Top (OTT) platforms in India. During 2013 and 2014, these platforms started offering the streaming of Premier League matches on smartphones [1]. OTT platforms have gained popularity in India, allowing users to access a wide range of shows and movies across various media channels, including Star [1]. The introduction of BIGFlix and the subsequent growth of OTT platforms have revolutionized the way Indians consume entertainment content. Netflix, the American streaming giant, officially launched its service in India in January 2016. It marked the entry of international OTT platforms into the Indian market, providing Indian audiences access to a wide range of global content. Amazon Prime Video entered the Indian market in December 2016, further intensifying the competition among OTT platforms. It offered a mix of international and local content. Disney acquired 21st Century Fox in 2019, leading to the integration of Disney's content into the Hotstar platform. In April 2020, it was rebranded as Disney+ Hotstar, offering a blend of Disney and Hotstar content. Some mergers and acquisitions took place in the Indian OTT market, such as the acquisition of MX Player by Times Internet and the merger of Zee5 and SonyLIV. These developments aimed to strengthen the platforms' positions in the highly competitive market. The history of OTT platforms in India shows their rapid growth and adaptation to the diverse preferences of Indian audiences. The market continues to evolve, with increasing investments in original content production and the introduction of regulatory frameworks to shape the industry's future.

## Technological Advancement of OTT

Advancement can be seen in almost every kind of OTT platform so first, see what kind of OTT is available in the Indian market. The platform of OTT can be made in two types according to its quality. These can also be separated by calling them audio or video (Tiwary et al., 2020).

1. **Audio OTT:** Whenever you listen to any music such as a song, story, bhajan, ghazal, etc., it all comes in the audio type of OTT. In a way, in the OTT platform where you can only listen and see nothing, all that audio will come in OTT. In this, you can see images or photos, but no video will be able to play in it. The audio OTT platform is called AOD. The full form of AOD is Audio on Demand i.e., Audio on Demand. There are also two types of types in this which can be called free or paid.
    a. **Ad-based Audio on Demand OTT–** It is like free streaming services where you don't pay any money for the content. Whenever you listen to any music such as a song, story, bhajan, ghazal, etc., it all comes in the audio type of OTT. In a way, in the OTT platform where you can only listen and see nothing, all that audio will come in OTT. In this, you can see images or photos, but no video will be able to play in it. The audio OTT platform is called AOD. The full form of AOD is *Audio on Demand* i.e., Audio on Demand. Gaana App is an example of this kind of App.

In this, you can listen to any of your songs, or you can listen to any song by searching on the internet. Are. Apart from this, you will also find many other apps or websites where you can listen to songs for free. For example, some famous AAOD OTT platforms are Gaana, Wynk, Pocket FM, etc. In this, you do not have to pay any kind of charge to listen to the songs, but in this, you must see many types of ads. That's why this app earns money only through these ads and makes songs available to you.

    b. **Subscription-based Audio on Demand OTT–** It also works like a paid application where you pay the fee for watching the content. In this, to use these OTT apps, you will have to take their subscription. Taking a subscription to any app means buying it for some time. It depends on you which plan you want to take such as weekly, monthly, yearly, or anything else. Just like you recharge on mobile, you will have to recharge some of these apps. After this, you will be able to use these apps without ads. In this only Gaana and Wynk apps are famous. The only difference will be that you will not have to listen to any ads in between or after the songs. Also, you will have opportunities to listen to unlimited songs (Hutchins et al., 2019)

## Audio OTT Apps are in high demand.

According to recent FICCI-EY research, (EnY Report, 2023) 200 million individuals would be streaming music monthly in 2021, up from the current level of 100 million. According to the Indian Music Industry's (IMI) Digital Music Study Report 2021 (Manuel, P. 2014). Indians listened to music an average of 21.9 hours per week last year, which is more than the 18.4 hours per week that the rest of the world did. The music business rejoices as streaming income of Rs 1,134 crore in 2020 accounting for 85.1% of the overall revenue of Rs 1,332 crore, according to the IMI report. It is going to increase at an exponential rate when 5G will be rolled out at the pan-India level. The 5G roll-out is expected to greatly benefit the audio OTT segment, as 10 times faster internet and up to 50 percent latency reduction will be expected." This will also help more apps stream high-quality music.

The growth of regional content will be the largest driver for the audio OTT market going ahead, according to "Regional content is already a significant deal. With over 70% growth in streams for Wynk, it is the fastest-growing sub-segment. Reaching a diversified and large audience has helped non-film,

*Figure 1. Pie chart showing categorized users of OTT*
*(Vidooly Report 2020)*

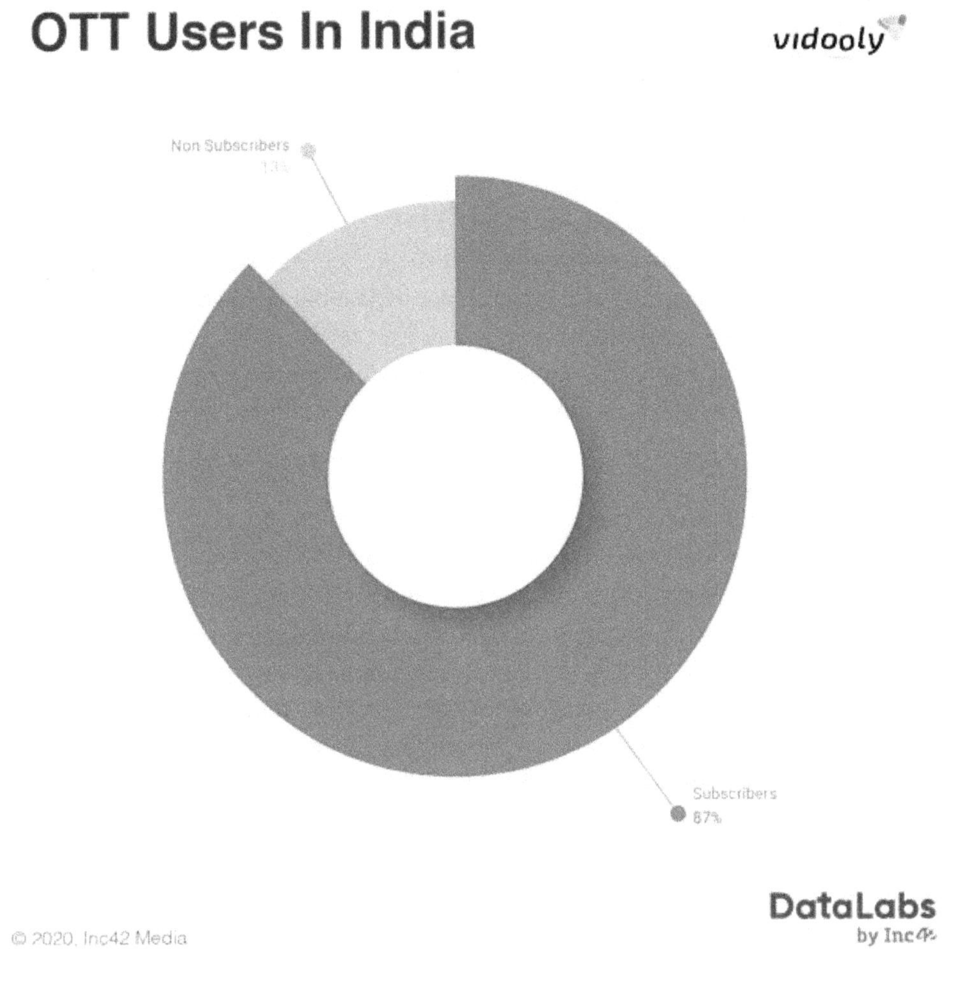


non-regional, and non-Hindi artists achieve significant success. Songs from regional languages, like Bhojpuri, Haryanvi, Punjabi, Tamil, Telugu, Kannada, etc., consistently make it to the YouTube rankings, demonstrating high consumer demand for music in their home tongue (Thomas et al., 2022). For instance, Renuka Panwar's Haryanvi song *52 Gaj Ka Daman*, which has received over 1 billion views to date, led the global rankings on YouTube for 13 weeks. But the biggest challenge is that do we have any mechanism available to filter this regional content? Probably not (Bhattacharya M., 2020). The digital database of Indian languages is not yet online. When the words of the main language are not in the online digital database, then it is a difficult task to get hold of their synonyms and misspellings.

2. **Video OTT:** The other name for this OTT platform is VOD or video on demand. This may be used to infer a wide range of things, such as whether you like to watch a serial or a movie. If you choose, you may watch any program, activity, festival, educational film, etc. All of it is available

on this OTT platform. VOD platforms come in three different varieties (Capgemini 2020). There are 3 types of this:

a. **Ad-based Video on Demand (AVOD OTT)**: It is also similar to AAOD to a great extent, that is, its features are similar to it, but its quality is different. It means to say that in AAOD you were listening to something with an ad, then in AVOD you will also see an ad in the middle of that video. You won't, however, be charged anything for this.
b. **Transaction Video on Demand (TVOD OTT):** You may only view a movie or television show on a site that offers video on demand. You don't see many films or TV shows, especially new ones, on the free or ad-supported OTT platform.
c. **Subscription Video on Demand (SVOD OTT):** It comes last in the video on demand and nowadays its trend is increasing very fast. This is because, in the first two, you will be able to watch only some free or medium movies or series, but if you want to watch famous movies or series, then you will have to spend money on them. Will have to get it done. Many types of apps and websites are not only famous in India, but some are famous all over the world. Such as Netflix, Amazon Prime, Hotstar, Zee5, Altbalaji etc. These are some of the apps that you must have heard about.

## Indian OTT Platforms and Their Growth

India's video streaming market is expected to be worth INR 12.97K Cr by 2024, according to Price-waterhouseCoopers (PwC) in its research titled "Global Entertainment & Media Outlook 2019-2023 (TRAI Report, 2023). The Over-the-Top (OTT) market in India has experienced significant growth over the past decade, and it continues to be one of the fastest-growing segments of the country's media and entertainment industry. Indian OTT platforms are growing at many levels either at the financial axis or at content, as many OTT platforms have invested heavily in producing original content, including web series and films. These productions have gained popularity not only in India but also globally. Shows and movies like "Sacred Games," "Mirzapur," and "Delhi Crime" have drawn international acclaim and at the same time, OTT platforms have recognized the importance of regional content. Many platforms offer content in various Indian languages, including Hindi, Tamil, Telugu, Marathi, Bengali, and more, allowing them to tap into regional markets.

Many OTT platforms have adopted a mobile-first approach, recognizing the significance of mobile devices in India. They offer user-friendly mobile apps for streaming content. While making OTT available on mobile devices, OTT platforms offer various monetization models, including subscription-based and ad-supported models. This provides revenue opportunities for both platforms and content creators. The Indian OTT streaming video market is currently in its second expansion phase and is projected to generate revenues of approximately $3 billion. The market is fiercely competitive, with companies investing heavily in content development and customer experience. The average revenue per user is projected to reach $8.81 in 2023. India's OTT market is valued at $2.92 billion in FY 2021-22 and is projected to reach $7 billion by FY 2027. The market could be worth $11-13 billion by 2030. Monetization of short-form video content alone could generate $8-12 billion by 2030. These figures highlight the immense financial potential and growth opportunities in the Indian OTT industry. [TRAI Report, 2023]

## Content Quality and OTT Players

India is a linguistically diverse country, and OTT platforms have recognized the importance of regional content. Many platforms offer content in various Indian languages, including Hindi, Tamil, Telugu, Marathi, Bengali, and more, allowing them to tap into regional markets.

The introduction of more than 30 streaming platforms in a matter of barely three years says eloquently about the enormous potential this business has, even though the stats indicate that it is still in its early stages of growth. The quick expansion has given brands, advertisers, and advertising in the digital consumer market a broad berth in addition to content owners and producers (Panjiar et al.,2023).

*Figure 2. OTT players and monetization model*
*(IFF Report, 2023)*

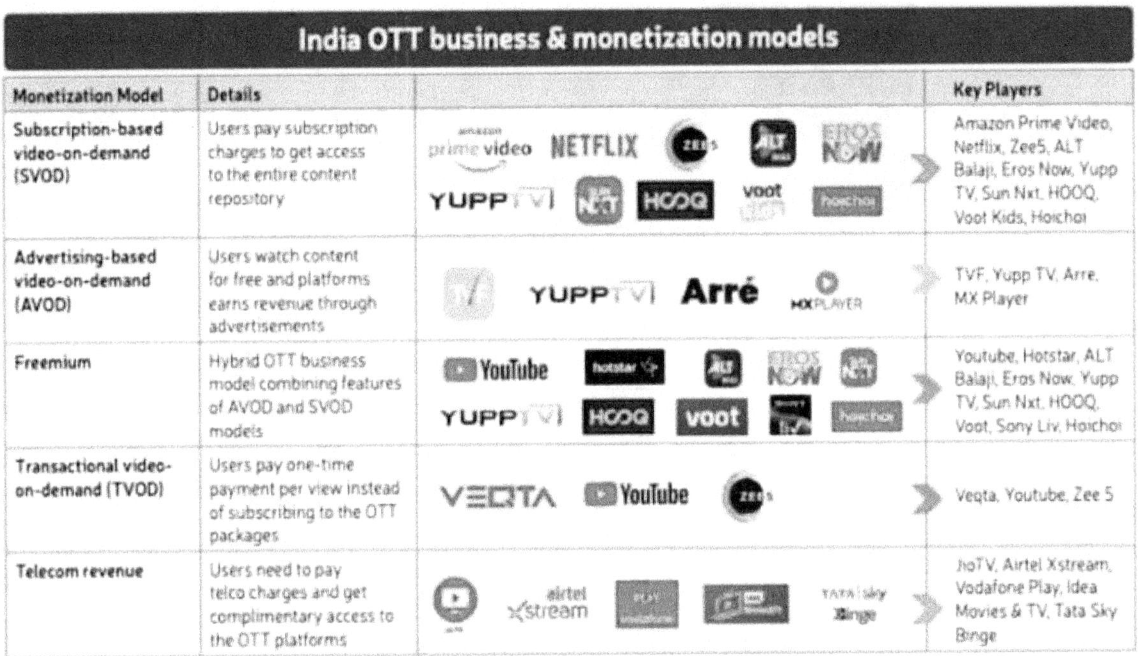

**India OTT business & monetization models**

| Monetization Model | Details | | Key Players |
|---|---|---|---|
| Subscription-based video-on-demand (SVOD) | Users pay subscription charges to get access to the entire content repository | | Amazon Prime Video, Netflix, Zee5, ALT Balaji, Eros Now, Yupp TV, Sun Nxt, HOOQ, Voot Kids, Hoichoi |
| Advertising-based video-on-demand (AVOD) | Users watch content for free and platforms earns revenue through advertisements | | TVF, Yupp TV, Arre, MX Player |
| Freemium | Hybrid OTT business model combining features of AVOD and SVOD models | | Youtube, Hotstar, ALT Balaji, Eros Now, Yupp TV, Sun Nxt, HOOQ, Voot, Sony Liv, Hoichoi |
| Transactional video-on-demand (TVOD) | Users pay one-time payment per view instead of subscribing to the OTT packages | | Veqta, Youtube, Zee 5 |
| Telecom revenue | Users need to pay telco charges and get complimentary access to the OTT platforms | | JioTV, Airtel Xstream, Vodafone Play, Idea Movies & TV, Tata Sky Binge |

There are currently many kinds of OTT platforms in India where users may watch films, TV shows, web series, etc. by subscribing on a monthly or annual basis (Tah S. 2022). In addition to this, there are other OTT sites where you may access a variety of services for nothing in exchange for repeatedly watching advertising (Wayne M. 2017). The following is a list of major OTT platforms available in India right now: Amazon Prime, Hotstar, Netflix, Zee 5, Voot, MX Player, SonyLIV, etc.

- **Hotstar:** It is an OTT platform owned by Disney that presently has more than 400 million monthly active users and dominates the Indian market. Live sports programming and blockbuster programs like Game of Thrones are two of its most well-known offerings. This platform allows users to watch information even without a membership service.

*Figure 3. Graphical representation of different OTT platforms*
*(Vidooly Report 2023)*

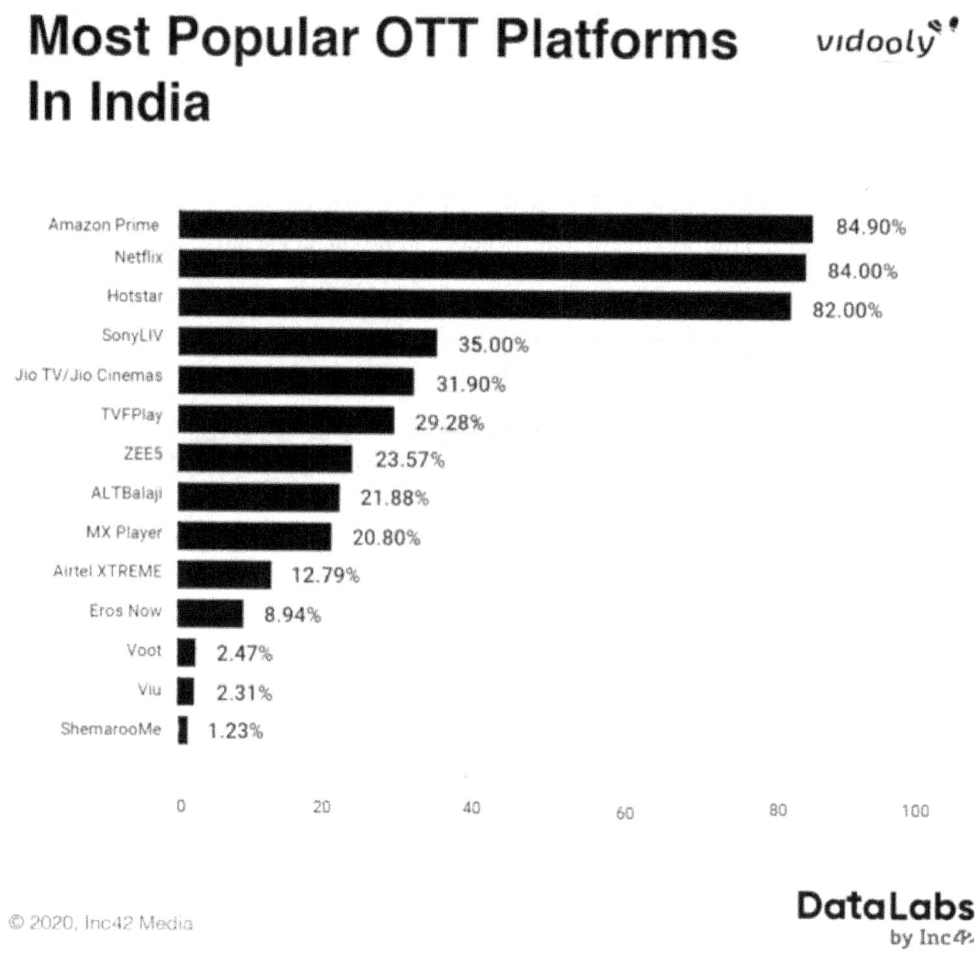


- **Netflix:** The most well-known streaming video company made its way to India in 2016. Four months after its debut, the original Indian television series Sacred Games was made available in it. Following this, low-cost internet plans, original TV programs, reality TV series, and films were introduced in India, and the country's ability to take subscribers quickly grew.
- **SonyLiv:** Sony joined the streaming space in 2013 by launching its own OTT service, following the lead of Disney-backed Hotstar. Then it began expanding quickly. This software had over 100 million downloads last year. To expand in India, SonyLIV, like Hotstar, broadcasts live spots for many international football competitions. Additionally, it includes gaming and a few minigames.
- **Zee5:** The OTT platform Zee5 was introduced in 2018. Over 80 Live TV stations and 100,000 hours of on-demand entertainment are available there. The platform's key selling point is its compatibility with different languages, along with its voice search functionality, live TV, and other features. By the end of this year, Zee5 will also release the short-form multimedia app Hypershots to compete with ByteDance's TikTok.

- **Voot:** Viacom18's 2016 debut alongside a number of its TV channels, Voot has amassed more than 45,000 hours of content. The following fall under this category: MTV, Nickelodeon, Viacom 18 Motion Pictures, Colours TV, and MTV Independent. It also provides information in a variety of languages.
- **MX Player:** After being named the most popular video streaming app in India, MX Player introduced an in-app gaming option to their video streaming platform. Five original web series were developed by MX Player in February 2019 with the help of Times Internet.
- **AltBalaji:** This company, managed by Ekta Kapoor's Balaji Telefilms, is a home video streaming service that focuses on Tier 2 and Tier 3 cities in India. This business app has the endorsement of Mukesh Ambani's Reliance Industries Limited.

## Growth Factors of OTT in India

- **Urbanization and Westernization:** User interest in customized OTT interfaces has increased as a result of migration to large cities and cultural shifts in media consumption.
- **Access to digital services:** It has given OTT a competitive edge thanks to affordable high-speed mobile internet, a doubling of internet users, and an increase in the usage of digital payments, among other factors.
- **Media democratization**: small content producers and accomplished artists benefit from India's OTT sector. Additionally, it aids in allowing access to regional films both domestically and internationally.
- **Features:** Fewer advertisements, the ability to pause and resume playback, and the ability to watch movies from anywhere at any time (for example, when traveling) Opportunities and other factors have all contributed to the lucrative expansion of the OTT sector in India.
- **Popularity:** The immense popularity of OTT in India is attributed to the rapid adoption and development of Internet infrastructure. These components enable OTT platforms to deliver content directly to audiences, bypassing traditional distribution and media networks.
- **Isolation in Pandemic:** The COVID-19 pandemic played an important role for OTT platforms to gain acceptance and popularity as people were spending more time at their homes during the lockdown. And it has helped people overcome boredom.
- **Easily Available being Portable:** Streaming media on the go is not the only advantage provided by OTT platforms. One can stream ad-free services at a low cost and download videos in multiple languages to their device for offline consumption.
- **Premier Content from Selected Viewers:** Some TV shows and movies are presented on streaming platforms even before they are aired on television. It is the perfect medium to take the excitement of first-time movie or web series fans to new heights.
- **One device many Providers:** A seamless viewing experience across multiple devices is an essential feature for OTT platform owners looking to build a stable following and customer base. A multi-screen OTT experience is another important factor that has helped the streaming industry garner an ever-growing customer base.
- **Fresh Content:** People are attracted to these platforms for quality and fresh content as they are tired of traditional Indian TV serials which go on forever. OTT opened up the world to creators and artists who had lost their relevance due to the change in the type of entertainment. Many have made their way back into the limelight.

An independent report (Farooqui, M. 2021) in the year 2020 predicts that OTT platforms in India would generate significant revenues. The Indian OTT industry is expected to expand by 45% by 2023, which would help these platforms grow to 130 billion rupees in revenue. The market for OTT platforms in India now has a value of Rs 250 crore. According to a forecast by Ernst & Young, there will be 500 million OTT users in the nation by the end of this year. After the US, India will have the second-largest OTT market (Joshi, S. 2015).

*Figure 4. Different OTT platforms in India*

**Regulatory Framework:** Currently, there are no specific regulations for OTT platforms in India. However, the Indian government has recognized the need for regulation and is taking steps to implement it. The Information Technology (Intermediary Guidelines and Digital Media Ethics Code) Rules, 2021 (IT Rules) was introduced to regulate the previously unregulated OTT sector. These rules include a Code of Ethics for platforms to self-classify content based on age appropriateness. However, the IT Rules have faced criticism for potentially limiting media freedom. Draft amendments to the IT Rules released in 2022 have raised concerns about transparency and censorship. The goal of content regulation should be to ensure appropriate audiences rather than censorship. Two main aspects are there in this scenario –

1. **Content Classification and Age Ratings:** Following self-regulation and the government's guidelines, OTT platforms have been using age ratings and content classifications to ensure age-appropriate viewing. These labels help users and parents make informed choices about the content they access.
2. **Self-Regulation Codes:** Before the finalization of government regulations, many OTT platforms in India had established self-regulatory codes to classify and label content based on its suitability for different age groups. These codes were meant to provide viewers with information about the content's nature, making it easier for them to make informed choices.

## CHALLENGES OF OTT IN INDIA

### First Problem: Network Usage Charges

#### Problem Area Identified: Lack Of Country Detailed Regulatory Framework

The majority of OTT platforms are controlled and run by sizable foreign companies, and they gather and store subscriber data from Indians abroad to boost the GDP of the country where they are located rather than directly boosting the Indian economy or telecom development. There would have been enormous benefits for the Indian exchequer if AGR had been used in this situation (Sharma, K.S. 2020).

Telecom firms claim that OTTs should pay an "Access Charge" since they are using their infrastructure "free of charge." In this situation, it is necessary to go in this direction to ensure the proper operation of the developing media and entertainment sector. Although OTT communication services are now free, the platforms generate enormous indirect income and profits through adverts and other monetization techniques using subscriber data acquired across the Indian subcontinent. The majority of OTT platforms are controlled and run by sizable foreign companies, and they gather and store subscriber data from Indians abroad to boost the GDP of the country where they are located rather than directly boosting the Indian economy or telecom development.

#### Reason: Lack of Appropriate Law

Since TSPs are Indian corporations, the earnings they generate are subject to taxes under 25-year-old rules in India. They spend a lot of money setting up, maintaining, and updating their networks technologically, yet OTTs are allowed to use these networks for free. The vast resident databases of subscriber data are also prohibited from being commercialized by law. OTTs are free to do that. Similar communication services are offered by TSPs and OTTs to the same Indian population in the same environment, region, and application layer, BUT under different laws that are detrimental to TSPs and the Indian nation. This and other issues should be fixed under the Telecom Act in the future (Upadhye, A. 2019).

### Second Problem: Censorship Concern

#### Problem Area Identified: Absence of Direct Regulation

OTT services are not directly regulated by any organization or law. They are under the exclusive jurisdiction of the Ministry of Electronics and Information Technology (MeITy) (KPMG Report 2022).

**Case Study -** An Indian court bench informed the Centre that establishing a legal framework was necessary in addition to guidelines for control before granting temporary protection from arrest to Aparna Purohit, the head of Amazon Prime Video India Originals, in connection with the FIR lodged against her regarding the Tandav web series.

On February 27, 2021, the Ministry of Electronics and Information Technology published the Information Technology (Intermediary Guidelines and Digital Media Code of Conduct) Rules, 2021. (Rohit Jacob Jose, 2020).

**Reasons:** The Suitable Boy on Netflix, Tandav on Amazon Prime, which stars Saif Ali Khan, and several sequences disrespecting Hindu deities in ALT Balaji are the most recent issues involving OTT platforms. XXX gave a few current instances. This argument has been sparked by the entire situation.

## Third Problem: Impact On Telecom Revenue

### Problem Area Identified: Complex and Decentralized Financial Framework

The demand for regulation of over-the-top ("OTT") communication services (such as WhatsApp and Telegram) has been around for many years. This demand is based on the logic of "same service, same rule" in which telecom service provider companies want that as many companies providing OTT services are using telecom services to provide their services and earning revenue from lag. Part of that revenue should also be distributed to the telecom companies as the license from the government has been taken by the telecom companies and not by the companies providing OTT (Bhattacharya et al., 2020).

OTT content providers argue that apart from concerns related to net neutrality, the costs incurred by OTT content providers are not necessarily traffic-sensitive, making payment based on the number of users or the amount of traffic transmitted is unfair. There has been a drastic change in the numbers in the industry after the entry of Jio. As anticipated, with the merger of Vodafone and Idea, the industry saw a similar increase in revenue (Q-o-Q from 18Q3) (TRAI ARPU Report 2022).

*Figure 5. Graph showing average revenue per user w.r.t different telecom players*

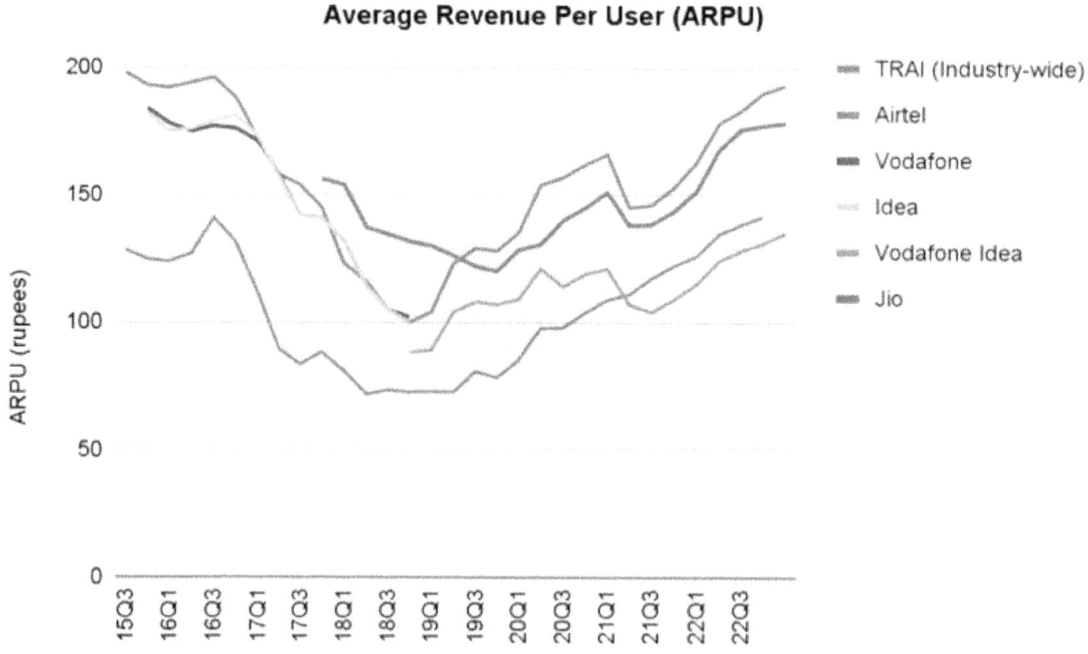

**Reason -** In the analysis, we find that this growth is positively correlated with the change in voice and data usage across the region – that is, the usage of both services is also increasing, especially since the entry of Jio in the market, and not just Vodafone – increased further with the integration of Idea. Data thus implying an increase in data usage - and therefore services availed using such data - cannot be blamed for decreased or negatively impacted revenue streams. The goal of regulation should be to serve the best interests of the public, rather than to protect company profits (Robert and Ravi, 2015).

## Fourth Problem: The Threat of Cybercrime

### Problem Area Identified: Lack of IPR Policies and Identification of Breach of Trust

In the process of subscribing to an OTT platform, people share their confidential information (like bank details, credit card details, etc.) which can be misused or where the threat of cybercrime exists. Piracy is a concern for both ad-supported and subscription-based video streaming platforms. The yearly revenue loss for these services is 30%. Within an hour and a half of their debut, well-liked episodes like "Scam 1992" on the SonyLIV app and "Ashram" on MX Player were pirated.

This is a significant issue for the OTT industry. In India, there is still a ton of room for growth in this industry, but the issue of piracy might potentially deal a serious blow.

## Case Study One: Content Theft

There are many Apps on many websites which steal the original content from these paid OTT platforms and make it available to their subscriber free of cost. These apps are not available on Google Play Store, instead of that, you have to download these Apps from their website only. A few of them are Pikashow, Momix, Popcorn Flix, etc. The complete program appears on these pirated apps just a few hours after it premieres on the country's biggest OTT platforms. Some of these platforms are Pkashow, Momix, T TV, Zephy, Oho, Jaguar, Snowflake, Beebalm, Velvet, Mr. TV, MovieBox, HD Sinema, and Momdro. Thop TV etc.

## Case Study Two: Link Sharing

**Telegram -** Telegram App has been able to gain a lot of popularity in a short period but its use has been huge in the last few years for other reasons and one of them is that many companies have been cheated on this app irrespective of the number of OTT platforms. Be it Netflix or Prime Video and Disney Plus Hotstar, the movies on all the app's content can be watched here for free, due to which these OTT platforms are facing a lot of loss, people use Telegram to watch movies and download and share them fiercely (Dhiman, 2023).

**Twitter -** On Telegram, people used to look for connections to get pirated or leaked movies. But those times when this software offered URLs to watch films or series for free are long gone when Telegram's trick came to an end, individuals created a new technique on Twitter. This work now makes use of Twitter. Twitter is being used more and more by users to get films for free. Many films have recently been uploaded on various microblogging sites. They might be downloaded by anybody.

Twitter is growing more and more well-liked for obtaining free television shows and films. Elon Musk's social networking site featured videos from films like Shrek and Evil Dead. At the same time,

Figure 6. Two examples of live streaming Apps which allow you to see any content for free

Figure 7. A sample screenshot of telegram channels providing OTT content freely for download

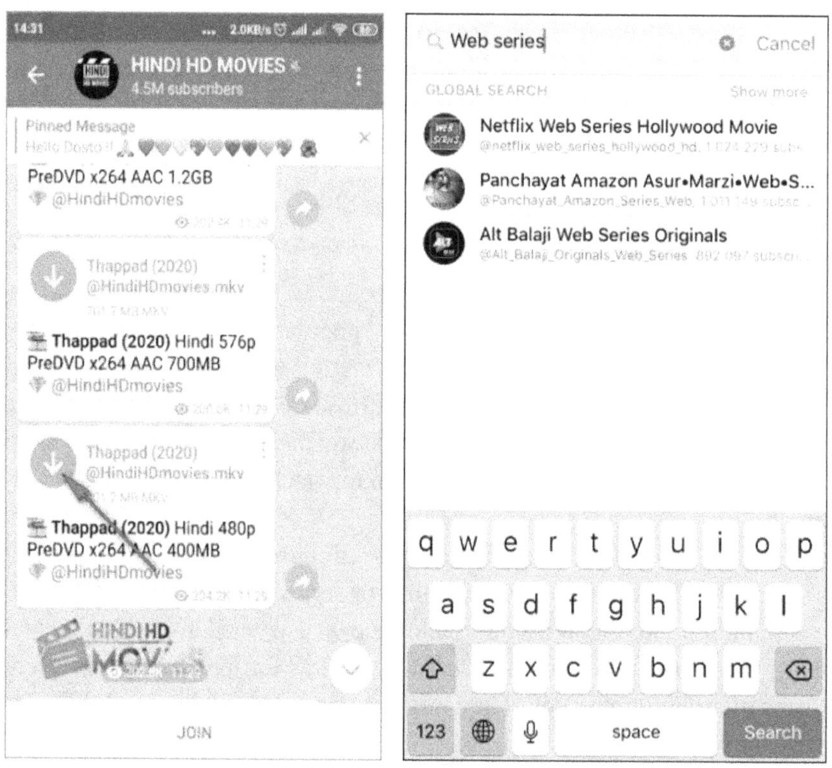

*Figure 8. Screenshot of a few telegram channels providing movies of different platforms unethically*

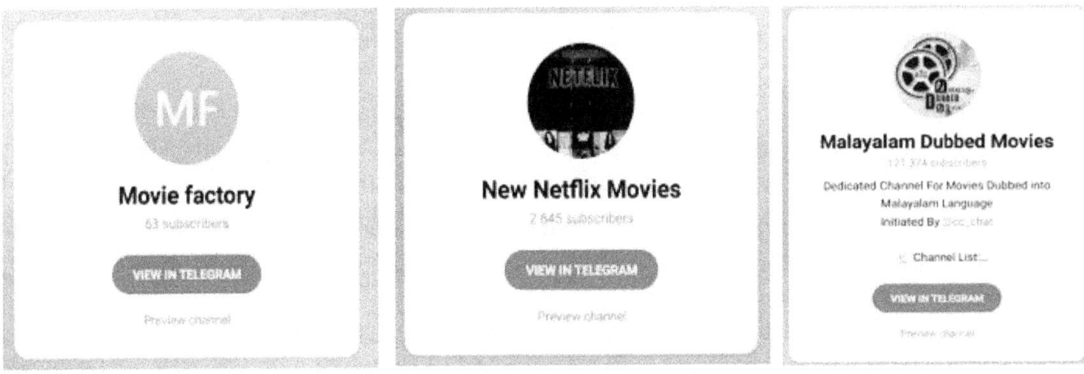


Keanu Reeves' newest film is titled John Wick Chapter 4. The movie was offered for free download going against the rule of the movie industry. Elon Musk's has unveiled this new feature. Only Blue Tick subscribers have been given access to this feature. Blue tick members have a 2-hour or 8GB video upload limit. Users shared a lot of movies as soon as this option was available. Visit the Twitter video downloader page to get these films. By visiting the website and copying the URL from the tweet, videos may be downloaded quickly. However, it is unlawful to distribute pirated content, however later on this feature enforced the approval of the Twitter team from a copyright point of view, the challenges of reporting every video and its removal from the platform remain not upto the mark as the user base is very high so people can download before it gets removed promptly.

*Figure 9. Announcement of the video feature and its copyright issue*

## Case Study Four: Terror Funding

### Problem Area Identified: Lack of Control and Immunity of Social Media Websites

According to news portals, the Indian OTT platform providing video streaming services is losing 30%-35% of revenue every year. Security agencies have a suspicion that this money can also be used against India i.e., for terror funding. The 1993 blast accused Dawood Ibrahim has also been involved in pirating Indian films for many years. According to sources, people associated with the International Mafia Gang leader Dawood gang have joined the piracy of OTT content and are collecting thousands of crores of rupees every year (Rohit Jacob, 2020).

This is a serious risk of loss of personal data. This stolen personal data may be available on the dark web for sale. This is a serious issue that has to be handled and to be addressed as early as possible.

**Reason -** Indian piracy occurs for two reasons. First of all, this product is completely unavailable in India, which leads to this particular content being pirated. Unexpectedly, the product doesn't arrive in India until much later than it does elsewhere.

Piracy harms OTT platforms' business. They may lose their potential as a result of this issue of data theft or piracy.

## Fifth Problem: Effect On Users And Experiences

### Problem Area Identified: Challenges to the Morality of Society

Critics have long argued that the prevalence of vulgarity and obscenity in the content on these platforms harms the youth and poses a threat to the social fabric. The social cohesion and moral foundation of society may be harmed by the content of OTT platforms because of the absence of censorship.

India now has the greatest global per capita mobile data use, according to Ericsson Mobility Reports. The average monthly data use is currently 10.022 GB, and by 2024, it is predicted to increase to 18 GB. Since young people make up the majority of OTT users, the government must have control over what they watch and how it is presented to them in the interest of innovation. Pornographic websites and videos are outright forbidden by the government on the one hand, while they are permitted to be broadcast freely via OTT platforms on the other. This has led to an increase in violent and rape episodes, which is the effect. Remarkably, even major production companies and well-known performers are already participating in such scenarios. In this case, OTTs like porn sites need to be strictly enforced (Abid Ali et al., 2021).

In a recent study on OTT content, it was observed that 9 out of 10 online series or OTT shows currently air in a way that is not fit for family viewing. These television programs and online series are chock-full of sexual content and vulgarity. Numerous popular television programs and online series that portray profanity and sexuality in the name of innovation can be found on Amazon Prime Video, Netflix, Disney Plus, and Hotstar. A lot of these programs have also generated controversy.

## Case Study One: Amazon Prime

The OTT-based web- series *Tandav*, which debuted on Amazon Prime Video in the year 2021, generated a great deal of criticism. This online series features Saif Ali Khan, Dimple Kapadia,

Tigmanshu Dhulia, and Sunil Grover. The Central Government had to step in as a result of the growing controversy surrounding this program. Hindu gods have even been made fun of in a Tandav scenario. A member of the BJP (a political party in India), has requested in a letter to the minister of information and broadcasting that this web series not be allowed to play on Amazon Prime Video because of this scene. Additionally, the show's director, producer, and actors were requested to sincerely apologize for offending Hinduism.

Another web series Mirzapur on Amazon Prime Video has also generated controversy. It has been claimed that this series is harming Uttar Pradesh's civilization and culture. This series' controversy grew to such a height that the Supreme Court heard the case. The third season of this online series is set to arrive, although it is still currently airing on Amazon Prime Video. In addition, this web series contains several explicit scenes that you cannot view with your family. Not only that, but both seasons of this online series contain a lot of foul language.

Abuse and violent images in the web series named Patallok on Amazon Prime Video have also caused criticism. In this series, Jaideep Ahlawat and Abhishek Banerjee portrayed the leading parts. The creators of this program have been sued by the Sikh and Gorkha communities. Both of these communities have claimed that this series has offended their feelings.

## Case Study Two: Netflix

Sacred Games on Netflix has also generated criticism. Actors like Saif Ali Khan, Pankaj Tripathi, and Nawazuddin Siddiqui have also worked in this web series. This web series is based on the 2006 best-selling novel by Vikram Chandra. A joke regarding the past prime minister was made in this series, which caused controversy. In addition to this, this web series is loaded with vulgarity, sensual scenes, and offensive language. This web series has a lot of unedited moments that you cannot watch with your family. Netflix has been streaming this web series since 2018.

## Case Study Three: MX Player

Aashram, a web series by MX Player, has also generated controversy. In this series, Bobby Deol portrayed the lead role. The fact that a Hindu saint was portrayed in this series as a drug dealer, rapist, and corrupt person caused criticism. According to rumors, Baba Ram-Rahim of Hisar is the inspiration for the Prakash Jha-directed television series. It has several sensual sequences that are intended to propagate indecency. Additionally, it contains a lot of profanity.

## Case Study Four: Alt Balaji / HoiChoi

In addition to these contentious online series, Ekta Kapoor's Balaji Telefilms company's OTT platforms Alt Balaji and Bengali OTT platforms Hoichoi and Ullu also aired many pornographic web series and programs. Erotic scenes are prevalent in web series like "XXX" and "Gandii Baat" on ALTBalaji. Furthermore, the obscenity threshold has been broken in the MX Player online series "Mastram." These web series and programs are all still available for internet streaming.

## Sixth Problem: Obscenity Means Business In Ott

### Problem Area Identified: Content Monitoring and Discretion of User's Choice

Many times, police have registered cases against Alt Balaji, Hot-Shot, Flimzmovies/Fenio/Kuku/Neo Flix, Ullu / Hotmasti, Cheeku Flix, Kuku, Prime Flix, apart from the website X-Videos and Porn Hub website, which broadcast obscene content. People are accessing such obscene material very easily through mobile. This can have fatal consequences for the youth. The Internet is growing rapidly in India. According to the Ministry of Telecom, there were 391.50 million internet users in the country as of 31 December 2016, it is expected that this number will increase to 900 million in 2025. Internet is getting cheaper in India and with the arrival of Jio, there is competition to make net packs cheaper in the market. The example of the Internet is evident from the fact that no matter how much TRP Kapil Sharma's show is on TV, the number of people watching Kapil's show on YouTube reaches crores. According to the report of the news, portal Let's OTT, in July 2022, the highest number of streaming (11 million) for an adult comedy show was done on the online OTT platform Max Player in a single day.

According to the report of Let's OTT (Let's OTT Report 2022), Alt Balaji's May 2022 viewership has increased by 70% as compared to last year. Its monthly active users have also increased by 33%. There is more adult content on Alt Balaji's OTT.

## Seventh Problem: Promotion Of Drugs

### Problem Area Identified: Freedom of Expression and Dummy Advertisement

Made in Heaven is a web series in which in almost every scene the main characters or actors smoke, drink, do drugs, and use profanity. If we look closely at many web series, we see this as a trend. Sometimes they are placed unnecessarily just to please the audience and glamorize the story. Made in Heaven, the main character is upset and then another character gives him cigarettes and alcohol and he thanks him for his 'much-needed support' and is told to be normal. Such scenes are very common in other web series as well.

### Sociological Analysis: Impact of OTT on Society in the Indian Context

#### *Promotion of Criminal Mindset*

How the web series is bringing practical changes in society and how society is copying it indiscriminately can be easily understood from many incidents. Today, the way crime is being shown in the web series is directly affecting the mental state of the person and the incidents of crime are also increasing rapidly. Looking at the web series, nowadays it is understood that it is easy to escape from the police and how to escape from the police after committing a crime. Crime is being glorified in web series which is not right—currently playing a very deadly role in web series. In the script for the web series, it is shown as if you and I are living in it. Web series run because negative thoughts keep running in our minds. When those negative thoughts appear to result in an incident, then we feel that we can also do it, which is directly reflected in society and the incidents of crime are increasing. In a reverse manner, many real-time crime incidents have been made a story in many OTT-based web series, and in those web series,

*Figure 10. True crime documentaries series on OTT*
*(Instagram)*

the crime is presented in a very heinous manner to make it more realistic and crunchier so that more and more viewership can be assured.

**Case Study -1 -** Inspired by the crime-based web series Money Heist, a 27-year-old youth in Hyderabad formed a gang in which some youths, including a woman, were recruited to carry out the kidnapping incident. In police interrogation, the main accused said that he was inspired to commit the kidnapping incident after watching the web series. In which the main character used to recruit people to carry out many criminal incidents.

**Case Study – 2 -** In the Jahangirpuri area of India's capital, Delhi, three minors killed a man in film style after being impressed by the crime-based web series. On being caught, during police interrogation, he told that he wanted to become famous in the world of crime and he committed the crime after getting influenced by the crime-based web series 'Pushpa'. The accused told during interrogation that they were influenced by the lifestyle of gangsters shown in films and web series like 'Pushpa' and 'Bhaukaal'.

**Case Study – 4 -** The Dancing on The Grave web series premiered in April 2023 on Amazon Prime Video. The story of this series shows the story of a high-profile murder that took place in 1991 in Bangalore. A husband murdered his wife to grab her property and buried her in a luxurious house. However, in the OTT movie murder was done in a very heinous manner later on it was glorified.

**Case Study – 5 -** Netflix started a documentary series based on true incidents of crime called Indian Predator some time ago, in which some such cases are shown, which shook the country and were in discussion for a long time. There have been four documentary series so far under the title Indian Predator.

**Case Study – 6 -** An FIR was lodged under section 295A against the makers of Netflix's other web series 'A Suitable Boy'. It was alleged that the kissing scene between a Hindu girl and a Muslim boy shot against the backdrop of a Hindu temple was hurtful and insulting to Hindus. *Dolly Kitty Aur Woh*

*Chamakte Sitare*, produced by Ekta Kapoor, was about a married Hindu woman with two children. She sleeps" with a Muslim delivery boy and his cousin, who works in a call center selling 'gift items' using 'sex chat'. The same. Disney Hotstar's web series Parivar made fun of rituals like Yagya (ritual worship) by making the Pandit (priest) do some incoherent nonsense in the name of mantras. It also had scenes where a character insulted a priest performing a havan. Insulting one's mother and hurling insults at the priest called to perform the ritual is considered a comedy and then one of the characters calls the ritual a "nautanki" (drama) before pouring water on the havan-kunda (sacrificial fire).

**Case Study – 7 -** This trend is not limited to the Hindi industry only and has spread to regional industries as well. Zee5's Tamil web series Godman was also controversial in that, unlike any Tamil devotional film, instead of glorifying Hindu deities, it pitted Jesus against the local deities (Siru Dhivam), declaring Jesus to be a friend of Amman Devi. Has been done. Navaras, a Netflix web show, has also been accused of insulting the Quran. Islamic organizations in India announced that the poster contained a recitation of Quranic verses and should not be published in newspapers or posters. Raza Islamic Academy in India posted a poster of a series of Quranic verses in a local Tamil newspaper on Netflix Accused of publishing, hurting the sentiments of Muslims, and insulting their Quran.

Reason –

1 – **Absence of Law -** Absence of Censor Board These web series do not have to face any kind of censorship anywhere. Because of this, he can put the scene dialogues according to his own, and every key is exactly as he wants it to be. They can show the same, although some people in India are also taking wrong advantage of its lack of censors and are serving people by making web series only and only with lewd sex scenes. But there is only a special audience who likes this type of web series, their number is very small.

Many OTT content providers have made a grouped and self-signed rule of law to regulate the content however there is no significant improvement observed. About 15 OTT platforms, including big players like Netflix, Disney Plus Hot Star, Voot, Alt Balaji, etc. Together, they have made a code for content regulation this year, in which five types of content have been agreed not to be shown on OTT platforms. given. However, the government did not consider it sufficient. In this regulation code, OTT players had decided that no content insulting national sentiments would be shown. Also, content promoting child pornography, hurting religious sentiments, and promoting terrorism in any form will not be shown. Also, no such content will be shown, which has been banned by the court in any way. The government did not accept the regulation code of OTT players.

2 - **Incompetent Existing Law -** In addition, the Indian Penal Code (IPC) tries to prosecute anybody who is or has ever distributed or sold pornography as that term is defined in Section 293 of the code. The next objective is to damage religious sentiments willfully and maliciously, which is prohibited under section 295A. Any publication of false information is considered defamation under section 499 of the IPC, as is any insult to the modesty of women under section 354 of the IPC. The Indecent Representation of Women (Prohibition) Act, 1986, is one of the several Acts relating to this matter. The Act places a strong emphasis on achieving a total prohibition of the indecent portrayal of women in ads,

Further, there are provisions, as mentioned in Sections 67A, 67B, and 67C of the Information Technology Act (IT), 2000, which mention and punish any person broadcasting or publishing obscene material of any kind, Providing for fine and imprisonment. This includes any sexually explicit material as well as material where children are depicted in sexual acts, however, there is no central body to report that these acts are not fully implementable in the case of OTT. In many cases, OTT content providers have skipped citing various reasons and lacking clear guidelines that can be enforced on them as these acts were primarily designed for computers not for the media being streamed through communication devices. Strict laws defining all terms and conditions is the need of the hour to control this menace in OTT which is being presented in the name of liberty of creativity.

3 – **Absence of Censor Board -** Absence of Censor Board These web series do not have to face any kind of censorship anywhere. Because of this, he can put the scene dialogues according to his own, and every key is exactly as he wants it to be. They can show the same, although some people in India are also taking wrong advantage of its lack of censors and are serving people by making web series only and only with lewd sex scenes. But there is only a special audience who likes this type of web series, their number is very small.

**A Comparison between Global OTT platforms and Indian OTT platforms:** Depending on the Zener, choice, and interest OTT platforms are available across the world and in every language, ir does not matter if you are a Marvel movie lover or just a melodramatic binge watcher. Many of them are Hulu, Netflix, and Amazon.

There are many criteria to review OTT while considering both Indian and Global players –

1. **Categorical Content:** Global OTT players have their presence in each category of age group. In the USA, the usage of Over-the-Top (OTT) platforms varies across different age groups. According to Statista, 75% of 18 to 34-year-olds, 72% of 35 to 44-year-olds, and 56% of 45 to 54-year-olds use OTT platforms. The growth of OTT and streaming video consumption is evident across age groups, with a shift in consumer demographics driving this growth. While younger consumers initially led the adoption of OTT, older audiences are now the main drivers of growth, emphasizing the emerging potential of older consumers in the OTT space.
2. **Code of Ethics:** The Indian government has issued new guidelines for digital media content, mandating OTT platforms like Netflix, Disney Hotstar, and Amazon Prime to provide age-related classifications for their content as part of a mandatory Code of Ethics for online curated content. This move is aimed at ensuring age-appropriate content consumption and requires a prominent display of content warnings related to themes like violence, nudity, and substance abuse. Additionally, streaming platforms will be required to implement parental locks for certain content categories and establish a self-regulatory body to oversee adherence to the guidelines and address grievances whereas The U.S. Department of Energy's Office of Technology Transitions (OTT) seeks subject matter experts to review applications for upcoming funding programs related to the Bipartisan Infrastructure Law. Reviewers play a crucial role in providing unbiased evaluation and are compensated for their services. The department is looking for reviewers with expertise in areas such as energy technology, innovation, economic development, and diversity. Interested individuals can apply by sending their resumes and registering on DOE Exchange, and selections will be made based on program needs while assessing conflicts of interest at the point of selection.

3. **Guidelines to regulate the content provider:** The Indian government has introduced new guidelines for digital media content, mandating that OTT platforms like Netflix and Amazon Prime provide age-related classifications for their content and this is similar to other countries to regulate the content.
4. **Feature-Rich Content and Quality of the Content:** Global OTT content is normally full of features, and it is available in many formats i.e., HD, Full HD, etc. Indian content is also available in similar levels and types however it is not that available up to that level. Global players are making their content available to different platforms i.e., mobile, browser, App-based, etc. while only a few OTT players in India are doing well in this direction.
5. **After Effect of OTT:** In India, few OTT players have been reported for malicious funding and few are under scanning by investigations agencies. The effect of OTT and their implications, what they wish to represent is a serious matter of concern where in a global scenario this is less prevalent because of strong regulations.
6. **Implication to social fabrics:** Implications of OTT on society are a global concern these days not only in India. The impact of unregulated online content on society has prompted discussions about new regulatory frameworks in the USA. In Latin America, countries like Uruguay, Costa Rica, Colombia, and Argentina are also facing regulatory challenges due to the rising popularity of Over-the-top (OTT) services [2]. Additionally, India has been grappling with the regulation of OTT platforms, which has sparked concerns about censorship and freedom of speech. These global efforts aim to ensure age-appropriate content and protect audiences from potential harm.
7. **Due Caution and Discretion:** According to the new draft of Indian OTT regulations, OTT platforms must take into account India's diverse racial and religious context and use "due caution and discretion" when depicting the activities, beliefs, practices, or views of any racial or religious group. Similarly, stringent regulations are present in Western countries when evaluating viewership and content quality.
8. **Parental Lock and Child Safety:** Under the new regulations, OTT platforms must incorporate a parental lock feature into their services. In addition, they are required to assign viewer grades and ratings based on parental control, a practice already implemented by OTT providers worldwide.
9. **Educational Content:** Global OTT players have a variety of content and series specifically for kids and school-going children such as Discover+, Animal Planet, Discovery, TLC, and Discovery Turbo, etc., while this type of categorical content availability is very little in the Indian scenario.
10. **Grievance Redressal Mechanism:** The Indian OTT regulation act includes appointing a Grievance Officer and setting up a three-tier grievance redressal mechanism, with the government overseeing the Level III mechanism and maintaining a grievance portal. It also mandates the classification of content and display of ratings, as well as the observance of journalistic conduct norms and program codes by news publishers on digital media, and a similar type of mechanism is available in other global OTT platforms too. The difference between those two entities lies in terms of transparency and compliance in a democratic manner.

## CONCLUSION

The majority of the problems center on the content and the government's control over it. A large number of these platforms have consented to self-regulation. IAMAI, the Internet and Mobile Association of

India published a document under the codename "Code of Best Practices for Online Curated Content Providers." This code's single and primary goal is to create a foundation for open disclosure. As long as the material is described or represented and the maturity rating is mentioned, it is also considered to be open and disclosed. The code was split into three versions, with the most recent one being released in September 2020. The Supreme Court of India addressed a notice to IAMAI in November 2020. Additionally, after this, a notification said that the Ministry of Information and Broadcasting will now be in charge of overseeing all online curator content providers, or OCCPs, including OTT platforms. A recent update stated that the government released rules titled Information Technology (Intermediary Guidelines and Digital Media Code of Conduct) Rules 2021 on February 25, 2021, including a three-tier regulating structure for OTT and digital news media.

## Future Scope

- **Establishment of a Regulatory Body:** Given the existing circumstances, it is necessary to use an impartial regulatory body to control material on OTT platforms.
- **Formation of Strict Regulation:** In addition, light-touch laws should be set up for OTT communication platforms like WhatsApp, Signal, and Telegram. The government should implement tight criteria for content development on OTT while keeping in mind consumer interest and cyber fraud risks.
- **Viewers' Responsibility:** If a stringent access and regulatory policy is not put in place to restrict access to underage content, viewers must ensure that minors in households do not have unrestricted access to OTT content.

The media and entertainment segment are one of the fastest-growing sectors in India. However, it is challenging to keep in mind the evolving content trends and ever-changing consumer demands. OTT platforms have helped maintain sanity in these difficult times by being a major source of entertainment. Almost every streaming service in the world is offering a free trial and users are taking advantage of the trial. However, the real test is customer retention which is achieved not only with great content but also with great technology, it is expected that artificial intelligence will be used in the future to create neat and long-lasting content.

## REFERENCES

Ali, A.. (2021). Impact of Smartphone: A Review on Negative Effects on Students. *PalArch's Journal of Archaeology of Egypt / Egyptology, 18*(4), 5710-5718. Retrieved from https://www.archives.palarch.nl/index.php/jae/article/view/7164

Ani. (2020, August 5). *LetsOTT the innovative Digital Streaming (OTT) Search Engine*. BW Businessworld. https://www.businessworld.in/article/LetsOTT-the-innovative-Digital-Streaming-OTT-Search-Engine/05-08-2020-305314/

Badlani, G. (2022, July 25). OTT censorship – through the lens of freedom of speech and expression. *Lucknow Law Review*. https://www.lucknowlawreview.org/post/ott-censorship-through-the-lens-of-freedom-of-speech-and-expression

Bhattacharya, M. (2020). Growth of OTT in India the Covid Factor. E-cineIndia. (ISSN)2582-2500.

Bhattacharyya, R., Kohli, S., & More, R. (2020). Change In Consumption Habits Of Entertainment During Lockdown And The Role Of OTT Platforms. *Elementary Education Online, 19*(4), 4686–4695. https://bibliomed.org/?mno=98027

Capgemini. (2020). Impact of the Covid-19 Outbreak on Media & Entertainment. Capgemini. https://www.capgemini.com/wp-content/uploads/2020/06/Covid-19-Impact-on-Media-Entertainment-industry_VF.pdf

Chan-Olmsted, S. M., & Park, J. T. (2000). From On-Air to Online World: Examining the Content and Structures of Broadcast TV Stations' Web Sites. *Journalism & Mass Communication Quarterly*, *77*(2), 321–339. doi:10.1177/107769900007700207

Dhiman, B. (2023) Key Issues and New Challenges in New Media Technology in 2023: A Critical Review. *Journal of Media & Management. SRC/JMM-184*.

Farooqui, M. (2021). OTTs set to launch more shows in 2021, invest more in regional content. *Moneycontrol*. https://www.moneycontrol.com/news/trends/entertainment/otts-set-to-launch-more-shows-in-2021-invest-more-in-regional-content-6326001.html, Accessed on July10, 2021

Hutchins, B., Li, B., & Rowe, D. (2019). Over-the-top sport: Live streaming services, changing coverage rights markets, and the growth of media sports portals. *Media Culture & Society*, *41*(7), 975–994. doi:10.1177/0163443719857623

Jose, R. J. (2020). Factors influencing the shift from traditional TV to OTT platforms in India. *International Journal of Advanced Science and Technology*, *29*(7s), 4044–4051. http://sersc.org/journals/index.php/IJAST/article/view/22888

Joshi, S. (2015). Impact of Over The Top Services on Telecom Service Providers. *Indian Journal of Science and Technology*, *8*(S4), 145–160. doi:10.17485/ijst/2015/v8iS4/62238

KPMG. (2017). The 'digital first' journey- how OTT platforms can remain 'on-demand' ready. *KPMG Report*. KPMG. https://assets.kpmg/content/dam/kpmg/in/pdf/2017/10/The-Digital-First-journey.pdf

Mansouri, N. (2023). *Money, magic, and machines: International Telecommunication Union and liberalisation of telecommunications networks and services*. London Review of International Law. doi:10.1093/lril/lrad011

Manuel, P. (2014). The regional North Indian popular music industry in 2014: From cassette culture to cyberculture. *Popular Music*, *33*(3), 389–412. doi:10.1017/S0261143014000592

Naik, A. (2023b, March 2). Regulation of OTT Platforms: Challenges and solutions. *WION*. https://www.wionews.com/opinions-blogs/regulation-of-ott-platforms-challenges-and-solutions-567573

Novak, T. P., Hoffman, D. L., & Yung, Y. (2000). Measuring the Customer Experience in Online Environments: A Structural Modeling Approach. *Marketing Science*, *19*(1), 22–42. doi:10.1287/mksc.19.1.22.15184

P, N. S., & Thomas, P. C. (2022). Understanding the Factors Lead to the Adoption of Content Localization Strategy by OTTP in India: A Review. *International Journal of Scientific Research and Management, 10*(05), 19–23. doi:10.18535/ijsrm/v10i5.com01

Panjiar, T., Chaudhuri, R., & Waghre I, P. (2023, June 12). *A Public Brief on Demand on Fair Share.* Internet Freedom Foundation. https://content.internetfreedom.in/api/files/divco3ywedt9rpe/sce88g-bwixr5ro4/fair_share_analysis_brief_1_zq0MuI6hrX.pdf/?ref=static.internetfreedom.in

*Performance Indicators Reports.* (n.d.). Telecom Regulatory Authority of India. https://www.trai.gov.in/release-publication/reports/performance-indicators-reports

Rohit Jacob Jose. (2020). Factors influencing the shift from traditional TV to OTT platforms in India. *International Journal of Advanced Science and Technology, 29*(No. 7s), 4044–4051.

Sharma, K. S. (2020). Are OTT platforms taking over theatrical experience? *Economic Times.* https://brandequity.economictimes.indiatimes.com/news/media/are-ott-platforms-taking-over-theatrical experience/73887372

Sujata, J., Sohag, S., Tanu, D., Chintan, D., Shubham, P., & Sumit, G. (2015). Impact of Over-the-Top (OTT) Services on Telecom Service Providers. *Indian Journal of Science and Technology, 8*(S4), 145. doi:10.17485/ijst/2015/v8iS4/62238

Sukwadi, R., Lasiman, S. C., Inderawati, M. M. W., & Suprata, F. (2019). Niche theory within video platform competition: Traditional vs modern. *IOP Conference Series. Materials Science and Engineering, 567*(1), 012003. doi:10.1088/1757-899X/567/1/012003

Tah, S. (2022, May 19). *Indian OTT platforms get box clever to become sustainable.* Voice&Data. https://www.voicendata.com/indian-ott-platforms-get-box-clever-to-become-sustainable/

Tiwary, I. (2020). Amazon Prime Video: A Platform Ecosphere. In Global transformations in media and communication research (pp. 87–106). Springer International Publishing. doi:10.1007/978-3-030-44563-8_5

Upadhye, A. (2019b, April 4). Regional Content - The next big thing for digital in India. *Reputation Today.* https://reputationtoday.in/regional-content-the-next-big-thing-for-digital-in-india/

Vidooly. (n.d.). *Industry Insights Archives.* Vidooly. https://vidooly.com/blog/category/video-industry-insights/

Wayne, M. (2017). Netflix, Amazon, and branded television content in subscription video-on-demand portals. *Media Culture & Society, 40*(5), 725–741. doi:10.1177/0163443717736118

## ADDITIONAL READING

Alam, T. (2020). Why OTT Platforms Towards Regional Contents Other Than Hindi. *Entrackr.* https://entrackr.com/2018/06/ott-regional-content/Accessed on 1 July 2023.

Babar, K., & Laghate, G. 2020. Netflix India: Netflix Leases Space For India Headquater In BKC. *The Economic Times*. https://economictimes.indiatimes.com/tech/internet/netflix-leases-space-for-india-headquater-in-bkc/articleshow/67280626.cms?from=mdr 1 July 2023.

Banerjee, A., Rappoport, P. N., & Alleman, J. (2014). Forecasting video cord-cutting: The bypass of traditional pay television. In *Demand for Communications Services–Insights and Perspectives* (pp. 59–82). Springer US. doi:10.1007/978-1-4614-7993-2_4

Mitter, S. (2020). *[App Fridays] Unacademy Wants To 'Spark Your Imagination' With Interactive Storytelling App Graphy*. YourStory.com. https://yourstory.com/2020/08/unacademy-spark-imagination-interactive-storytelling-app-graphy.

# Section 4
# Societal Impact and Inclusion

# Chapter 12
# Bridging the Digital Divide:
## Exploring the Role of OTT Platforms in Advancing Digital Inclusion in the Global South

**Nitesh Behare**
https://orcid.org/0000-0002-9338-8563
*Balaji Institute of International Business, Sri Balaji University, Pune, India*

**Suraj Sharma**
*Ramachandran International Institute of Management, India*

**Varada Inamdar**
*Arihant Institute of Business Management, India*

**Swapnali Bhosale**
*Arihant Institute of Business Management, India*

**Shrikant Waghulkar**
https://orcid.org/0000-0002-3767-3765
*Arihant Institute of Business Management, India*

## ABSTRACT

*The chapter will give an in-depth exploration of role of OTT platforms in facilitating digital inclusion in the Global South, which will cover various aspects like economic impact, digital inclusion, challenges, recommendations, and future prospects. It will also explore how OTT platforms have contributed to bridging the digital divide, fostering access to information, entertainment, and educational content in regions with limited infrastructure and resources. The chapter will analyze initiatives and strategies employed by OTT platforms to reach underserved populations, impact of affordable pricing models, localized content, and innovative approaches to overcome connectivity challenges. The challenges in the Global South are also examined in this chapter including infrastructure limitations, connectivity barriers, privacy concerns, regulatory frameworks, and ensuring equitable access for all populations. Additionally, it will discuss the implications of OTT platforms for socio-economic development, empowerment, and creation of digital opportunities in the Global South.*

DOI: 10.4018/979-8-3693-3526-0.ch012

*"OTT is a key to bridge the digital divide and Global South's digital inclusion..."*

...Unknown

## INTRODUCTION

In today's digitally connected world, access to the internet and various free & paid technologies have become an important factor in terms of gaining socioeconomic and educational opportunities. Nonetheless, there is a significant disparity exists in between those who have access to and can be benefited from these incredible technologies and those who do not. This disparity, basically known as *"digital divide"*(DiMaggio P., 2004). This digital divide poses various challenges in accomplishing equitable development, predominantly in the Global South, which encompasses developing countries which are facing socio-economic obstacles to digital inclusion (Warschauer M., 2010).

Over-the-top (OTT) platforms, which provide video streaming services over the internet, have emerged as dominant tools in bridging the digital divide. These platforms can be easily accessible using smart devices which offer a wide range of digital services like video streaming, e-commerce, messaging, and many more. Increasing popularity of OTT and its influence have transformed the way people communicate digitally, access information and engross in various phases of their daily lives (Napoli P. M., 2019).

On the other hand, the Global South, with its vast geographical areas, diverse populations, and infrastructural limitations, faces exceptional challenges in accomplishing digital inclusion. Conversely, the emergence of OTT platforms has created new opportunities for addressing these challenges and narrowing the digital divide (Koskinen, et al., 2019). By understanding the role of OTT platforms in advancing digital inclusion in the Global South, people can gain insights into their probable impact on education, digital connectivity, cultural exchange and entrepreneurship (Das, 2023).

In our increasingly interconnected world, the accessibility of internet connectivity and vigorous infrastructure plays a significant role in ensuring digital inclusion. OTT platforms, with their vast reach (amongst their audience) and technological capabilities, have potential to considerably enlarge internet access in the Global South. These OTT platforms have been influential in increasing penetration of the internet, specifically through mobile internet, overcoming conventional barriers which are related with physical infrastructural limitations. By offering innovative technologies and their services, OTT platforms can encompass connectivity to formerly underserved areas, allowing individuals and communities with access to information, services and opportunities (Kulkarni, 2019). OTT platforms have a transformative impression on digital inclusion in the Global South by giving access to education, endorsing entrepreneurship & economic empowerment and also nurturing cultural exchange. OTT platforms empower individuals to impart various skills and knowledge, empower small production houses to reach an extensive customer (viewers) and showcase varied cultures and stories to the global audience (Koskinen, et al., 2018). By bridging the digital divide and augmenting digital inclusion, OTT platforms contribute to a more connected, comprehensive and economically vibrant society in the Global South. While OTT platforms hold enormous potential in evolving digital inclusion, there are many challenges like band width of internet, affordability and accessibility of OTT platforms and regulatory framework in global south (T & I, 2020) (Pertierra R., 2018). Collective efforts between industry stakeholders, policymakers and communities are crucial to navigate these challenges and ensure that the advantages of OTT platforms are harnessed effectively by their users. In advancing digital inclusion in the Global South, this chapter delves into the multifaceted role of OTT platforms. (Andy Nguyen, 2022) By leveraging the capabilities of OTT platforms and fostering inclusive strategies, businesses can empower individuals and communities, drive sustainable development and unlock the full potential of the Global South in the digital era.

## Definition of Digital Inclusion and Its Significance for the Global South

According to many intellectuals, digital inclusion refers to the *"goal of ensuring that all individuals, especially those from disadvantaged groups, have equal access to and the necessary skills to use digital technologies"*. It comprises of bridging the gap between all those who have access to digital resources and those who do not. Digital technologies have potential to encourage social inclusion by removing barriers and offering real-time services. Conversely, there is a need to apprehend and establish clear links between digital technologies and social inclusion. Accomplishing digital inclusion needs to address various barriers to access the technology, promoting digital literacy and creating an inclusive digital ecosystem. By doing so, more equitable and participatory society can be formed where everyone can benefit from the advantages of digital technologies (Nguyen, 2023).

Various authors and organizations have provided key definitions of digital inclusion. Some of them are mentioned below.

According to World Bank (The World Bank, 2023), digital inclusion refers to *"the affordable and effective use of information and communication technologies (ICTs) for all members of society to enable individuals and communities to access and use digital services, participate in social and economic activities, and enhance their well-being."*

European Commission (Commission, 2022) in their policies defined Digital inclusion as *"the ability of individuals and groups to access and use information and communication technologies. It encompasses a range of skills, knowledge and attitudes required to confidently and effectively engage with digital technologies and services."*

The International Telecommunication Union (ITU, 2022) in their mandate described Digital inclusion as *"the process of ensuring access to and use of digital technologies for all individuals and communities, particularly the disadvantaged and vulnerable groups, to improve their social, economic, and educational opportunities."*

Professor Ellen Helsper (J.A.M, et al., 2016) who had a great contribution in the field of digital inclusion had developed *'Digital Inclusion Scale'* to understand the gap in digital inequalities. Professor Ellen Helsper defines digital inclusion as *"the extent to which individuals can access, use and benefit from digital technologies in ways that promote positive outcomes in terms of their well-being, life chances and capabilities."*

(UN, 2021) in their article defined Digital inclusion as *"equitable, meaningful, and safe access to use, lead and design of digital technologies, services, and associated opportunities for everyone, everywhere"*.

(Rachelle Bosua, 2021) in their article stated that *"Digital inclusion is social inclusion in the 21st century which ensures individuals and disadvantaged groups have access to, and skills to use, Information and Communication Technologies (ICTs) and are therefore able to participate in and benefit from today's growing knowledge and information society"*.

According to (Ragnedda & Mutsvairo, 2018), *"digital inclusion is the ability of individuals and groups to access and use information and communication technologies. Digital inclusion encompasses not only access to Internet but also the availability of hardware and software; relevant content and services; and training for the digital literacy skills required for effective use of information and communication technologies".*

(Sue Bennett, 2012) stated that *Digital inclusion refers to mind-sets, strategies and initiatives that seek to ensure that all people in society have equitable access to technology regardless of their personal circumstances. It is underpinned by the belief that access to technology and the ability to use it effectively are important to citizenship and social cohesion.*

Over all, *Digital inclusion refers to effective and equitable access to information and communication technologies (ICTs), ensuring that individuals and communities can contribute in the digital society and benefit from digital technologies, irrespective of their personal circumstances.*

## Significance of Digital Inclusion for the Global South

In the developing countries from Asia, Africa, Latin America, and Caribbean zones, Digital inclusion holds immense significance for the Global South. It plays a crucial role in driving economic and social development by empowering small businesses and entrepreneurs to access international markets, signifying and promoting innovation in various fields, fostering job creation and entrepreneurship (Heeks, 2022). Furthermore, digital inclusion enhances educational opportunities and skill development by offering access to the online resources and various e-learning platforms, bridging educational gaps and empowering the workforce of the developing countries (Bastion & Mukku, 2020). Digital inclusion improves healthcare and public service accessibility through telemedicine, digital health records and online government portals, benefiting individuals in underserved areas and augmenting efficiency and equity. Additionally, digital inclusion promotes financial inclusion, allowing individuals to access various financial services, make transactions and build financial flexibility (Hamdar, 2021). It also fosters social inclusion, empowering marginalized communities to voice their perspectives, access information and participate in decision-making processes. Many intellectuals and socio-economic entities have emphasized that digital inclusion is highly important for the Global South as it drives economic growth, enhances education and healthcare, promotes financial inclusion, fosters social empowerment and preserves cultural diversity (DISD/DPIDG, 2021). By bridging the digital divide and ensuring equitable access to digital technologies, the Global South can unleash its potential, overcome barriers, and actively participate in the global digital economy and information society (ITU, 2022). Lastly, digital inclusion facilitates the preservation of cultural heritage and enables cultural exchange, showcasing the richness of the Global South's diverse cultures to a global audience (Seward, 2020). Conversely, lack of access to ICT plays a significant role in social and economic outcomes for people in both developed and developing countries. Moreover, digital inclusion plays a major role in achieving digital equity, when every individual has the digital skills and resources needed to fully participate in the social, economic and civic aspects of society. (IEEE, 2019)

## Introduction to OTT Platforms as Facilitators of Digital Inclusion

The acronym OTT stands for Over-the-Top. This convenient little term explains the new delivery method of film and TV content over the internet, across many different devices, without the need for traditional broadcast, cable, or satellite pay-TV providers. In simple terms, OTT streaming means paying an internet provider, for internet access to access various OTT platforms like Netflix, Hotstar etc. to watch streaming contents without paying for cable TV. (Endavo, 2019). OTT platforms play a substantial role in promoting digital inclusion by leveraging their technological advancements and user-friendly interfaces, thereby enhancing accessibility and inclusivity on a global scale. To elevate streaming quality and deliver seamless user experiences, these platforms often leverage innovative technologies (Chakraverty, 2020). These OTT platforms invest heavily in designing user friendly interface, ensuring that their platforms are intuitive and easy to navigate, even for individuals with limited technical knowledge. This user-friendly approach increases the reach of digital content to individuals who may have previously felt hesitant or overwhelmed by technology (Endavo, 2019). OTT platforms continue to minimize barriers of digital divide and enable individuals to engage with digital content and services by continuously evolving and adapting to their user needs (Lin, 2016). Additionally, by reaching to remote and underserved areas where traditional broadcast infrastructure may be limited, but OTT platforms have potential to bridge the geographical divide. With increasing internet accessibility, individuals in these regions can seamlessly access digital content through OTT platforms, empowering them to participate in the digital world and benefit from educational resources, cultural exchange and entertainment. For various content producers and content providers, OTT platforms also serve as a platform to showcase their work and talent. These OTT platforms offer number of opportunities for artists, independent filmmakers and storytellers who may have faced barriers in traditional media distribution channels. By intensifying diverse voices and perspectives, OTT platforms contribute to a more inclusive media landscape which reflects the richness of human experiences. Overall, accessibility, advancements, user-friendliness, and opportunities for diverse content on OTT platforms expressively contribute to digital inclusion by making digital content accessible to their users, engaging the audiences and relevant contents to a wide range of individuals and communities.

## UNDERSTANDING THE DIGITAL DIVIDE IN THE GLOBAL SOUTH

The digital divide refers to the gap in accessing the digital technologies and the internet between various communities, regions and socioeconomic groups. However, digital divide is a worldwide problem, it is predominantly noticeable in the Global South which include countries from Asia, Africa, Latin America, and the Caribbean (Ayanso, 2010) (Reddy, et al., 2022). In these regions, internet connectivity, lack of affordable devices, disparities in digital infrastructure, lack of consistent supply of electricity and digital literacy create unequal opportunities and hamper socioeconomic development. Moreover, socioeconomic factors like inequality, poverty, linguistic barriers and rural-urban divides intensify disparities in digital access (Reddy, et al., 2022) (Hussain, 2014). In Global South, for Bridging the digital divide, it requires a comprehensive approach that considers socioeconomic conditions, policy frameworks, cultural contexts and technological advancements. It is important for encouraging inclusive economic growth, empowering marginalized communities and ensuring that no one is left behind in the digital era. Stakeholders can frame effective strategies and interventions to nurture digital inclusion and create a more equitable digital landscape for all by understanding the dynamics and challenges of the digital divide. (PennState, 2019)

### Examination of the Digital Divide in the Global South

Examining digital divide in Global South entails considering additional factors which contribute to inequalities in access and usage of digital technologies. These disparities include rural-urban divides, gender inequalities and various barriers faced by marginalized communities. Out of the mentioned disparities, gender disparities can limit women's access to digital tools and opportunities, need targeted initiatives to promote gender equality in the efforts taken for digital inclusion (Mahmudul Hasan Laskar, 2023). Another disparity, i.e. rural-urban divide, which highlights the challenges faced by population residing in rural areas with very limited basic infrastructure and internet connectivity options, imposing personalized solutions like community networks and mobile connectivity. Marginalized communities which includes ethnic minorities, people with disabilities and aboriginal populations, always face unique barriers caused by limited resources, discrimination and lack of representation. Addressing such interconnecting inequalities entails comprehensive policies, personalized programs and community engagement to ensure everyone is benefitted (OECD, 2018).

Collaborative efforts from industries, various agencies of developed countries and governments must involve to bridge the digital divide in the Global South. International organizations, development agencies and donor countries can offer technical and financial support for infrastructural development and capability building initiatives. In ridging the digital divide, governments play a crucial role in creating and facilitating policy environment, establishing regulations and setting targets for connectivity that promote healthy competition, affordability and data protection. Private organizations, comprising telecommunications companies and technology providers, can invest in infrastructure, offer affordable connectivity solutions and partner with local communities. NGO's, Civil society organizations and initiatives by communities are influential in executing grassroots projects, providing digital literacy training and encouraging rights of marginalized communities. Factors that are responsible for digital divide can be addressed by creating a more inclusive digital landscape in the Global South. This includes investing in infrastructure, reducing costs for required services, promoting digital literacy and skills training, encouraging localized content, bridging rural-urban divides, addressing gender disparities and endowing marginalized communities. Considering digital inclusion as an essential element of development agendas, OTT industry can unleash the complete potential of Global South by fostering social advancement, economic growth and sustainable development for everyone.

## Factors Contributing to Limited Access and Connectivity Challenges

Several factors contribute to limited access and connectivity challenges in the Global South, exacerbating the digital divide. These factors include:

- *Infrastructure limitations:* Many regions in the Global South lack the necessary digital infrastructure, such as reliable broadband networks, cellular coverage and electricity. Remote and rural areas are particularly affected, as the cost of extending infrastructure to these areas is often prohibitive. Insufficient investment in infrastructure development hampers connectivity and restricts access to digital services (Reddick, et al., 2020) (Antonio & Tuffley, 2014) (IEEE, 2022).
- *Affordability and cost barriers:* The cost of internet services and digital devices can be prohibitively high for individuals in the Global South, especially those living in poverty. Limited income and high data costs, relative to income levels, pose significant challenges. The affordability of smartphones, computers and other devices also presents a barrier to access, making it difficult for individuals to own the necessary tools to engage in the digital world (Reddick, et al., 2020) (Antonio & Tuffley, 2014) (IEEE, 2022).
- *Socioeconomic inequalities:* Inequities in income and wealth distribution contribute to disparities in digital access. Low-income communities and marginalized groups face significant challenges

in affording internet services and digital devices. Additionally, rural-urban disparities exacerbate the digital divide, with urban areas typically having better connectivity and access compared to remote and rural regions (Reddick, et al., 2020) (Antonio & Tuffley, 2014) (IEEE, 2022).

- *Digital literacy and skills gaps:* Limited digital literacy and lack of technological skills hinder effective engagement with digital technologies. Many individuals in the Global South lack the necessary knowledge to navigate the internet, use digital tools and evaluate online information. This skills gap restricts their ability to fully benefit from the opportunities provided by digital technologies (Reddick, et al., 2020) (Antonio & Tuffley, 2014) (IEEE, 2022).
- *Language and content barriers:* Language diversity in the Global South poses challenges in accessing relevant digital content. A significant portion of online content is primarily available in major international languages, limiting access for individuals who do not speak or understand these languages. The lack of localized content in native languages restricts the relevance and inclusivity of digital services (Antonio & Tuffley, 2014) (IEEE, 2022) (Reddick, et al., 2020).
- *Government policies and regulations:* Inadequate policy framework, limited regulatory frameworks and bureaucratic processes can hinder the development of digital infrastructure and the provision of affordable connectivity. Restrictive policies and regulations may also stifle competition, limiting the entry of new players into the telecommunications market and impeding innovation.

Addressing these factors requires comprehensive strategies and collaborative efforts involving governments, private sector entities, civil society organizations and international stakeholders. By investing in infrastructure development, promoting digital literacy and skills training, implementing affordable connectivity initiatives and fostering localized content creation and connectivity in the Global South can be gradually overcome, promoting digital inclusion and fostering equitable participation in the digital world. (Reddick, et al., 2020)

## Implications of the Digital Divide on Socio-Economic Development

The digital divide has significant implications for socio-economic development which results in perpetuates economic disparities by limiting access to digital tools and opportunities, deterring employment prospects, entrepreneurship endeavours and access to e-markets. The education and skills gap in digital divide also limits educational opportunities and the acquisition of digital skills, deterring development of digitally literate workforce and impeding overall human capital development. Social exclusion and marginalization in global south are aggravated as individuals without digital access are unable to engage in social interactions, participate in civic activities and access government services. Limited access to digital health resources and telemedicine services leads to health disparities, with individuals facing challenges in receiving timely and appropriate medical advice and care. The digital divide also deepens cultural and knowledge gaps, as limited access to digital content and online information resources restricts individuals' capability to participate in cultural exchange, access diverse perspectives and benefit from digital knowledge repositories (Molala & Makhubele, 2021) (A.G.M & Dijk, 2006).

On the other hand, Innovation and entrepreneurship are stalled by limited digital access, preventing individuals and communities from harnessing the power of technology to drive innovation, start their businesses and contribute to overall economic growth of the country. Financial inclusion also gets impacted mainly in areas with limited access to digital banking and payment systems, as individuals face challenges in accessing affordable financial services and participating in formal economy. Limited digital access hinders democratic processes and citizen engagement, limiting transparency, accountability and citizen participation in governance. Rural areas are unreasonably affected by the digital divide, impeding their development potential by restricting access to markets, agricultural information, and e-commerce platforms. Moreover, the digital divide restricts global connectivity and collaboration opportunities, isolating communities and hindering their ability to participate in global networks, collaborations and knowledge sharing (Gamble, 2023) (Sharma & Banerjee, 2022) (Swalehin & Nasir, 2011).

## THE RISE OF OTT PLATFORMS IN THE GLOBAL SOUTH

The Global South has witnessed significant rise in online streaming contents which have gained enormous popularity and become catalysts for digital content consumption. This phenomenon has revolutionized how people in the Global South access and engage with information, entertainment and cultural content. Penetration of affordable smartphones, expansion of internet connectivity and the accessibility of diverse content have boosted the growth of OTT platforms and offering a wide range of selected digital content. This explores the rise of OTT platforms in the Global South, highlighting their impact on media consumption, cultural exchange and the overall digital transformation in the region.

### Overview of the Growth and Adoption of OTT Platforms in the Region

The Global South has experienced significant growth and adoption of OTT services, including platforms like Netflix, Amazon Prime Video, and Disney+, have gained popularity in emerging economies within the Global South, leading to increased subscription numbers (GVR, 2023) (Intelligence, 2023). To cater to local preferences, OTT platforms have invested in creating and acquiring regional content, attracting a wide range of audiences. The extensive use of smart devices (mobile, tablets, etc.) has fuelled the growth of OTT platforms in the Global South. OTT Platforms have prioritized mobile optimization, developing user-friendly mobile applications and embracing a mobile-first approach. This strategy of OTT platforms has heightened convenience and accessibility, contributing to the widespread adoption of OTT platforms. With telecom operators, strategic partnerships, played a critical role in intensifying the reach of OTT platforms and subscriber base too. These partnerships always encompass bundling OTT services with mobile data plans and making OTT content reachable to a larger audience.

Because of such strategic decisions of OTT platforms, in recent years, the OTT market has experienced substantial growth, with a projected CAGR of 28.19% (Intelligence, 2023). By 2025, the market is expected to reach USD 2,830 billion, with a CAGR of 29.55% (Straits, 2023) (Intelligence, 2023). Factors like flexible packaging, diverse content selection, wider device availability and the increasing adoption of Subscription Video on Demand (SVOD), services in emerging markets contribute to this growth. However, the key challenges like video content piracy remain unanswered. The COVID-19 pandemic has also augmented the demand for streaming services and the launch of new platforms has reformed the media landscape.

### Analysis of Key OTT Players and Their Impact on Digital Inclusion

As per the report of (Research, 2022), global OTT market observed substantial growth, reaching a value of US$ 150.51 billion in 2021. It is estimated to continue increasing and is likely to reach around

US$ 1241.6 billion by 2030, with CAGR of 26.42% from 2022 to 2030. This growth in OTT industry, indicates a considerable increase in the adoption and popularity of OTT platforms globally which are driven by factors such as increasing internet penetration, growing demand for on-demand content, and advancements in streaming technology. Following infographics of (Research, 2022) give clear indications about future of OTT market across the globe.

*Figure 1. Future of OTT*

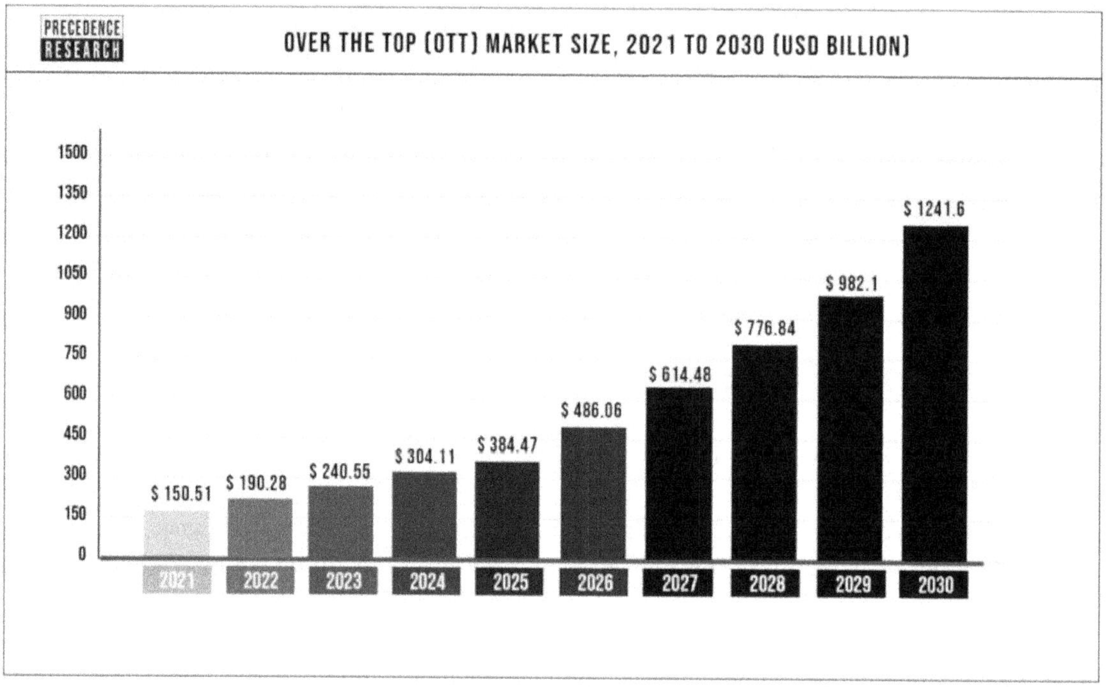


Source: https://www.precedenceresearch.com/over-the-top-market

OTT platforms, like Netflix, have been instrumental in promoting digital inclusion wherein it focuses on affordability and accessibility by offering different pocket friendly subscription plans, comprising mobile-only option in the selected markets. The platform also makes sure its service is accessible on a wide range of devices, catering to users' preferences and economic capabilities. For serving to the diverse audiences and promoting cultural inclusivity, Netflix has heavily invested in producing localized content (Patrick Frater, 2023) (Levy, 2018). By implementing such initiatives, Netflix has already taken a step forward for bridging digital divide and promoting digital inclusion. To generate more revenue, other OTT platforms have also started following Netflix and initiated huge investments in generating local but original contents which in-turn resulted in market growth. Further, COVID-19 pandemic has accelerated the demand for OTT services, as people were spending more time at home and seeking various entertainment options. Collaborations of OTT platforms and content creators have also played significant role in intensifying content libraries and enhancing user experiences. Looking ahead for more business opportunities from Global South, many leading platforms have started focusing on enhancing user experience with the help of technological advancements, adoption of video-on-demand services, and expansion into emerging markets. Amalgamation of technologies such as virtual reality, artificial intelligence, augmented reality machine learning, inclusion of deep learning in data science etc. will provide users with immersive viewing experiences with personalization and recommendation algorithms. Overall, the growth of the OTT market and the efforts of key players in promoting digital inclusion are reshaping the way people access and consume digital content, creating opportunities for a more inclusive and diverse digital landscape.

Various OTT platforms, like Amazon Prime Video, Disney+, iflix, YouTube, Hotstar, iQiyi, Viu, Showmax, HOOQ, MX Player, Zee5 etc. have made substantial contributions to digital inclusion through their initiatives and focused strategies. These OTT platforms have recognized the significance of catering to the unique needs and preferences of their target markets, resulting in increased accessibility and inclusivity. These platforms have played vital roles in intensifying digital inclusion and making digital entertainment more accessible to diverse audiences. By offering localized content, subtitles in multiple languages, offering foreign contents dubbed in local languages and affordable subscription plans, have ensured that digital entertainment is accessible to an extensive audience, including users in global south. By embracing digital inclusion, these platforms have not only expanded their user base but also fostered cultural representation, supported local talent, and contributed to the growth of the digital entertainment industry in their respective regions (Saraswat, 2023).

## Initiatives to Address the Digital Knowledge Gap in the Global South

To address the gap of digital knowledge of a global south, governments, non-profit organizations and private sector entities have taken various initiatives and recognized the significance of equipping individuals of global south with necessary digital skills and knowledge to participate in the digital age. Following are some notable initiatives:

a. *Infrastructure Development:* the very fundamental initiative taken was to expand internet connectivity in underserved areas recognizing the importance of robust digital infrastructure. These initiatives comprise building internet infrastructure, intensifying broadband coverage and deploying mobile networks to remote regions. By bridging the connectivity gap, individuals of the Global South can access online resources, digital tools and educational platforms. Infrastructural development has also enabled increased business opportunities in terms of local content development for the OTT industry. This progress has also played significant role in advancing digital inclusion and nurturing the expansion of the digital entertainment industry in the region.
b. *Digital Literacy Programs:* Various digital literacy programs like basic computer literacy, internet usage, online safety, digital communication etc. have been launched to deliver training and resources to individuals of the Global South with the aim to enhance digital skills. Marginalized communities, youth, women and rural populations, were often targeted to empower and navigate them to the digital world effectively which helped them in reducing socio-economic disparities and fostering social equality (UNICEF, 2019).
c. *Affordable Internet Access Programs*: for digital inclusion, affordability remains a significant barrier for global south. To address this, initiatives have focused on offering affordable access to internet services by subsidizing internet costs, offering low-cost data packages and promoting availability of affordable smartphones and tablets. Such programs enable individuals of the Global South to

access online educational resources, job opportunities, entertainment services and digital services (OET, 2022).

d. *Digital Skills Training for Educators*: Understanding the significance of educators in shaping digital literacy, some initiatives have also focused on training teachers and educators in digital skills. These programs equip instructors with essential knowledge and tools to assimilate digital technologies into the classroom, promote digital literacy among students and prepare them for the digital future (Gautam, et al., 2023).

e. *Public-Private Partnerships:* Collaborations between private sector entities, non-profit organizations and governments have been influential in driving initiatives to address the digital knowledge gap. Partnership between public-private entities can leverage the expertise and resources of different stakeholders to implement large-scale digital inclusion programs. These partnerships often focus on skill-building, community engagement and infrastructure development to ensure a holistic approach to address the digital divide.

f. *Digital Community Centres*: Community-based digital centres have developed as an essential hub for digital learning and accessing technology. These centres provide training facilities, internet connectivity with digital devices to individuals of the underserved communities where they can learn, explore, and acquire digital skills, fostering digital inclusion and empowering communities. These centres provide access to technology, training, mentorship, and networking opportunities.

g. *UNESCO's Mobile Learning Week:* UNESCO, every year organizes such annual event which explores innovative ways to leverage mobile technology for education. The event brings together policymakers, educators and experts to discuss and share best practices in digital learning (Unesco, 2022).

h. *Africa Code Week:* It's a significant initiative meant at promoting digital literacy and coding skills among African youth with the key objective of empowering African individuals with coding and computational thinking skills and enabling them to become active participants in the digital economy. It is a public-private partnership that brings together non-profit organizations, governments, educational institutions and private sector companies. Under Africa Code Week initiative, various programs on coding as well as training sessions across various African countries along with providing hands-on coding experiences to the participants have been organized. By fostering development in the digital skills, Africa Code Week seeks to bridge the digital divide, promote innovation, and create opportunities for future employment and entrepreneurship in Africa (SAP, 2022) (ACW, 2022).

i. *ITU Digital Transformation Centres:* The International Telecommunication Union (ITU) has established Digital Transformation Centres in various countries with the aim of providing training, capacity building, technology transfer, knowledge sharing and technical assistance in areas related to information and communication technologies (ICTs). These centres aim to empower individuals with digital skills, promote digital inclusion and empower individuals and communities with the essential skills and knowledge to fully participate in the digital economy (ITU-D, 2022) (ITU, 2020).

j. *Google's Internet Saathi Program:* Google launched this program in India in collaboration with Tata Trusts, which aims to train rural women to become "Internet Saathis" or digital mentors. Through the program, Internet Saathis organize training sessions in villages, teaching women how to use smartphones, access internet and utilize various online services. These women then educate their communities about the benefits of internet and provide digital literacy training (Trust, 2020).

k. *One Laptop Per Child (OLPC):* OLPC is a non-profit organization that provides low-cost laptops and educational resources to children of underserved areas. The program aims to enhance digital literacy and educational opportunities for children in the Global South.
l. *Digital Bridge Initiative*: This initiative, led by the World Bank, aims to connect unconnected communities in developing countries. It focuses on building digital infrastructure, expanding internet access and providing digital skills training to promote inclusive growth and development.
m. *Digital Empowerment Foundation (DEF):* DEF works towards bridging the digital divide in India and other countries through various programs, including digital literacy training, community digital centres and mobile-based interventions. They focus on empowering marginalized communities, women and rural populations.

## AFFORDABLE PRICING MODELS AND ACCESSIBILITY

For bridging the digital divide in the Global South, Affordable pricing models and improved accessibility are concern areas wherein various initiatives have been instigated to address affordability by lowering the cost of internet access and required digital devices. These initiatives also include subsidized internet services, promoting low-cost data plans and penetration of affordable smartphones or tablets in the marginalized areas. Subsequently, to shorten the gap other key efforts also been taken which focuses on augmenting accessibility by intensifying internet infrastructure in remote areas and implementing community networks (Gamble, 2023). Aim of these initiatives is to make digital services more affordable and accessible to individuals and communities, which are ensuring that cost does not deter digital inclusion. OTT platforms can cater to different income levels and multiply their user base by offering affordable pricing models and flexible subscription options. (Krick, et al., 2021)The ability to adapt and iterate on pricing strategies is essential for a successful transition to OTT 2.0. Research, subscriber demand and flexibility, and adaptability to accommodate different pricing plans are important for the success of a subscription based OTT business. To attract and retain customers, OTT business are offering different pricing packages such as:

a. *Freemium Pricing:* Freemium pricing models have gained popularity in the Global South by offering free access to a limited range of content whereas through paid subscriptions customer can access premium contents. Freemium pricing models allow their users to experience the desired platform and its benefits without upfront financial commitments. It facilitates them to evaluate the value and utility of the service before determining to upgrade to premium features. This approach reduces the financial barrier by allowing users to enjoy some content for free while providing an incentive to upgrade for enhanced content and features (Kumar, 2014) (Fuchs, 2022). By providing a basic, no-cost tier, OTT platforms entice users to explore their content and experience the value they provide. This strategy serves as a powerful marketing tool, enticing users to upgrade to the premium tier, which offers enhanced features, exclusive content and overall superior experience. The freemium model allows OTT services to generate revenue through a combination of subscription fees, targeted advertising, and upselling opportunities while maintaining a strong user base through an accessible entry point.
b. *Promotional Pricing:* Providing time-based or volume-based discounts which helps OTT businesses to convert prospects into customers. Promotions can be customized based on factors like geography,

customer type or usage. While promotions can boost OTT's customer acquisition rates, relying too heavily on them may erode value and create an expectation for continual discounts (Origin, 2020) (Chand, 2021).

c. *Cross-Platform Packaging and Bundling:* this pricing strategy have emerged as one of the effective strategies to enhance accessibility and promote digital inclusion in the Global South. By clubbing multiple OTT platforms in a single package or bundle, users can access a diverse range of content from various OTT platforms through unified interface which ultimately expand their viewers. By offering bundled services, OTT platforms can enhance the value proposition for users, increase engagement and encourage greater adoption of digital entertainment in the Global South. Moreover, these initiatives contribute to the growth of the digital entertainment industry by fostering collaboration and competition among OTT providers, ultimately benefiting both consumers of global south and the industry as a whole (Zuora, 2021) (ReelnReel, 2022).
d. *Pay-As-You-Go Charges:* Some customers prefer not to commit to a subscription and only pay for what they use. OTT platforms should offer this level of customization and manage detailed usage and billing accordingly (Zuora, 2021) (ReelnReel, 2022).
e. *Consumer Choice:* Consumers desire the freedom to choose the programs they want, when and how they want them. OTT content providers offers such package and price fresh content in compelling ways to attract and retain subscribers (Zuora, 2021) (ReelnReel, 2022) (Origin, 2020).
f. *Ad-Supported Model:* Some OTT platforms offer free access to content supported by advertisements. Advertisers pay to reach the platform's audience, allowing the platform to provide free or discounted subscriptions to users. This model can make OTT services more affordable for consumers while still generating revenue for the platform (Zuora, 2021) (ReelnReel, 2022) (Origin, 2020).
g. *Hybrid Model:* This model combines elements of different pricing models to provide a flexible and affordable offering. It could include a combination of free content, tiered pricing and optional add-ons or rentals for specific content. The hybrid model allows users to customize their experience and pay for what they value most (Zuora, 2021) (ReelnReel, 2022) (Origin, 2020).
h. *Pay-Per-View:* OTT platforms can offer pay-per-view options for live events, exclusive content, or premium releases. This model allow users to access specific content for a one-time fee, making it a cost-effective choice for occasional viewers or those with specific content preferences (Origin, 2020) (Zuora, 2021) (ReelnReel, 2022).

## LOCALIZATION AND CULTURAL RELEVANCE

In the diverse regions of the Global South, localization and cultural relevance are important in terms of the success of OTT platforms. By adapting local content, fostering user experience and proper marketing approaches to local cultures and preferences, OTT platforms can efficiently engage wider audience of global south. This includes partnership with local creators, translating content in regional languages, considering cultural nuances and personalizing recommendations to make a sense of familiarity and authenticity. Furthermore, adding social features, aligning affordable pricing with local economic realities and leveraging cultural festivals and events further augment the cultural relevance and accessibility of OTT platforms. Customizing content to local languages, cultures and local customer preferences is important for successful OTT platforms in the Global South (Kaushik Das, 2023). This includes adapting the local content library to reveal diverse cultural backgrounds and languages, breaking down language obstructions and offering culturally relevant adaptations like dubbing or subtitling. Promoting local content industries and diverse narratives is also crucial, as it amplifies the voices of local creators and contributes to industry growth. By featuring diverse narratives, platforms enhance inclusivity and attract a broader audience (Farooqui, 2021).

It is proven by some researches that localization and cultural relevance create familiarity and resonance with the audience of global south and ultimately leading to higher user engagement and loyalty. Incorporating nuances of culture, references and sensitivities towards global south is important to avoid misunderstandings. It also allows OTT platforms to tap into cultural events and traditions, aligning their content releases and promotional activities with local audiences. Furthermore, cultural relevance increases accessibility by offering relevant content and eliminating barriers to entry. It nurtures a sense of ownership and connection with the platform, encouraging active participation and advocacy. The promotion of local content industries empowers creators and contributes to the region's economic growth. OTT platforms can apply various approaches to serve global south which includes investing in localization efforts, acquiring and licensing local content and leveraging data analytics to understand audience preferences. Featuring local content supports underrepresented voices and cultures, while localization creates an immersive experience in the preferred language (Toppan, 2021). Data-driven insights help platforms curate personalized recommendations and meet specific demands. By showcasing local content and embracing diversity, OTT platforms offer visibility and recognition for underrepresented voices. They offer a wide range of content options that reflect the cultural richness and diversity of the region, fostering cross-cultural dialogue and understanding. Additionally, they contribute to the economic empowerment of local communities and create lasting connections with users.

## INNOVATIONS IN CONNECTIVITY AND INFRASTRUCTURE

Innovations in connectivity have revolutionised worldwide access to information and services. Telecommunications, e-commerce, education, healthcare, and entertainment have all been transformed by improved internet and mobile networks. Bridging the digital divide in the Global South is a key focus to overcome connectivity challenges. Governments collaborate with private players for infrastructural development and deploy high-speed internet, encompassing access to remote areas. Investments in installing cellular towers, fibre-optic cables and satellite technologies, can bridge the gap and promote social development and economic growth. High speed mobile data with affordable prices make connectivity more accessible, whereas offline viewing options allow content consumption without continuous internet access (Review, 2023). Innovative solutions like mesh networks and community-based networks empower local communities to establish their own networks. By combining these approaches, governments, organizations and technology companies are working towards affordable and inclusive connectivity, enabling individuals and communities in the Global South to participate in the digital world and its opportunities.

## SOCIO-ECONOMIC DEVELOPMENT AND EMPOWERMENT

For any prospering society, socio-economic development and empowerment of marginalized communities are vital components and in today's digital age technology plays a critical role in driving progress and transforming lives of individuals. Availability and accessibility to digital tools, information and opportunities can have substantial impact on individuals & communities and empowering them to advance their livelihoods, inflate their knowledge and participate in economic and social activities. From education and healthcare to entrepreneurship and civic engagement, the digital revolution has the potential to foster socio-economic development, bridge inequalities, and empower individuals to shape their own futures. In this context, leveraging technology and promoting digital inclusion have become key strategies for promoting socio-economic development and empowering individuals and communities worldwide (IT for Change, 2013) (OECD, 2018).

## The Entrepreneurial Potential Facilitated by OTT Platforms in the Global South

In the Global South, by providing a low-entry barrier, OTT platforms have democratized entrepreneurship for content creators and small businesses. Earlier, traditional media channels and their distribution networks had narrowed the reach of entrepreneurs and content creators. Conversely, with the introduction of OTT platforms, anybody with a creative idea and internet access can generate and dispense their content globally. This paradigm shift has given excellent opportunity to the new wave of digital entrepreneurs who wish to leverage the power of OTT platforms to showcase their talent and ideas. OTT platforms facilitate entrepreneurs to tap into global markets, reaching mass audiences that were earlier inaccessible. For example, various artists, musicians and filmmakers from remote regions of Africa, Asia, and Latin America can now showcase their talent with the world, gaining exposure, recognition and monetization opportunities. This has led to the emergence of a vibrant creative economy in the Global South, fostering cultural diversity and encouraging innovation.

## Economic Impact on Job Creation, Revenue Generation, and GDP Growth in the Global South

Economic impact of OTT platforms in the Global South is substantial and multidimensional. Firstly, OTT platforms have generated new employment opportunities across several sectors. OTT industry had generated direct job opportunities for translators, dubbing artists, subtitlers and editors in the Global South. Moreover, it also geared up indirect job opportunities in technical infrastructure, content production, support services and marketing. Localization efforts also contribute to economic development, training programs, entrepreneurship and cultural exchange. Overall, localization enhances employment prospects, fosters industry expansion and promotes diversity and representation in the global media landscape. The demand for content production, data analysis, digital marketing and customer support had also amplified which also resulted in creating various direct and indirect jobs in these areas. Additionally, as OTT platforms inflate their operations and invest in infrastructure in such areas, they also generate different employment like logistics, cloud computing and customer service centres. Now, in the Global South, these platforms have also become key revenue generators. They empower content creators and businesses to monetize their offerings through well designed subscription models, advertisements and collaborations. Revenue generated from this contributes to the growth of digital economy which encourages local businesses and also attracts foreign investment. In addition, development due to OTT platforms has also resulted in amplified tax revenues for governments, further boosting economic development and created a multiplier effect on other sectors of the economy. The digital ecosystem created by OTT platforms improves productivity, fosters innovation and improves competitiveness. As a result,

Global South has experienced accelerated economic growth, improved living standards of their residents and reduced income inequality.

### Empowering Marginalized Communities Through Access to Digital Platforms in the Global South

Access to the digital platforms has immense potential to empower marginalized communities where individuals and businesses can participate and showcase their talent. This inclusivity breaks down conventional barriers and empowers individuals to overcome geographical constraints and societal biases. These platforms also empower marginalized communities by providing access to content related to education and skill development resources. Educational contents of OTT platforms allow individuals from remote areas of global south to access quality learning materials and bridge the digital divide. Whereas, skill development resources, like online courses and tutorials, can equip individuals with the essential skills to strive in the digital economy.

## CHALLENGES AND LIMITATIONS

In a global digital landscape, OTT platforms have revolutionized the way people access digital content, offering convenience to their viewers, choice of content they prefer and personalized experiences. Conversely, along with their numerous benefits, OTT platforms also faced many challenges and limitations which is essential to be addressed for their sustained success and positive impact. So, by deeply understanding and addressing these challenges, OTT Platforms can pave the way for a more robust and sustainable OTT industry which prioritizes user rights, fosters innovation and also contributes to socio-economic development of communities, predominantly in the Global South where these challenges are often more noticeable (The Tribune, 2021). Let's explore some key challenges faced by OTT platforms and their potential solutions.

### Addressing Regulatory and Legal Frameworks

One of the important challenges faced in the Global South by OTT platforms is navigating regulatory and legal frameworks. It becomes essential to establish clear and reliable regulations that balance the need for consumer protection, healthy competition, and regular innovation. It is expected that OTT platforms must work closely with policymakers and regulatory bodies to address issues like content censorship, licensing, taxation and intellectual property rights. Collaborative efforts between government entities and industry stakeholders are essential to create a favourable regulatory environment which promotes growth of OTT while ensuring compliance with legal requirements.

### Privacy Concerns and Data Protection in the Global South

The propagation of digital platforms, comprising of OTT services, increases concerns about privacy and data protection, especially in Global South where data privacy laws may still be evolving. OTT users assign their personal information to OTT platforms and it is imperative that vigorous data protection laws are in place. OTT platforms must adhere to strict data privacy policies, implement secure data storage and transmission protocols, and obtain informed consent from users regarding data collection and usage. Furthermore, providing users with control over their personal information and engaging in transparent data practices are indispensable for building trust and ensuring user confidence.

## Ensuring Equitable Access and Inclusion for all Populations

It is critical to address the issue of equitable access and inclusion for all populations as OTT platforms have immense potential to bridge the digital divide. In Global South, disparities in internet connectivity, affordability in terms of OTT and other digital subscriptions and digital literacy remain major barriers in accessing OTT services. It is important for private sector entities, governments and civil society organizations to collaborate in implementing various initiatives which can promote affordable internet access, digital skills training and infrastructural development in underserved areas. Customization strategies, like zero-rating programs, community network initiatives and public access centres, can help to overcome barriers and ensure that marginalized populations can benefit from the opportunities offered by OTT platforms.

## Overcoming Infrastructure Limitations

In serving marginalized communities, infrastructure limitations, such as poor network coverage, very limited bandwidth and unreliable connectivity, pose significant challenges to widespread adoption of OTT platforms in the Global South. Innovative solutions such as satellite internet, wireless connectivity, and mobile networks to reach remote locations are required to overcome these limitations and expand connectivity to underserved areas. Additionally, heavy investment in infrastructure development, including deployment of fiber-optic cables and expansion of broadband networks is necessary to improve the quality and reliability of internet services. Collaborative efforts between telecommunications companies, governments and tech-service providers are crucial in addressing these infrastructure limitations.

## Innovations in Connectivity Solutions for Underserved Areas

Innovative connectivity solutions are being developed to overcome the challenges of limited infrastructure in underserved areas. For instance, initiatives like “Internet.org” and “Project Loon” aim to offer access of internet to remote areas using high-altitude balloons or drones. Community-driven networks, where local communities build and maintain their own networks, are also emerging as a solution to improve connectivity in underserved regions. These innovations not only expand internet access but also generate opportunities for OTT platforms to reach previously untapped markets and engage with new audiences (Espinoza & Reed, 2018).

## Public-Private Partnerships in Expanding Internet Access

In order to overcome infrastructural constraints and increase internet access, public-private collaborations are essential. When the public, corporate, and non-profit sectors join together, they can pool resources, expertise, and networks to expedite internet infrastructure deployment, lower connectivity costs, and support digital literacy initiatives. These partnerships can leverage the strengths of each sector to address the problems and restrictions associated with internet access in the Global South, fostering an atmosphere that supports the expansion and long-term viability of over-the-top (OTT) platforms.

## Other Challenges

Other than the challenges mentioned above, for, "Bridging the digital divide in the Global South and leveraging the role of OTT platforms in advancing digital inclusion" face various challenges and limitations. These includes affordability and accessibility, content relevance and localization, data privacy and protecting consumer rights, digital skills development, regular supply of electricity, stability in connectivity, gender digital divide, local content production, cultural sensitivity, language barriers, digital divide in education, sustainable digital inclusion, digital security, bridging urban-rural divide, connectivity reliability in remote areas, limited local content creation infrastructure, addressing socio-cultural barriers, limited local technical support and content censorship and freedom of expression. Overcoming these challenges requires rigorous efforts from stakeholders across sectors to ensure that digital inclusion is prioritized where OTT platforms can play a transformative role in empowering individuals and marginalized communities in the Global South.

# CONCLUSION

In conclusion, for bridging the digital divide in global south and exploring the role of OTT platforms in advancing digital inclusion, it is important to have a collaboration between policymakers, industry stakeholders and OTT platforms. Policymakers should give priority to the infrastructure development in terms of internet, must invest in digital literacy programs for marginalized and remote communities and also to create a regulatory environment that enables social development. Industry stakeholders should invest in R&D for affordable technologies, work together with local communities and support local content industries. OTT platforms should encourage local contents, collaborate with local telecommunications providers, and prioritize user privacy and data protection.

Looking ahead, there are favourable opportunities to further enhance digital inclusion in the Global South. By exploring innovative solutions like satellite internet and community networks, internet connectivity can be extended in underserved areas, thereby bridging the digital divide. Moreover, adapting and localizing digital technologies to cater to the needs and preferences of diverse populations will also advance the accessibility and engagement of the consumers. Similarly, investing in digital skills development, predominantly targeting marginalized communities, will empower individuals to participate in the digital ecosystem. By encouraging local content industries with diverse narratives will contribute to a more inclusive and representative media landscape.

To realize these prospects, industry stakeholders, sustained collaboration and partnerships between governments and OTT platforms are very crucial. Public-private partnerships can play significant role in bringing individuals and communities of global south in main stream by driving initiatives to increase internet access, invest in infrastructure and support digital literacy programs. With dedicated efforts and a customer-centric approach, the potential for digital inclusion in the Global South can be fully harnessed, social development, leading to economic empowerment and improved well-being for individuals and communities. By embracing the recommendations and grasping the opportunities that lie ahead, the Global South can truly unlock the transformative power of digital inclusion.

## REFERENCES


ACW. (2022). *Africa Code Week.* ACW. https://coderina.org/africa-code-week/ [Accessed 12 Jun 2023].

A.G.M, J. & Dijk, V. (2006). Digital divide research, achievements andshortcomings. *Poetics, 34*(5), 221–235.

Aishwarrya,C. (2020). *Enhanced User-Experience – A key differentiator for OTT platforms in 2020?* Media Samosa. https://mediasamosa.com/2020/11/11/enhanced-user-experience-ott-strategy/

Nguyen, A. (2022). Digital Inclusion: Social Inclusion in the Digital Age. In *Handbook of Social Inclusion, Research & Practices in health and social care* (pp. 1–13). Springer Nature Switzerland.

Antonio, A. & Tuffley, D. (2014). The Gender Digital Divide in Developing Countries. *Future Internet, 6*(4), 673-687.

Ayanso, A. &. C. D. &. L. K. (2010). The Digital Divide: Global and Regional ICT Leaders and Followers. *Information Technology for Development, 16*(04), 04-319.

Chand, S. (2021). *Sales Promotion Tools: Consumer-Oriented and Trade-Oriented Sales Promotion.* Your Article Library. https://www.yourarticlelibrary.com/sales/sales-promotion-tools-consumer-oriented-and-trade-oriented-sales-promotion/32305 [Accessed 22 Jun 2023].

European Commission. (2022). *Digital inclusion.* EC. https://digital-strategy.ec.europa.eu/en/policies/digital-inclusion#:~:text=Digital%20inclusion%20is%20an%20EU,with%20the%20COVID%2D19%20pandemic. [Accessed 15 Jun 2023].

DiMaggio, P. H. E. C. C. &. S. S. (2004). Digital inequality: From unequal access to differentiated use. Social Inequality, 355-400.

DISD/DPIDG. (2021). *Leveraging digital technologies for social inclusion.* UN. https://www.un.org/development/desa/dspd/2021/02/digital-technologies-for-social-inclusion/

Endavo. (2019). *Everything You Need to Know About OTT.* Endavo. https://www.endavomedia.com/what-is-ott/ [Accessed 14 Jun 2022].

Espinoza, D. & Reed, D. (2018). Wireless technologies and policies for connecting rural areas in emerging countries: a case study in rural Peru. *Digital Policy, Regulation and Governance, 20*(5), 479-511.

Farooqui, J. 2021. *Regional OTTs on the rise as native audiences demand local language content.* [Online] Available at: https://www.exchange4media.com/digital-news/regional-platformsdriving-the-second-ott-wave-in-india-114698.html [Accessed 01 Jul 2023].

Fuchs, J. (2022). *Freemium Pricing: How to Leverage It & 4 Popular Models.* Hubspot. https://blog.hubspot.com/sales/freemium-pricing-strategy

Gamble, C. T. (2023). *Bridging the Digital Divide: The Imperative of Digital Inclusion.* LinkedIn. https://www.linkedin.com/pulse/bridging-digital-divide-imperative-inclusion-tournis-gamble/

Gautam, P., Upadhyay, A. K., & Deepmala, S. (2023). Digital Literacy among Educators. Noida, UP, India, 11th International Library Information Professionals summit (1-LIPS) on new and Innovative Libraries in Digital Era: Services and Practice, (pp. 413-418). Noida.

Geraldine de Bastion, S. M. (2020). Data and the global South:Key issues for inclusive digital development. In: D. Heinrich-Böll-Stiftung Washington, ed. Data and the global South:Key issues for inclusive digital development. Washington, USA: Heinrich-Böll-Stiftung Washington, DC.

GVR. (2023). *GVR Report coverOver The Top Devices And Services Market Size, Share & Trends Report.* GVR. https://www.grandviewresearch.com/industry-analysis/over-the-top-ott-devices-services-market

Himabindu Reddy, S. J. A. J. V. W. (2022). A Critical Review of Global Digital Divide and the Role of Technology in Healthcare. *Cureus, 14*(09).

Hussain, D. M. (2014). SOCIAL IMPLICATIONS OF DIGITAL DIVIDE IN GLOBAL CONTEXT: AN APPRAISAL. *ZENITH International Journal of Multidisciplinary Research, 4*(2), 271-281.

IEEE. (2019). *What Is Digital Inclusion? The Global Effort to Bring Everyone Online.* IEEE. https://ctu.ieee.org/what-is-digital-inclusion-the-global-effort-to-bring-everyone-online/

IEEE. (2022). *Solutions to the Digital Divide: Moving Toward a More Equitable Future.* IEEE. https://ctu.ieee.org/solutions-to-the-digital-divide-moving-toward-a-more-equitable-future/

Intelligence, M. (2023). *OTT MARKET SIZE & SHARE ANALYSIS - GROWTH TRENDS & FORECASTS (2023 - 2028).* Mordor Intelligence. https://www.mordorintelligence.com/industry-reports/over-the-top-market

IT for Change. (2013). *ICTs for empowerment and social transformation - A note prepared by IT for Change for ActionAid International.* IT for Change. https://itforchange.net/index.php/icts-for-empowerment-and-social-transformation-a-note-prepared-by-it-for-change-for-actionaid-0 [Accessed 03 Jul 2023].

ITU. (2020). *International Telecommunication Union.* ITU. https://dig.watch/actor/international-telecommunication-union [Accessed 21 Jun 2023].

ITU. (2022). *Digital inclusion of all.* https://www.itu.int/en/mediacentre/backgrounders/Pages/digital-inclusion-of-all.aspx#:~:text=Overview,and%20services%2C%20equally%20and%20equitably. [Accessed 04 JUn 2023].

Koskinen, K. (2019). Digital Platforms in the Global South:Foundations and Research Agenda. In: *Information and Communication Technologies for Development. Strengthening Southern-Driven Cooperation as a Catalyst for ICT4D.* Switzerland: Springer Nature Switzerland.

Koskinen, K., Bonina, C., & Eaton, B. (2018). Digital Platforms in the Global South: Foundations and Research Agenda. *Development Implications of Digital Economies, 1*(1), 1–28.

Krick, K., Boswell, J. S. & Poretsky, S. (2021). *Bridging the digital divide for an inclusive digital economy.* Research Gate.

Kumar, V. (2014). *Making "Freemium" Work.* HBR. https://hbr.org/2014/05/making-freemium-work [Accessed 26 Jun 2023].

Levy, A. (2018). *Why Netflix Loves to Highlight Its Local Content.* Fool.com. https://www.fool.com/investing/2018/04/24/why-netflix-loves-to-highlight-its-local-content.aspx.

Lin, Y. W. M. Z. H. Z. R. S. Z.-J. M. (2016). *Content Promotion for Online Content Platforms with the Diffusion Effect.*, *21*(Jun), 49.

Laskar, M. (2023). Examining the emergence of digital society and the digital divide in India: A comparative evaluation between urban and rural areas. *Frontiers in Sociology, 12*.

Massimo Ragnedda, B. M. 2018. Digital Inclusion Empowering People Through Information and Communication Technologies. In: Digital Inclusion. An International Comparative Analyses. s.l.:Lexington Book.

Molala, T. S., & Makhubele, J. C. (2021). The Connection Between Digital Divide And Social Exclusion: Implications For Social Work. *Humanities & Social Sciences Reviews*, *9*(04), 194–201. doi:10.18510/hssr.2021.9427

Kulkarni, M. (2019). Digital accessibility: Challenges and opportunities. *IIMB Management Review, 31*(01), 91-98.

Napoli, P. M. N. (2019). Media as platforms: Definition, dynamics, and drivers. *Journal of Communication*, *69*(2), 277–301.

Nguyen, A. 2023. Social Inclusion in the Digital Age. In: V. H. V. College of Health Sciences, ed. Handbook of Social Inclusion. Springer Nature Switzerland.

OECD. (2018). *BRIDGING THE DIGITAL GENDER DIVIDE: INCLUDE, UPSKILL, INNOVATE.* Australian government.

OET. (2022). *Advancing Digital Equity for All Barriers & Strategies.* OET. https://tech.ed.gov/advancing-digital-equity-for-all/barriers-and-strategies/

Frater, P. (2023). *Netflix to Increase Asia Content Spending by 15% in 2023 as Revenue Growth Accelerates in the Region – Study.* Varriety. https://variety.com/2023/tv/news/netflix-content-spending-asia-pacific-1235543665/

PennState. (2019). The Digital Divide. In: INFORMATION, PEOPLE, AND TECHNOLOGY. Research Gate.

Pertierra, R., &. C. J. V. (2018). Digital inclusion in the Global South: An analysis of mobile and OTT services in the Philippines. *Telematics and Informatics*, *35*(03), 540–550.

Rachelle Bosua, M. G. (2021). Access to Flexible Work Arrangements for People With Disabilities: An Australian Study. In M. B. L. Queen Aigbefo (Ed.), *Anywhere Working and the Future of Work* (pp. 134–161). IGI. doi:10.4018/978-1-7998-4159-3.ch006

Reddick, C. G., Enriquez, R., Harris, R. J. & Sharma, B. (2020). Determinants of broadband access and affordability: An analysis of a community survey on the digital divide. *Elsevier - PMC COVID-19 Collection.*

ReelnReel. (2022). *OTT Content Pricing Strategies for OTT Platforms.* ReelInReel. https://reelnreel.com/ott-content-pricing-strategies-for-ott-platforms/

Research, P. (2022). *ICTOver the Top (OTT) Market.* North America: Precendence Research.

Review, T. (2023). *Kenya's Digital Future: A Journey of Connectivity.* Innovation and Prosperity. [Online]

Heeks, R. (2022). Digital inequality beyond the digital divide: conceptualizing adverse digital incorporation in the global South. *Information Technology for Development, 28*(04), 688-704.

Ruhiya Seward. (2020). *Understanding digital access and use in the Global South.* IDRC.

SAP. (2022). *Conquering Africa's Digital Divide: SAP Africa Code Week Empowers 2.6 million Youth in 2022.* SAP.

Saraswat, S. (2023). *Revolutionizing Entertainment: The Growth of OTT Platforms in India and Around the World.* LinkedIn. https://www.linkedin.com/pulse/revolutionizing-entertainment-growth-ott-platforms-india-saraswat/

Sharma, A., & Banerjee, A. (2022). Socio-Economic Determinants of Digital Divide in India. *Demography India, 51*(1), 78–92.

Sreekumar, T. T., &. A. A. I. (2020). OTT platforms in the Global South: Challenges and opportunities for digital inclusion. *Journal of Global Information Management, 28*(4), 50–70.

Straits. (2023). *Over-the -Top (OTT Market).* Strait Research. https://straitsresearch.com/report/over-the-top-market

Bennett, B. (2012). Digital Natives. In: Z. Yan, ed. Encyclopedia of Cyber Behavior. Research Gate.

Swalehin, M. & Nasir, M. S. (2011). Digital Divide in India and China: A Comparison. *Research J. Humanities and Social Sciences, 2*(4), 181-185.

The Tribune. (2021). Rise and Expansion of OTT Platforms: Making Quality Content Accessible to All. *The Tribune.* https://www.tribuneindia.com/news/brand-connect/rise-and-expansion-of-ott-platforms-making-quality-content-accessible-to-all-indian-355860

The World Bank. (2023). *Digital Development.* World Bank. https://www.worldbank.org/en/topic/digitaldevelopment/overview

Toppan. (2021). *Cultural Sensitivity: How to Market Your Brand to Other Culture.* Toppan. https://toppandigital.com/us/blog-usa/cultural-sensitivity-market-brand-cultures/

Trust, T. (2020). *Stories of Change.* Tata Trust.

UN. (2021). *Leveraging digital technologies for social inclusion.* UN. https://www.un.org/development/desa/dspd/2021/02/digital-technologies-for-social-inclusion/

Unesco, (2022). *Digital Learning Week.* UNESCO. https://www.unesco.org/en/weeks/digital-learning [Accessed 01 July 2023].

UNICEF. (2019). Digital literacy for children:exploring definitions and frameworks. UNICEF Office of Global Insight and Policy.

van Deursen, A. J. A. M., Helsper, E. J., & Eynon, R.J.A.M. (2016). Development and validation of the Internet Skills Scale (ISS). *Information Communication and Society*, *19*(6), 804–823. doi:10.1080/1369118X.2015.1078834

Warschauer, M., & Matuchniak, T. (2010). New technology and digital worlds: Analyzing evidence of equity in access, use, and outcomes. *Review of Research in Education*, *34*(1), 179–225. doi:10.3102/0091732X09349791

Hamdar, Y. (2021). *Digital Transformation in the Global South: Enhancing Citizens' Trust.* Data Appliance. https://datapopalliance.org/digital-transformation-in-the-global-south-enhancing-citizens-trust/

Zuora. (2021). *OTT CONTENT PRICING STRATEGIES.* Zuora. https://www.zuora.com/guides/ott-pricing-strategies/

# Chapter 13
# Business Transformation Projects:
## The Impact of the Over-the-Top

**Antoine Toni Trad**
https://orcid.org/0000-0002-4199-6970
*IBISTM, France*

## ABSTRACT

*This chapter presents the business transformation project (BTP) and the impact of over-the-top (OTT) based platforms and group work (OTTBGRW). The OTT supports employees' GRoup Work (GRW) in the context of complex collaborative professional or training-educational activities. A BTP can be supported by the OTTBGRW (BTPOTTBGRW) which uses the author's specific research project concept and artefacts like critical success factors and areas, authentic transformation framework, and an integrated decision-making system. The BTPOTTBGRW can be used in all BTP's levels, phases, and activities. The BTPOTTBGRW uses existing related standards, methodologies, cultural/technical specificities, innovation trends, and legacy business practices. A BTP includes domains, like information and communication systems (ICS), multimedia, business engineering, security, and many others. The OTTBGRW enables business organizations (simply Entity) to use agile cross-functional collaboration GRW based environments, and dynamic multimedia services'-based interaction/management concepts.*

## INTRODUCTION

*Project Group Work* is used in all BTP's phases, and uses existing technologies standards, methodologies, and best practices. The transformed ICS supports the OTTBGRW in organizations (simply *Entity)* to integrate agile GRW collaborations, integrational patterns, and OTT services' management. OTTBGRW's limitations are in general *Entities*' capabilities to transform their traditional system(s); and to integrate the OTTBGRW in the actual environment and to find skilled BTP personal (simply the *Staff*). *Project Group Work*'s integration depends on: 1) An In-House (IH) Enterprise Architecture (EA) Concept (IHEAC, simply *EA Concept*); 2) An Applied Holistic Mathematical Model (AHMM)

DOI: 10.4018/979-8-3693-3526-0.ch013

for OTTBGRW (AHMMx4OTT, simply *Applied Mathematical Model*) (Trad, & Kalpić, 2020a); 3) Services based architecture for BTP's infrastructure and platforms (Trad, 2015a; 2015b); 4) Holistic enterprise patterns (Trad, & Kalpić, 2022a; 2022b); 5) Distributed, cloud, common multimedia based platforms, and OTT Platforms (OTTP); 6) Agile Transformation Management (ATM) (Spencer, 2016); 7) Digital Transformations' (DT) successful termination; and 8) Offering and opening OTTP services to internal and external requests. The *Project Group Work* supports *Entity's* multimedia-based innovation which can be modelled by the *Applied Mathematical Model*. The *Applied Mathematical Model* uses its internal Rating & Weighting Concept (RWC) which includes: 1) Critical Success Areas (CSA); 2) Critical Success Factors (CSF); 3) Key Performance Index (KPI); and 4) A specific mixed-method (Trad & Kalpić, 2020a). The *Project Group Work* is supported by an internal Decision-Making System (DMS) for OTTBGRW (DMS4OTTBGRW) and Knowledge Management System (KMS) for OTTBGRW (KMS4OTTBGRW, simply *Intelligence*) that is based on a mixed-method developed by the author, and is based on Action Research (AR) based Learning Process (ARbLP). The ARbLP also manages Return On Experiences (ROE). A GRW concept forces the used transformation framework to synchronize all BTP activities, like the integration of the Polymathic OTT based Interdisciplinary GRW (PIGRW). Automated GRW can produce different types of problems, that can be proactively managed. The author's transformation framework is a pioneering one and its originality lies in that it can be used in any type of BTP or transformation project and in any project's phase. This framework delivers a ARbLP subsystem that can learn and persist ROEs/experiences from any type of tasks or encountered problem or a project solution (Trad, 2023d). The author's Transformation Research Architecture Development framework (*TRADf*) interactions include the following components: 1) ICS and media technologies; 2) *Intelligence* subsystem; 3) *Project Group Work*; and 5) Research and Development Project (RDP), which is the first CSA, and its heading are in fact the initial CSFs.

## THE RESEARCH DEVELOPMENT PROJECT

### A Different Concept and Gap

This chapter has multi-dimensional approach to RDPs because it has: 1) A dynamic adapted RDP approach; 2) Promotes an IH transformation framework; 3) Uses an organizational methodological approach that is based on *EA Concept*; and 4) The integration of PIGRW (Trad, & Kalpić, 2023a). This RDP uses an iterative and incremental approach and *TRADf's* recurrent evolution; in which the author which reuses his previous works. Such an unconventional approach can be misunderstood and that is the author's major concern; it also not aligned with standardized quantitative based academic concepts. The reuse of *TRADf*'s research components, concepts, and resources is necessary. The reuse of the author's existing resources, artefacts, and keywords can be viewed (by simplistic quantitative-oriented automated checkers) as duplications or similarities... Which is definitely not the case and added to the mentioned facts, the very high failures rates of BTPs, permits the search for other concepts and complex RDPs... Using defacto imposed research methods is not innovation, particularly in complex undertakings which need crucially renewed Polymathic concepts... The Researched Literature Review (RLR) and the resultant Gap have clearly showed that BTP's high rates of failure are the result of problems related to the integration of cross-functional fields. The RDP and BTP use the *Applied Mathematical Model* which includes the ARbLP based Heuristics Decision Tree (HDT) that in turn supports *Intelligence*. This chapter's

Research Question (RQ) is: "Which *Project Group Work* characteristics are needed to support BTPs and OTT based solutions?" This chapter uses author's previous works and existing *TRADf's* components, like: 1) The Business Transformation Project's Holistic Agile Management (Trad, & Kalpić, 2022c); 2) The Selection, and Training Framework selection and training framework (STF) for Manager's in Business Innovation Transformation Projects–Educational Recommendations (Trad, & Kalpić, 2014b); 3) Business Transformation Projects: The Role of Psychology-Based Resistance (RPbR) (Trad, 2023b); 4) Educational Transformation Project's Remote Group Work (ETPRGW) (Trad, 2022d); and others. This RDP's gap is mainly due to the fact that there are no similar: 1) Research concepts; 2) Very High failure-rates; 2) Inexistent framework; 3) The selection of accountants as transformation managers; and 4) There are not profile's description for transformation managers. The gap is due to the lack of: Efficient concepts; High failure-rates; PIGRW's integration; and *Project Group Work* similar concept. This RDP privileges Polymathism and the integration of PIGRW, and distributed ICS strategies. The *Project Group Work* combines: 1) The PIGRW, OTT and other online media technics; 2) *Applied Mathematical Model*, ARbLP and HDT based *Intelligence*; 3) Enterprise patterns; 4) To use the *EA Concept* to support BTPs; 5) To use PIGRW interfaces; and 6) *Project Group Work*'s statuses metrics, based on CSAs, CSFs, and KPIs (simply *Factor*).

## Factors' Management

*TRADf* and RDP are agnostic to the type of BTP (or transformation initiative) and APplication Domain (APD). The RDP selects the essential CSAs which include sets of CSFs. Each CSF includes a set of KPIs, where a KPI maps/relates to a unique BTP or *Project Group Work* requirement. A KPI links to a real ICS application variable, which implies that the evaluation processes receive realworld statuses' values. The defined requirements link to an HDT evaluation Table's a column (Putri, & Yusof, 2009). The RDP starts with the 1st phase (*Feasibility Phase*), which verifies the Factors' statuses' (that are managed by HDT evaluation Tables) and each CSA has one Table. To check if BTP's goals were fulfilled, the HDT returns a status value, and a set of next BTP actions. Based on RDP's RLR and the evaluated Factors using the following rules and constraints: 1) *Project Group Work* related resources' references that are credible and are estimated by *Intelligence;* and also comply with *TRADf's* classification storage; 2) RWC evaluation of Factors; 3) *EA Concept*'s modelling capabilities and management of KPIs' links with ICS variables; 4) The Architecture Development Method (ADM) is used to manage BTP's iterations, and KPIs' versioning; 5) The ADM manages also *TRADf's* iterations and Factors' evaluations; 6) In the case that RWC based Factors and all CSA Tables are positive, then the final results are stored in the *Final Table*; and 7) If columns of the *Final Table* exceed the defined minimum-constraint, then the RDP proceeds with PoC's execution and uses the selected Applied Case Studies (ACS).

## RDP's ACSs

This RDP combines various ACSs that are used mainly in the PoC and as references. One common ACS is related to *EA Concept* and originates from the Open Group. it covers BTP's ICS, *EA Concept*, modelling, integration of KPIs, and BTP common transformation scenarios. *Project Group Work*'s Factors are managed by the ARbLP based HDT; and the HDT uses the RWC that initiates BTP Factors in RDP's 1st iteration and afterwards these Factors are tuned in each of ADM iteration. BTP's and *Project Group Work*'s statuses are evaluated by *Intelligence* and which tunes Factors. The selected ACSs are: 1) The already mentioned common ACS; 2) The importance of PIGRW in interactive training environments that are optimal for mass training (Postle, 2003); 3) An ACS in which's aim is to combine all major OTTPs into a unified solution (Kher, 2022); 4) A streaming TeleVision (TV) enterprise, which offers services for subscribers, like cable channels supported by a common OTTP (InfoSys, 2023); 5) Marketing success ACSs related to OTT based media (UI, 2023); 6) Reimagining OTTPs' experience (Sinha, 2021); and 7) CSAs related to DTs in 10 known industries (Llewellyn, 2023). OTT accelerated DT's applications and integration's scuees (Patil, 2022). This chapter's CSAs are: RDP, *Applied Mathematical Model*, OTTP based APDs, ICS support, *EA Concept*'s integration, and *Project Group Work*'S main blocks. *Project Group Work* activities can be modelled using the *Applied Mathematical Model*.

## The *Applied Mathematical Model*

The adoption of a holistic and Polymathic OTTPs' integration approach can be enabled by the *Applied Mathematical Model*. The *Applied Mathematical Model* supports the *Project Group Work* by managing critical-risks to guaranty OTTP operations' coherency. For a *Project Group Work* requirement (or related problem type), *Intelligence* offers a set of Factors and adjacent sets of actions to be activated and used by the HDT. There is a lack of a Polymathic/holistic concept to support BTPs and that was confirmed by the RLR that used the following artefacts/resources: 1) Articles, artefacts, and resources related to BTPs, OTT/Media, ICS platforms, AI/Automation, …; 2) The author's previous RDP/RLR articles/works, and *TRADf*; 3) BTP's management and feasibility concepts; 4) Management and evaluation of Factors; 5) The application of the Empirical Engineering Research Model (EERM); and 6) OTTP's architecture and integration concepts. Polderman and Willems (Polderman, & Willems, 1998) have proved that a Polymathic-Holistic Mathematical Model (PHMM) is a subset of concrete real-world system like *Project Group Work*'s integration. PHMM's main facts are:

- A PHMM can deliver an abstract model of a real transformed business system (Hinkelmann, 2016).
- Architecture and modelling artefacts can be used for PHMMs (Sankaralingam, Ferris, Nowatzki, Estan, Wood, & Vaish, 2013).
- The *EA Concept*, Network modelling, AI, and HDT can be used by the PHMM.
- There is an immense gap between the PHMM based BTPs and archaic BTPs (Syynimaa, 2015).
- The PHMM and Factors based RWC/HDT are used to evaluate BTPs (Dogan, Çalgici, Arditi, & Gunaydin, 2015).
- Develop an IH PHMM, like the *Applied Mathematical Model* (Giachetti, 2012).
- A PHMM can describe an *Entity*, BTP, and *Project Group Work*, or other by using a mathematical language (Sankaralingam, Ferris, Nowatzki, Estan, Wood, & Vaish, 2013).
- A complex Factors concept for *Intelligence* is based on qualitative and quantitative methods, like the ARbLP based HDT (Zandia, & Tavana, 2011; Kim & Kim, 1999).
- A PHMM is optimal for EERM oriented RDPs (Easterbrook, Singner, Storey, & Damian, 2008).
- A PHMM can be used for video traffic and transmission in networks (Ming, & Nikolaos, 2009; Kokoreva, & Belezekovaf).
- Quality levels can be estimated mathematically (Ahmad, Floris, & Atzori, 2016).
- The first step is to define PMM's basic elements.

## Basic Elements

*Applied Mathematical Model*'s basic element are used in *Intelligence*, which is a proper to the author's approach; and the used nomenclature is shown in Figure 1, has the following rules:

- The symbol $\sum$ indicates summation of RWC's actions, showing the importance of members that are relevant. RWC's Ratings and weightings as integers ranging in ascending importance from 1 to 10.
- The symbol $\underline{\text{U}}$ indicates sets union.
- The *Applied Mathematical Model* defines the BTP and *Project Group Work* as models, using Factors, HDT, and RWC.

## The RWC

There many ways how to implement an RWC and *Intelligence;* it is possible to use an external/ standard RWC or an IH one like in the *TRADf's,* which has the following rules:

- CSAs' weightings $\epsilon$ { 1 … 10 } are integer values.
- CSFs' weightings $\epsilon$ { 1 … 10 } are integer values.
- CSFs' ratings $\epsilon$ { 0.00% … 100.00% } are floating point percentage values.
- A weighting is defined for each CSA/CSF, and a rating for each KPI.
- The *Applied Mathematical Model* uses an HDT that in turn uses the RWC.

External/standard RWCs can be:

- Object Management Group's (OMG) Decision Model and Notation (DMN), which can be applied for the specification of domain/business decisions and rules. DMN is optimal for BTP's decision-making and evaluations (OMG, 2022). The DMN can be also used for Tables' processing and HDT operations.
- RWC based on Weighted Criteria Matrix (WCM) which can evaluate BTPs based on defined evaluation of Factors. The WCM delivers KPIs with respect to defined criteria (Ouye).

The *Applied Mathematical Model* uses the HDT which is a qualitative construct that has specific calls to quantitative methods.

## A Quantitative-Qualitative Research Mixed Model

For a specific *Project Group Work* problem, RQ, or Factor, the HDT is used by relating breadth/depth tree and the acquired knowledge is stored in *Intelligence*. The set of BTP problems and the linked Factors are initialized in ADM's/BTP's preliminary phase (or initial iteration). Then the HDT inputs needed datasets that include: Raw data, Constraints, and Rules. Applying only quantitative analysis, is limited and superficial, therefore there is the need to use mainly a qualitative method, like the HDT. ARbLP is suitable for the BTP and *Project Group Work*, because AR is helpful for LPs and it provides BTPs with important knowledge on how to improve OTTPs' integration practices. AR support for BTPs is a systematic concept and offers solutions like: 1) Enabling professional development and efficiency;

*Figure 1. PMM's and AHMM's nomenclature*
*(Trad, & Kalpić, 2020a)*

**Basic Mathematical Model's (BMM) Nomenclature**

| | | |
|---|---|---|
| *Iteration* | = An integer variable "i" that denotes a *Project/ADM iteration* | |
| microRequirement | = (maps to) KPI | (B1) |
| CSF | = $\Sigma$ KPI | (B2) |
| Requirement | = (maps to) CSF = $\underline{\cup}$ microRequirement | (B3) |
| CSA | = $\Sigma$ CSF | (B4) |
| microMapping microArtefact/Req | = microArtefact + (maps to) microRequirement | (B5) |
| microKnowledgeArtefact | = $\underline{\cup}$ knowledgeItem(s) | (B6) |
| neuron | = action->data + microKnowledgeArtefact | (B7) |
| microArtefact / neural network | = $\underline{\cup}$ neurons | (B8) |
| microArtefactScenario | = $\underline{\cup}$ microartefact | (B9) |
| AI/Decision Making | = $\underline{\cup}$ microArtefactScenario | (B10) |
| microEntity | = $\underline{\cup}$ microArtefact | (B11) |
| Entity or Enterprise | = $\underline{\cup}$ microEntity | (B12) |
| EnityIntelligence | = $\underline{\cup}$ AI/Decision Making | (B13) |
| BMM(*Iteration*) as an instance | = EnityIntelligence(*Iteration*) | (B14) |

**The Generic AHMM's Formulation**

| | | |
|---|---|---|
| AHMM | = $\underline{\cup}$ ADMs + BMMs | (B15) |

**AHMM's Application and Instantiation for BTP**

| | | |
|---|---|---|
| *Domain* | = BTPOTTBGRW | (B16) |
| AHMM4(*Domain*) | = $\underline{\cup}$ ADMs + BMMs(*Domain*) | (B17) |

2) Increasing *Staff's* empowerment and collaboration; and 3) Synchronizing architecture, modelling, implementation, maintenance and ROE based practices (Hine, 2013). ARbLP collected knowledge are used to support BTP's applied transformational mathematical model.

## The Applied Transformation Mathematical Model

The *Applied Mathematical Model* is composed of: 1) A static view; 2) A dynamic (or behavioural) view; and 3) A dictionary/database of checked ROE based ARbLP instances. The *Applied Mathematical Model* is modelled and implemented using the BTP Transformation Mathematical Model (*BTMM*) that abstracts the BTP:

*Applied Mathematical Model*=Weigthing$_1$**AHMM*_Qualitative+Weigthing$_2$**AHMM*_Quantitative (B18).

*Applied Mathematical Model* = $\sum$ *AHMM* for an BTP iteration (B19).

*BTMM* = $\sum$ *AHMM* instances (B20).

Weigthing$_1$ and Weigthing$_2$ are processed and returned by the RWC. *BTMM's* objective/optimization function(s) uses constraints and (extra) ICS variables that need to be configured. These variable (for maximization or minimization estimations) can be, for example: 1) BTP's team Polymathic capacities; 2) Stability and robustness of OTTP's infrastructure; 3) OTTP's related costs, or another Factor. For *Project Group Work*'s PoC feasibility is the main constraint and success is quantified as a binary 0 or 1, where the objective function minimizes *BTMM's* risks and identifies *Project Group Work*'s feasibility. The *BTMM* is a mixture of existing methodologies ADM and *Applied Mathematical Model*, which are parts of *TRADf* (Kim, & Lennon, 2017).

## ADM's Usage, Intelligence and Managing ROEs

At BTP's start problem types and Factors are defined and a Scaled Agile Framework (SAFe) is selected to support PIGRW interactions (Scaled Agile, 2022). An *Entity* must implement its IH ADM, ARbLPs, and ROEs concepts, in order to manage *Project Group Work*'s complex problems. ROEs are used to enhance ARbLPs and for this goal the BTP needs: 1) A subsystem to manage ROEs; 2) ADM based *EA Concept* to support PIGRW activities; 3) OTTP's integration and agile interaction; 4) OTTP's infrastructure optimization; 3) Discovering new OTT business opportunities from stored ROEs; 4) Sharing experiences; 5) ADM based OTTP's implementation support ... The ADM or Unified Modelling Language (UML) are modelling methods that use BTP cycles. The ADM manages BTP's modelling, development, and maintenance activities; and ADM's main phases are: 1) Preliminary, which introduces *Project Group Work*'s goals; 2) Vision and domain architectures; 3) ICS' and PIGRW's architectures; 4) Innovative technologies' (like OTT) integration concepts; 5) BTP's requirements management; and 6) Managing *Intelligence*. *Intelligence* delivers solutions for BTP's problem types and these solutions have the form of recommendations on possible transformational activities, which minimize BTP risks. *Intelligence* launches the HDT to solve encountered problems and estimates *Project Group Work*'s risks (Hussain, Dillon, Chang, & Hussain, 2010). BTP's initial iterations transform legacy DMS and supports the adoption of HDT/ARbLP concepts. ARbLPs manage Decision or Knowledge Items (DKI), where DKIs and linked to ROEs. *Intelligence* supports *Entity's* ICS to manage ARbLPs and associated DKIs. The BTP interfaces *Intelligence* and stored DKIs, which are in turn linked to Factors. *Intelligence* supports BTP's enhancements and OTT based creative work.

## OTT Based Creative Work

OTTPs support professional collaborations, as shown in Fig. 2, where professionals use video infrastructure to interact and OTT is widespread and supports (FastEdge, 2023; Pappas, 2022): 1) Video collaboration and independent training; 2) Complex DTs; 3) Remote-tools collaboration; 4) Just-in-time feedbacks; 5) Tasks' management; 6) Quality of Service (QoS), specifically in Polymathic domains; 7) Modelling and implementation of BTP modules; 8) Management of ROEs and ARbLPs; and 8) Integration of cloud based ICSs.

*Figure 2. OTT based collaboration*
*(FastEdge, 2023)*

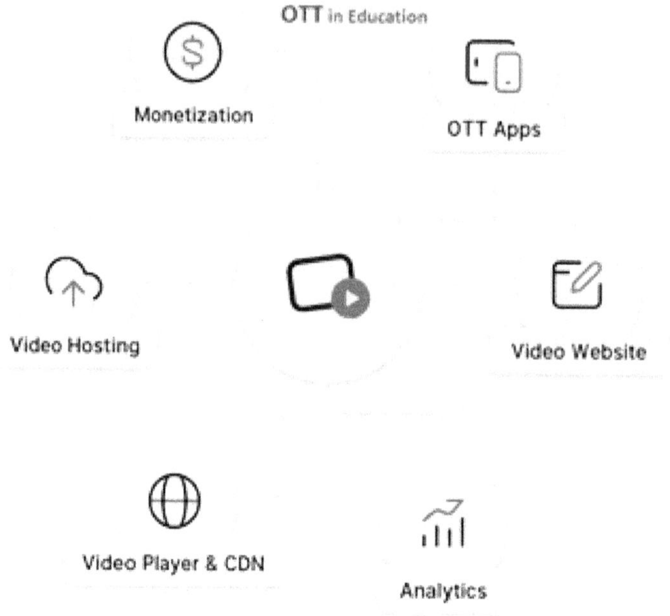


OTTPs supports DT's and BTP's implementations which use PIGRW and can offer OTT based PIGRW Patterns (OTPs) to establish video-based practices. OTPs support interactions using collaborative modelling and implementation (Botturi, & Belfer, 2014). Professional Modelling/design activities are important Factors for the *Project Group Work*, because such collaboration is the fundament of Polymathic BTPs. OTPs' support the PIGRW that is associated with BTP's design activities and is a central activity. An OTTP can support cognitive activities in design tasks. The OTP supports the PIGRW by providing (Han, Park, Hua, & Childs, 2021): 1) Access to *Entity* resources; 2) Support for architecture; 3) Media based interactive communication; 4) Optimal performance; and 5) Skills transparency for OTTP based APDs.

## OTTP BASED APDS

### Using Basic Artefacts

The *Project Group Work* uses Factors in different BTP parts and APDs, which needs a specific cross-sector interfacing. RDP, APD, and BTP Factors are identified in the RLR period and then optimized in various iterations (Thune, 2011). This RDP's APDs are the ones which intensively use OTTBGRW. The OTTBGRW is supported by different types of *EA Concept* artefacts, like EA patterns, services architectures, network components, and Building Blocks (BB), which can encompass common OTTP architecture related models, problems, and solutions. Solutions have consequent transitional state(s) which are managed by an agile concept like SAFe (Spencer, 2016).

## Using SAFe

BTP's complexities are due to continuous replanning and agile interaction between BTP teams who design and implement solutions. Therefore, BTPs make frequent changes that need concepts like SAFe to deliver QoS solutions. BTPs need *Entity's* commitment, which is *Project Group Work*'s strategic request and SAFe which is shown in Figure 3, supports the integration by applying the Development and Operations for OTTP which supports OTT's artefacts integration (Kersten, 2021).

*Figure 3. The relation between BTPs and stability*
*(Kersten, 2021)*

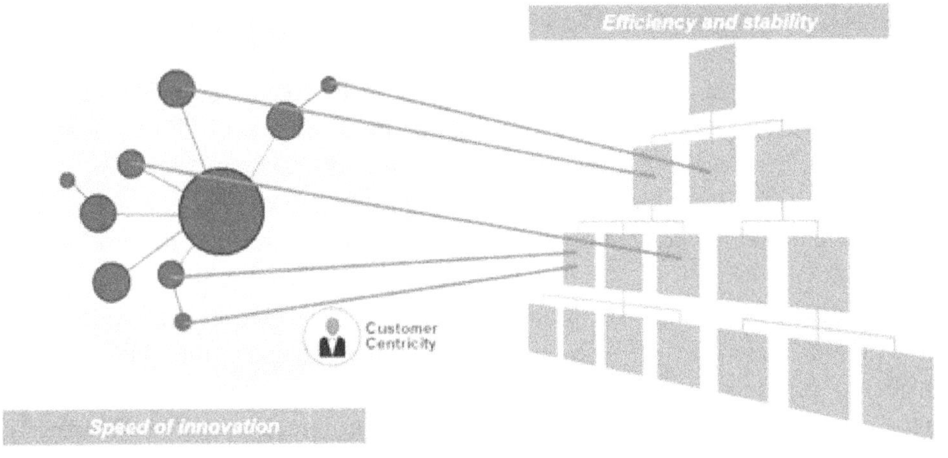


BTP's team *manages* agile transformation iterations and related OTTP's integrations by using SAFe, *EA Concept*, and the Cloud (Stojanov, 2015).

## Cloud's Integration

Clouds can support technical ICS and OTTP features like security, availability, performance, unreliable access, maintenance, and support. The Cloud based OTTP and hence *Project Group Work* can provide the following (Pappas, 2022): 1) Supporting the setup and maintenance activities; 2) Minimizing costs; 3) Enhancing employees retention; 4) Increasing productivity/availability and QoS; 5) Enables collaboration in distributed ICS/OTTP; 6) Supports scalability which permits growth of OTTPs; 7) Simplifying ICS/ OTTP's resources exchange; 8) Offers remote and mobile access; 9) Offers OTTP services to customers; 10) OTTP become robust and safe; and 11) Supports all types of OTTP activities, like offering OTT based Collaboration Pattern (OCP, simply *Pattern*).

## *Pattern*'s Usage

In OTTPs there is the need to integrate *Patterns* to enhance interactions and a successful use of *Project Group Work* is related to team-members' ability to work in virtual-teams. *Pattern* based groupwork forces teams to develop their version of this pattern. Interactions of teams includes design/modelling, planning, coordination, implementation, and other. Sets of (inter)actions are defined for teams using implemented *Patterns* and ROEs. *Patterns* can help in managing teams and enhancing communication, and organization, which enhances teams' engagement. There are differences in perceptions of *Pattern* based groupwork because team-members do not see same values, due to their skills. *Patterns* need the adoption of innovative OTTP technologies, business scenarios, and methodologies, but unfortunately here is a lack of skills; and *Patterns* need BTP's transition from an iterative model which requires continuous adjustments. *Patterns* are organized to accomplish BTP tasks with frequent communication, problem-solving approaches and is self-regulated groupwork. BTP teams must continuously improve *Patterns* to enable engineers focus on their tasks and improve existing ICS/OTTP infrastructure (Pisonia, Gijlersc, Nguyend, & Chene, 2021).

## Existing ICS/OTTP Infrastructure

OTTPs provide TV and film content over the internet where the content provider *goes over the top of existing internet services*. In the 90's, digital TV used the Internet Protocol (IPTV), as shown in Figure 4, to provide and then there was the need for: TV subscription; Contract, Set-top box… The content was distributed through private multicast-mpeg streaming networks; and this concept locked costumers to one service-provider and the content offered was limited to channels available in specific regions (TeleStream, 2023).

OTTP based video delivery supports (TeleStream, 2023): 1) A multitude of options, like viewing the content on various platforms such as: Smart TVs, computers, tablets, smartphones…; 2) Access to multiple distributors and view channels, offering more control on the content to purchase; 3) The delivery of content upon request, where a connected device has a unique connection to the provider; 4) As shown in Figure 5, OTTP transmission is over the internet; and 5) Overcoming a single operator Set Top Box (STB) technology required by IPTV.

*Figure 4. IPTV based environment*
*(TeleStream, 2023)*

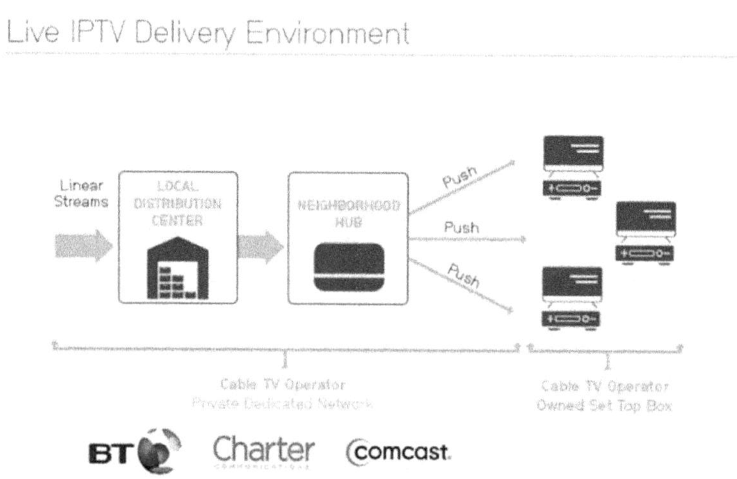

*Figure 5. OTT based environment*
*(TeleStream, 2023)*

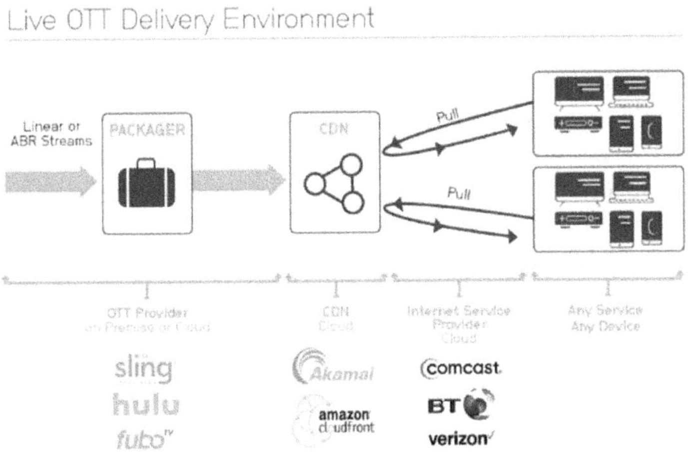


## ICS SUPPORT

In a BTP the first major change is succeeding the DT so that the *Entity* is fully digitized and can support OTTPs. There are many ways to achieve DTs, and the major way is to unbundle the legacy ICS into services.

### A Services' and Cloud Based OTTP Architecture

An important BTP phase is to unbundle the legacy ICS and offer a pool services, where a set of services is assembled in BBs which can be used in APD scenarios (from the functional perspective). BBs are used in *Application Services* which in turn are oackaged in Transformed Domain Components (TDC). BBs can contain *Business Services*, and a subsequent set of diagrams, (other) BBs, *Application Services…* Modelling diagrams use elements, BBs, and functionalities (Hosiaisluoma, 2022; Trad, 2023e), and these services can access and be managed by OTTP's Application Programming Interface (API) (Brightcove, 2023). APIs can be supported by the integration of a Cloud. The Cloud enabled an *Entity* Unified Communications (EUC) unifies the transmission of voice, data, and applications. Cloud services support the implementation and maintenance of an OTTP; these services include: Storage as a Service (SaaS); Infrastructure as a Service (IaaS); and Platform as a Service (PaaS). It supports also Unified Communication as a Service (UCaaS) that delivers collaborative voice, data and business applications to the end-user as a Cloud service, which also includes voice telephony, web-services, audio and video conferencing, and unified messaging; which are the basis of OTTP connectivity (NEC, 2015). OTTP based video streaming opens new business opportunities, because it is Cloud based, and is dynamic, which means that it can be personalized Diffusion to the Client (D2C) (AWS, 2022). When the unbundling process (that delivers services and APIs) and Cloud's integration terminates, it can be considered that the most important and critical part of the DT has been finalized. Especially in domains like Telecom (Deloitte, 2024).

## DT's Support for OTTP and Media

A DT supports OTTPs to remain *on-demand ready* and in the actual age customers access media massively. That needs seamless access to OTTP services which require intuitive client-interfaces. To stay competitive and efficient an *Entity* must terminate the DT, to harness all resources and achieve the *Digital First* and related optimal digital architecture. OTTP's content consumption is mass-based content and that requests large screens and investments in original contents' creation which is driving OTT base consumptions. OTTPs support live-streaming which leads to massive data usage. As shown in Figure 6, DT's process includes: Defining clear vision and strategy; Understanding OTT customers' proposition; and Precise business design and designing the plans (KPMG, 2017).

*Figure 6. DT's process*
*(KPMG, 2017)*

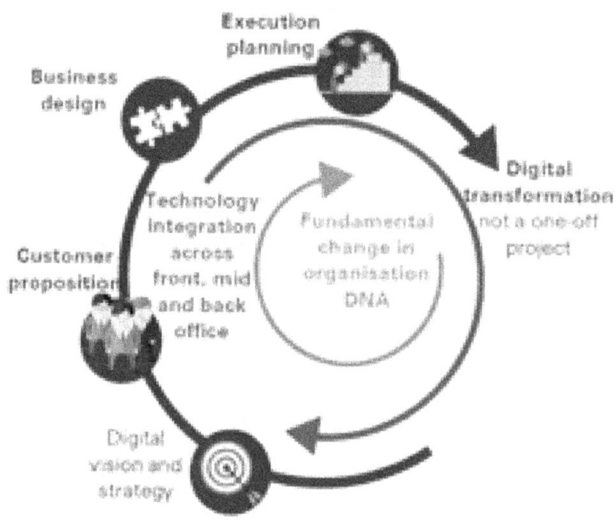


The hyper-evolution of ICS is promoting OTT's fast evolution; where DTs are revolutionizing the entertainment industry. People entertainment has been transformed based on (Koshy, 2022): 1) Enhanced user experience, where the OTTP has seamless viewing experience and has quality monitoring solutions; 2) Blending OTTPs create and distribute content; 3) All-in-one OTT subscriptions, are supported by aggregated platforms that provide a single-window to manage subscriptions; and 4) Data and analytics insights based on a data-driven content strategy that supports 360-degree customer view. DT based OTTP's evolution supports radio and TV related *Entities,* to deliver long-term strategies, like the case of Radio and Television of Montenegro that was transformed and innovative information exchange opened up new business possibilities and that has also enabled competition by smaller media *Entities*. Video On Demand (VOD) based platforms have enabled the evolution of major players like Netflix, HBO… Such platforms are: Available world-wide, Provide excellent user-experience, Personalization, and Availability and DT based OTTP based media usage (Janinović, 2022; Mohan, 2017). *The internet*

*disrupted the media and entertainment value chain. Business, however, is essentially still based on the fundamentals of a century ago: engaging and retaining audiences. What is changing dramatically is the way to monetize the products and services you offer them...* Destroying barriers and frontiers enabled service broadcasters to have new roles as global media *Entities*, offering digital output by OTTPs. The major changes are the pace and breadth, where scalability, globalization, and mass. Content is the major challenge and as OTT evolves, *Entity's* production increases. But there is no guarantee for costumers' loyalty, and therefore all their touch-points are critical for 360-degree customer management and OTT's delivery models which focus on content and customers' experiences. Real-time analytics and just-in-time payments are critical in a DT based OTTP and related operations (Starinieri, 2018).

## Supporting OTTP and Media Operations

DT based OTTP and media operations' support operations and services that fill the gap between OTT and ISPs, mainly for multimedia streaming services, which is BTP's major Factor. Actual integration of OTTs and ISPs are complicated and there is the need to define an interface-based solution based on the design of Business Processes (BP), which can depict the integration of OTTPs in ISP operations. BPs are scenarios which present the integration and include information and recommendations collected from ISPs. Their legacy BPs were based on long-term infrastructural investments that depended on subscriptions. Due to deregulation and competition, traditional BPs are the past. Data consumption and competitiveness is growing, so ISPs are transforming their platforms. BPs offers the possibility to remodel ISP services and that facilitates the integration of OTTP services on ISP platforms and the integration of Cybersecurity aspects (Castanheira, & Fernandes, 2019). A BTP can tailor OTTP's Cybersecurity aspects that takes into account customers' fragmentation across various media-platforms. Integrating OTTP's security requirements in early DT's phases needs the following controls (Atos, 2019): 1) DRM that enables digital management; 2) Encryption of data and communication; 3) Securing e-Transactions; and 4) Protecting customers' privacy (data, connections…). OTT based video piracy needs *private property protection in the form of copyright infringement*, because *sharing* on the internet made pirating of videos easy, because of the use of attack vectors used to retrieve and redistribute content. Piracy on the internet is a threatening subject and results in significant financial damage (Akamai, 2020). OTTPs' integration is extremely complex and that needs *EA Concept* based OTTBGRW.

# EA CONCEPT'S INTEGRATION

## Multimedia Basic Characteristics

Multimedia interactive internet-based video environments have enormous transformative capacities; especially for enhanced online-video activities. The *Entity* has to implement a platform to support online-video collaborations. For that there is a need to identify the optimal *Patterns*, which can also be useful in design processes and activities. Design activities for groupwork cultivate collaborative design sessions. OTTP based groupwork is significant for problem-solving, and implementing innovation initiatives. BTP's problems and frequent changes in Business, ICS, and infrastructure, need aggregated solutions. The OTTBGRW can support team-members' autonomy and independent thinking, specific characteristics that are lacked in traditional concepts. The OTTBGRW supports such virtual groupwork and ensures: Access to information, Complex design, Manage tasks communication, and enforce skills related to architecture and modelling (Han, Park, Hua, & Childs, 2021).

## Basics, Architecture, Modelling, and Integration

The ADM based *EA Concept* is generic and manages all BTP's phases and iterations; it also supports the design of its ICS and OTTP parts and integrates the *Project Group Work*. The BTP can use existing frameworks like TOAGF and/or SAFe to manage changes. The *EA Concept* supports the *Project Group Work* and aligns: 1) Vision and principles; and 2) Various existing standards. *Project Group Work*'s instances can be used to automate OTTP's services and to optimize interaction with ISPs. *Entities* must transform their infrastructure to link to ISPs and thus improve OTTP's performances. The OTTBGRW needs to define the Unit of Transformational Work (UoTW) that is supported by *TRADf*BB's granularity definition, which in BTPs is a very complex task (Kim, & Lennon, 2017). The BTP uses the *1:1* mapping and classification concept, which supports that its online resources map to OTTP services. The *EA Concept* supports *Project Group Work*'s activities, which use BB(s) bundles in the form of UoTW, that facilitates the integration of the OTTP and that facilitates Streaming Architectures (SA).

## SA's Application

SAs for data, is OTTP's core component and due to the fulgurant growth of data-volumes (that includes data from non-traditional data-sources like IoT sensors, security logs, and internet applications) OTTPs must be scalable. SAs integrate open-source frameworks like Spark/Hadoop to provide end-to-end ASs built on scalability of cloud Datalakes. An SA is a blueprint that includes software components, processes, and data streams... SAs manage data streams' which tend to generate massive data that are mainly semi-structured. Data-streams' processing is a complex task and needs adapted SA solution that includes BBs which offer functions to transform these streams into analytics-enabled datasets; and as shown in Figure 7 an SA must include (Levy, 2022): 1) A Message Broker/Streaming-Processor;

*Figure 7. An SA interaction*
*(Levy, 2022)*

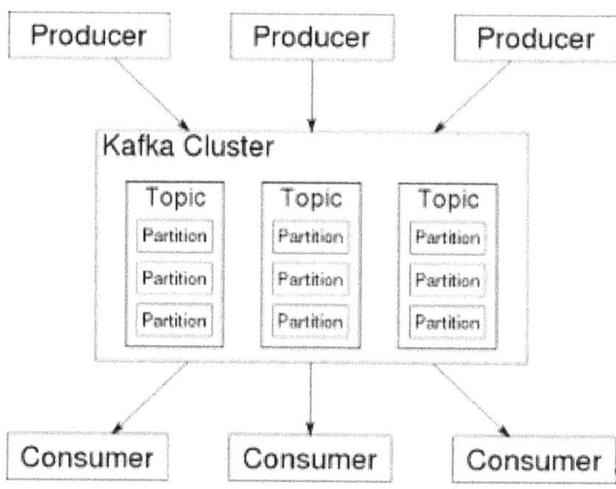

2) Batch/ETL support; 3) Data Analytics/Serverless Query Engine; and 4) Streaming data-storage. Streaming Data Analytics (SAD) is a process of extracting insights from data-streams; in which data is transformed into a tabular format. SAD main challenges are: 1) The nature of data which is constantly modified; 2) Data-volumes need to implement distributed architectures; and 3) Use streaming Structured Query Language (SQL). As shown in Figure 8, OTTP oriented SA and SAD includes deployments to various *Entities* and different ICS and media technologies, where the OTTP is *business-centric*. Value-chains are not ICS-centric but is a *self-service* environment/platform which automates SAD's actions like: Message ingestion; Batch and streaming Extract Transform Load (ETL) actions; Storage management; and Preparing SADs. SA's main advantages are to: Eliminate massive data-engineering projects; Improve OTTP performances, availability; Supports Cloud-based ICS/OTTP; Support agility; and Establish a Reference SA (RSA) (Levy, 2022).

*Figure 8. An SA overview*
*(Levy, 2022)*

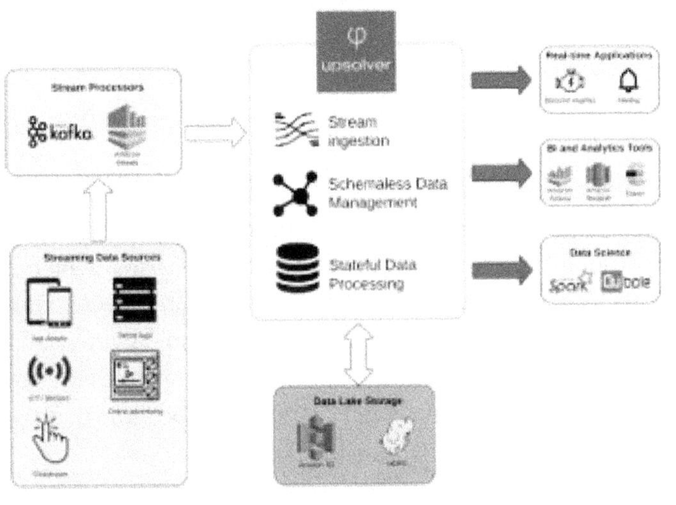


## RSA's Usage

A Quality of Experience (QoE)-centric service-delivery improves the collaboration between OTTs and ISPs. Quality assurance for customers is the OTTP's ultimate goal for the deployed multimedia-services and that requires the collaboration of various partners, like OTTs and ISPs. For that goal there is the need for a QoE driven approach that is based on RSA. An RSA requires modules and vertical interfaces for OTTs and ISPs collaboration. This collaboration model can be based on revenue's maximization which drives such a collaboration. Revenue depends on user-churn, which is related to QoE and an optimal collaboration strategy for the OTTPs' interactions. As shown Figure 9, the RSA includes: 1) Management of functional requirements, architecture and blocks; 2) Pricing modelling that defines economic rules, these rules show how much customers should pay and possible OTTP's congestion created by possible Smart Data Pricing (SDP); and 3) Optimal *Intelligence*.

*Figure 9. An RSA for the collaboration between ISPs and OTTs*
*(Ahmad, Floris, & Atzori, 2016)*

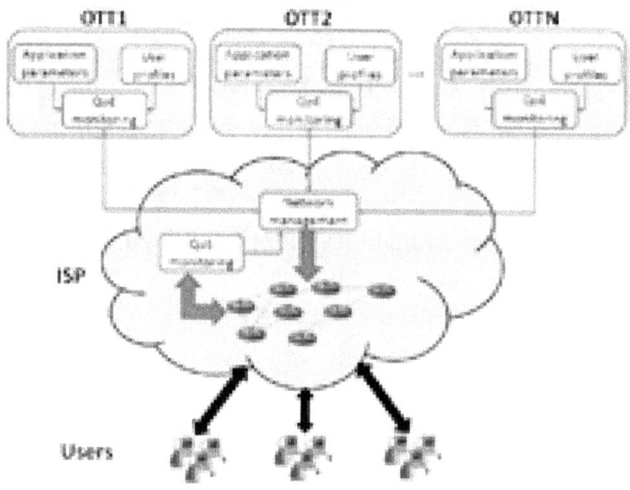


## Intelligence

*Project Group Work* is supported by *Intelligence* which delivers solutions to defined BTP and OTTP problems. The solutions recommend changes and optimize risks. *Intelligence* solves problems by using the HDT and manages different types of OTTP integration scenarios, which supports *Project Group Work* activities. *Intelligence* is supported by *Applied Mathematical Model*'s formalism and uses a holistic concept for delivering a set of *Project Group Work* recommendations. The *Project Group Work* interfaces *Intelligence*, in which various OTTBGRW solution templates are stored, then they are used by the HDT. The OTTP based *Entities* have today, efficient environments to support *Intelligence* like approaches and can enable *Project Group Work*'s blocks interactions

*Figure 10. An OTTP*
*(Calvello, 2023)*

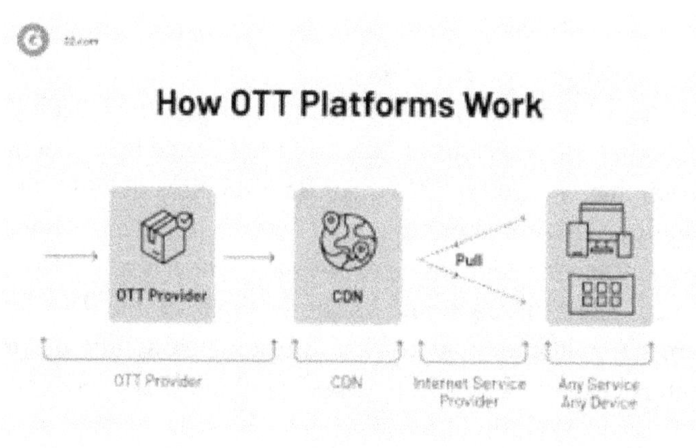

## PROJECT GROUP WORK'S MAIN BLOCKS

### OTTPs' Integration

An OTTP delivers films or TV content over the internet, using different types of devices. Brands select OTTPs to offer complete control over the content, user experience, and related data. Due to evolving use, more customers are leaving traditional TV services and are moving to OTT medias. There is a difference between OTTPs for streaming and traditional video-streaming. OTTP streaming is a direct-to-consumer video content platform which streams on various devices, while video-streaming the content is streamed across devices like computers, smart TVs, mobile devices, and OTT devices. The choice of an OTTP can be based on: Content delivery; VOD capabilities; Streaming functionality; Video monetization; Security tools; Data-analytics; Video library management; Customer support… OTTP viewers can access multiple distributors to have contents they're most interested in. IPTV based TV is no longer vital and it has performance issues and OTTPs must: 1) Organize content releases; 2) Offer filtering; 3) Integrate its own content delivery network; 4) Permit customers to create profiles; 5) Provide search functionality; and 6) Enable support in various APDs like OTTP based Education (OTTPbE) (Calvello, 2023).

### OTTPbE's Evolution

OTTPs has opened new cognitive horizons and youngsters spend about nine hours/day using streaming-medias. The enormous volume of content offers persons access to rich and diverse media and ISP/OTT providers are working on a strategy to offer full-accessibility. This has positive and negative side-effects on peoples' behavior, because of their exposure to massive-data content that can affect their attitude; and the main evolutions are: 1) Increased access to OTTPs which offer various instructional content and are remote; 2) Enhances creativity and social skills, where people can create content like videos, animations… It also provides the possibility to interact with others; 3) Exposure to inappropriate content like violence, sexually material… So there is the need for controls and disclaimers; 4) Increases sedentary habits which leads to decreased intellectual/academic performance, socialization, and physical activity; 5) Effects mental status because of exaggerated time spent on viewing content results in mental disorders like despair, anxiety… 6) Cyber-bullying and other online misdeeds; and it is difficult to monitor. Misdeeds can be cyber-stalking, exposure to inappropriate content, online predators… (ATC, 2023). OTT Video-on-demand (OTTV) has expanded beyond the entertainment industry in education and OTTPbE has proven to be feasible and that videos can be watched many times. The OTTPbE provides access to top educators and offers latest techniques and this interaction enhances memory retention. It also provides global high-quality education and enables to meet students' specific needs; but there are some resistances. OTTPbEs improve learning processes by using positive concepts like: Interactivity, Scalability, Costs reduction, Accessibility, and Efficient assessment (STEM, 2023). A OTTPbE oriented-system can handle massive students' activities, like in the case of India in which 260 million students enroll in more than 1.5 million schools, which need important quality improvements. These improvements need adapted ICS/OTTPs that include avantgarde tools/methods, OTTPbEs that support educators and students. Such a transformation of the education sector makes it more efficient and OTTPs can provide optimal OTTPbEs by offering (Mishra, 2022):

- Accessibility and affordability, especially in remote areas, because an OTTPbE doesn't depend on physical presence. A student can manage related materials and interact with educators that makes courses more affordable and makes education inclusive.
- Online support and efficient evaluation, that is due to improved OTTPbE's tools which offer possibilities to record classes, share presentations/voice/text, virtual attendance, transferring assignments, receiving feedback, and assessment tests are more objective.
- Gamification offered by OTTPbEs supports key gaming principles and application-oriented learning and is optimal for elementary and primary cycles.
- Placement is supported by preparing students for the professional-world and OTTPbEs offer upskilling courses in various fields; and it also supports Blended OTTPbE (BOTTPbE).

In BOTTPbEs, students choose the learning techniques that are best for them and BOTTPbEs enhance students' performance, achievements, and is the best strategy for the future of education. OTTPbE's investments are rising exponentially and that ensures future-focused learning methods because of the following reasons (Nasir, 2021):

- A one stop solution offers interactivity and a centralized repository, in which OTTPbEs support students and educators in *two-way interactions* and offer access to educational content.
- Data and feedback are offered by interactive tools communications in which collective data/resources are managed by the OTTP. Educators create and offer content, monitor students' participation and assess students' performance.
- Flexibility is supported by an adaptable, scalable, and integrated OTTP which enables educators' make learning interesting.
- Reducing costs in various ways like refocusing students towards online (or on-demand) features of the OTTPbE, reduces the use of infrastructure, such as classrooms, libraries…

*Figure 11. Advantages of OTTPbEs*
*(Mishra, 2022)*

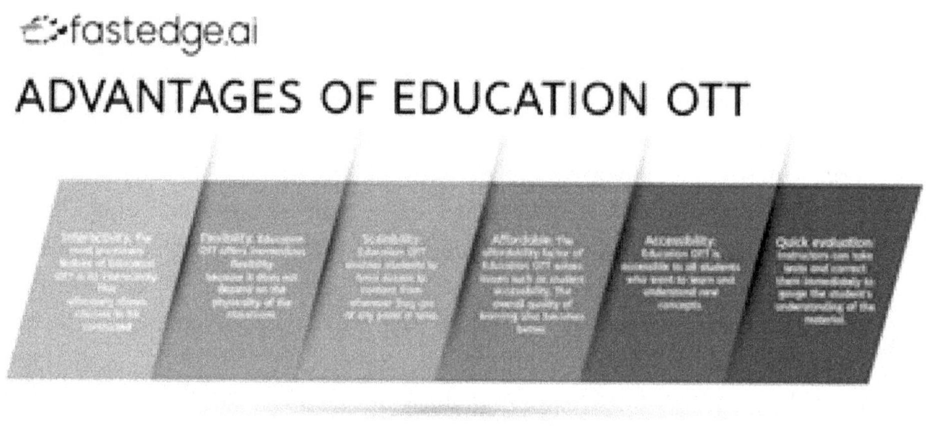

- OTTPbE's integration is feasible, because it requires that an ISP offers on-demand implementation capabilities for OTT based applications; but these facts depend on Telecom companies' (Telco) capacities and evolution.

## Telco's Evolution

In digitized *Entities* OTT based services are used and massive interactions have to be handled. Telcos offer products/solutions to deliver high-quality OTT services that are changing current business keytrends and opportunities in the media domain like in entertainment and OTT based sports industries. OTTPs are a growing trend in various traditional broadcasters and especially sports media industry. Customers have various ways to entertain as opposed to conventional TV which was the standard. Actual *on-the-go lifestyle* is a handcuffed to TV and a limited OTTP usage as this domain is transformed. Thus, versatility that is due to OTTP's content delivery capabilities imply that customers' requests can be largely fulfilled. Even if conventional TV will last for a while but the evolution of OTTPs has evident benefits because of their flexibility to comply with customers' needs, have a richer interactivity level, and offer inclusion. OTTPs' high-level of personalization capabilities and an ever-enhanced user experience, will make the difference. These facts will be difficult to follow by conventional TV systems, especially in APDs like sports, making this APD an important component for growth in the streaming services' industry. OTTPs let users watch-use content when they need to and from wherever they are. This type of specialized ADP content provider understands their customers and their experiences and that can make an OTTP for an APD exponentially grow. Sports content delivery delivers immersive viewing experience and is growing exponentially. OTTP providers will grow and the competition is tough, whereas conventional TV providers will continue to exist, adapt and incorporate limited OTT features and components in order to keep market-shares and are committed to deliver customers with quality content. The sports content APD and industry is currently aware of ever rising customers' requests. Added to these facts there are

*Figure 12. Growth by device*
*(IAB, 2016)*

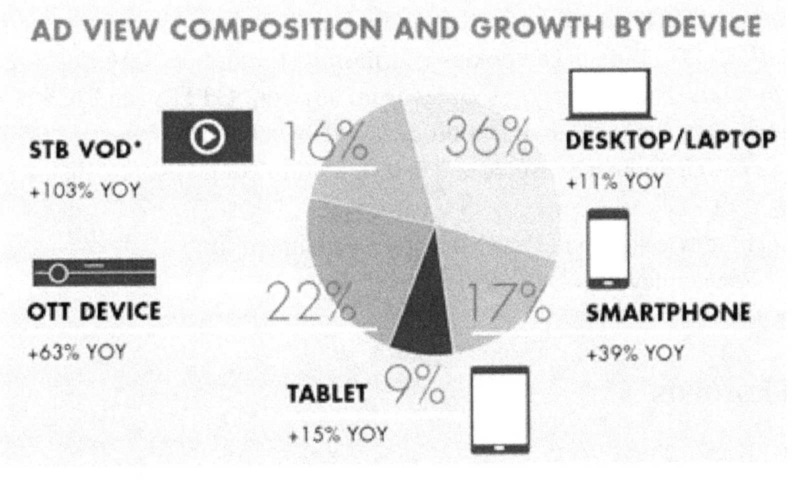

quickly rising numbers of OTTP players competing for customers which will push forward this industry and its evolution (Mullan, 2020).

## OTTP's Evolution

OTTPs' evolution is fulgurant, but nevertheless when choosing one the *Entity* or BTP must query the provider following questions (IAB, 2016):

- What's is the *Entity's* OTTP audience or clients in the sense of reach, frequency, and financial conditions ?
- On what type of OTTP is the *Entity's* content compatible and eventually available?
- Does the *Entity* offer or sell OTTP inventory independently of cross-platform package solutions?
- Does the *Entity* have OTTP audiences and other types of video audiences experiences and knowledge?
- Besides video media, does the *Entity* have any type of OTTP products or solutions that it can offer like video displays, sponsorships… ?
- What is the *Entity's* definition of and vision on fluidity (that is mainly related to TV publishers transforming inventory from one display/screen to the following on an individual action) ?
- How does the *Entity* manage and execute fluidity?
- How does the *Entity's* manage pricing and the differences between traditional and digital offerings and services?
- Does the *Entity's* guarantee and optimize target demos ?
- Can the *Entity's* support the management and measurement of cross-platform portability and de-duplication?
- What kind of measurement, integration, reporting and verification tools and environments does the *Entity* have ?
- How will OTTPs extend and enhance *Entity's* reach?
- What are *Entity's* real targeting capabilities and defined goals ?
- Can *Entity's* personnel access OTTP's inventory by using API channels or does it need external interfaces?
- Does the *Entity's* OTTP have the same capabilities as an internet web-browser ?
- Does the *Entity's* OTTP manage cookies and flashes; because of similar characteristics of OTT devices, digital ads, and 3rd party resources from adjacent OTTPs (and ICS components) which can generate various types of incompatibilities, which can create important OTTP problems.
- Is the *Entity's* OTTP audience diverse and is it limited to young people, males, or games' console audiences ?
- How much OTT devices (of video ad views) are accounted for ?
- How much video ad views on OTT devices has grown ?
- Define the set of OTT formats.

## OTTP's Used Formats

The OTTP has to define the set of possible media formats to be used and the issues which can play a role. Video formats are important because elements like codec and containers decide which video format the OTTP will use. Combining codecs and containers generate video formats and a codec is an encoder/decoder or can be a compressor (co)/decompressor (dec). Codec inputs a digital file-media and compresses it (or encodes it) for storage and decompresses it (or decodes it) for usage (viewing or transcoding). After codec's processing containers processing follows. Containers contain videos and also contain all the information related to the video and possible formats are (Banerjee, 2022):

- Matroska Video File (MKV) is a type of container that supports unlimited pictures, videos, and subtitle tracks, thus delivering complete package to *Entity's* customers.
- MP4 is the most common video container format related to uploading videos on OTTPs, and has many benefits, like video compression, encoding, and localization. MP4 videos can be used on multiple platforms.
- AVCHD is an optimal format for recording high-definition videos which can be persisted on multiple devices like DVDs, Hard-Disk (HD)… It is a format that is developed for HD videos only.
- The Audio Video Interleave (AVI) format was developed by Microsoft and was a part of its Video for Windows Technology. AVI format contains both audio and video data in a file container; and permits synchronous audio-and-video playbacks.

OTTP providers need strict regulations, policies and tax conditions, in order to protect business sustainability.

## Regulators, Policy Makers, and Tax Authorities

OTT based applications have become pervasive in improving communications, interactions, and groupwork; and digitization changed working concepts by using OTT based applications, like Zoom, Skype.... BPs have also been digitized and have become Cloud-based. ISPs are focusing on providing high-speed data accesses to support OTTPs which are at the epicenter of DTs and innovations, but stakeholders from different jurisdictions have concerns related to ICT/OTTP policies and regulations for OTT based applications. Figure 13 summarizes some important myths against evidence-based facts that are based on ACs within the context of the Internet Value-chain. OTTPs need clear acknowledgement and fundamental knowledge of Value-chains in Telcos and assessments of possible impacts. Governments and regulators receive demands to regularize OTTs, especially concerning voice and SMS revenues, which are causing an decline in sector revenues. ISPs are calling for regulators to (Stork, Nwana, Esselaar, & Koyabe, 2020):

- To share revenue with governments and ISPs.
- Review declining tax revenues due to declining sector-revenues from OTTs. Governments are looking for ways of tracking revenues derived from their citizens by OTTs.
- Broadcasters complain that streaming-services have unfair advantages, because streaming-services are not subjected to local broadcasting regulations
- Assess the status of OTT legislation and regulation, like the case of the Commonwealth ICT Ministers Forum mandated the Commonwealth Telecommunications Organization (CTO) that conducted a study on OTT applications to implement policy and regulatory mechanisms.
- ICS Value-chain has expanded and that requires regulatory tools and to consider regulations from other institutions and jurisdictions.

- Regulatory mechanisms concerning OTTs should not change
- New organizations and legislative frameworks are entering the market to address online customer protection, data privacy, and cybersecurity concerns.
- Cross-institutional interaction is critical because they are controlled and regulated by multiple organizations and laws.
- Global data-volumes usage.

*Figure 13. Myths vs facts in the OTT based applications debate*
*(Stork, Nwana, Esselaar, & Koyabe, 2020)*

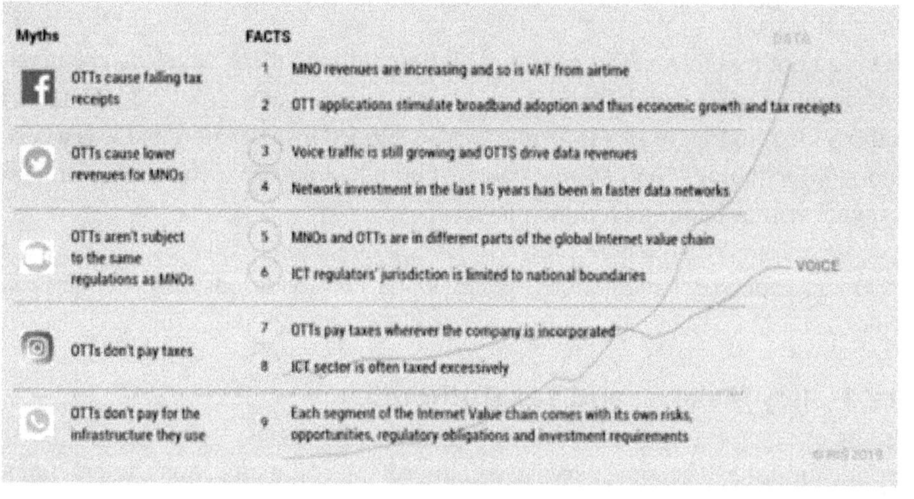


## Data Usage and Issues

Transforming OTT business needs data environments like Big Data & Analytics (BDA), which can support OTTPs and that includes (TCG, 2021; Wong, 2015; Kochar, 2022):

- OTT APDs' spending forecasts.
- UX's will enhance OTT based applications.
- Finds new business opportunities and gives the ability to differentiate customers.
- Finding new consumption patterns by fostering unmet audience needs.
- Support addressable advertising is less intrusive and relevant to customers.
- Real-time information is available from the OTT based applications on various devices
- Customers do not require direct involvement of Telcos and OTTPs are able to generate real-time reports on consumption.
- Manages information that is analyzed and fed in the BDA that constantly ingests data, cleans and enriches it in order to make it credible and usable.
- Supports Data-Science (DS) that delivers insights into customers' experiences which help ISPs adapt to increasing requests.
- Localizes customers' churn rate and creates detailed customers' viewpoints.
- Analysis of customers' market value by knowing customers better

- Segmenting customers on common traits such as demographics, geography, ….
- Classifying audiences into groups like people who like comedy series.
- Promoting high customer engagement by focusing on marketing campaigns.
- Improving advertising revenue by using personalization, advertisers can establish good relationship with customers.
- Using data clean rooms, which help share customer data with the advertising bodies to protect customers' data privacy.

## THE PROOF OF CONCEPT

This PoC is based on: 1) CSAs that include: RDP, *Applied Mathematical Model*, OTTP based APDs, ICS support, *EA Concept*'s integration, and *Project Group Work*'s main blocks; 2) The mentioned ACSs that are based on ICS/OTTP, various types of OTTP services; and 3) Proving how to implement *Project Group Work* using OTTP and BPs based application portfolio rationalization scenario.

*Figure 14. Order to payment BP*
*(Castanheira, & Fernandes, 2019)*

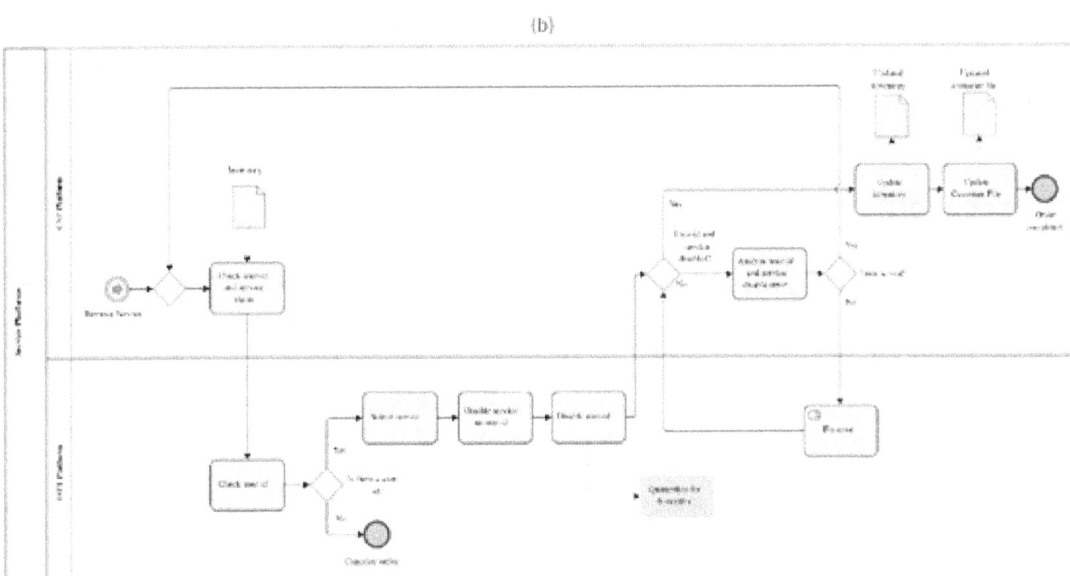


### Online Services Portfolio Rationalization Scenario, ICS Unification and CSFs

This PoC selects *Project Group Work*'s sets of Factors to satisfy BTP's requirements and implements needed services and BPs, as shown in Figure 14. The ACSs have common *Entities'* strategic BTP goals as shown in Figure 15.

*Project Group Work*'s executed operations in ADM phases, are: 1) In *EA Concept*'s Phase A (or the Architecture Vision) sets up an OTTP based BTP roadmap; 2) In Phase B, sets up *Entity's* target archi-

*Figure 15. BTP goals*
*(Jonkers, Band & Quartel, 2012)*

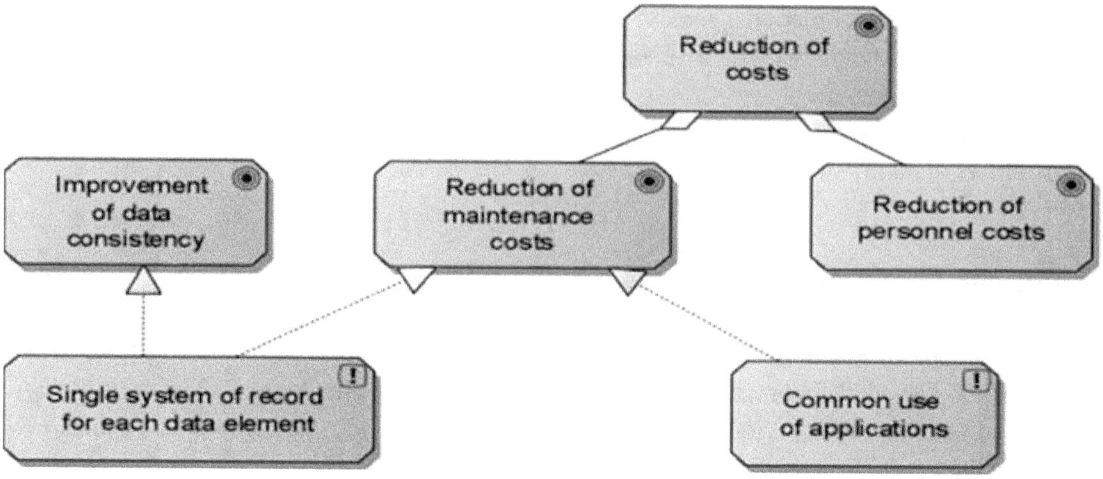


tecture and the set of OTTP's requirements definition(s); 3) In Phase C (or the Gap Analysis phase) that models the target OTTP based application landscape; 4) Phase D (or the Target Technology Architecture and Gap Analysis phase) implements OTTP's infrastructure and design models; 5) In Phases E and F phases (or implementation and migration planning) defines the transition architecture and intermediate situation, it also evaluates BTP's status. This PoC focuses on BTP's capabilities to support the PoC's execution process.

## PoC's Execution

The PoC is implemented by *TRADf* and uses an *Applied Mathematical Model* instance, where The *Project Group Work* interfaces *Intelligence* that uses Factors. Factors evaluation is presented in Table 1. The *Project Group Work* requires integrational methodologies and artefacts like Factors, *EA Concept*, SAFe, and *TRADf*. The *Project Group Work* also defines relationships between the OTTBGRW artefacts like the set of OTTP requirements and ICS BBs. The PoC was implemented using *TRADf* client's interface, where the starting activity is to setup *Project Group Work* Factors. Once the implementation-interface was activated, the scripting interface was launched to implement the needed BBs and BPs to process defined CSAs. After starting *TRADf's* client, the sets of Factors were selected and linked to HDT nodes and sets of BBs. The developed scripts are linked to the *Applied Mathematical Model* instance. The *Applied Mathematical Model*-based HDT uses BBs that are called by *Intelligence's* actions. The *Project Group Work* and its Factors, actions, were setup to be used, then the scripts were executed. This chapter's decision-table and its results concludes PoC's 1st phase, as shown in Table 1. Table 1 shows clearly that the *Project Group Work* can be used in BTPs. *TRADf* and *Applied Mathematical Model*'s constraint to implement the *Project Group Work* is that CSAs having an average result below 8.5 will be ignored. In the case of Factors' evaluation has an average result below 7.5, it will be ignored. Table 1 has the result of 8.60 which implies that *Project Group Work*'s integration is possible.

*Table 1. The project group work RDP's outcome is 8.60*

| CSA Category of CSFs/KPIs | Transformation Capability | Average Result |
|---|---|---|
| RDP's Integration | Usable-Mature | From 1 to 10 9.0 |
| OTTP Based APDs | Transformable-Possible-Mature | From 1 to 10 9.0 |
| ICS Support | Transformable-Possible | From 1 to 10 8.5 |
| IHEAC'S Integration | Transformable-Possible-Complex | From 1 to 10 8.00 |
| BTPOTTBGRW'S Main Blocks | Transformable-Possible | From 1 to 10 8.5 |

Evaluate First Phase

## CONCLUSION AND RECOMMENDATIONS

In the author's previous RDPs, the most significant recommendation was that the BTP must be managed by a Polymathic and holistic Architect of Adaptative Business Information Systems (AofABIS). The recommendations for the *Project Group Work* are based on the processing of Factors to assert its feasibility, coordination, and effectiveness. In this chapter the focus is on OTTPs and *Project Group Work* which try to transform *Entities'* multimedia capacities and reduce costs. There has been much developed in the domain of multimedia, but the author recommends the implementation of a *Project Group Work*. The main barriers rely on the fact that actual ISP and OTT providers deliver silos' approach and *Entities*' lack experiences in OTTPs and generic BTPs. This RDP delivers a list of recommendations on the manner on how to proceed with BTPs where the OTTBGRW proposes a virtual groupwork concept. The integration of the *Project Group Work* needs that ICS, ISP, and OTTP platforms are ready to be used to glue the various *Entity's* media services. There are many resources, formats, and solutions on media platforms' transformation, but the author recommends to verify OTTBGRW's capacities to improve media platforms. This chapter's resulting recommendations can be used to support BTP teams help *Staff* deliver optimal media platforms/services and use the OTTBGRW capacities. *Project Group Work*'s recommendations, and an IH transformation framework like *TRADf*, round up the proposed concept needed for dynamic media platforms, and the needed roadmap for integrating *Project Group Work* with: OTTPs, *Patterns*, *EA Concept*, OTTBGRW, SAFe. The RDP for verifying the *Project Group Work* is based on Factors' processing which resulted from RLR's various sources, and RLR's inputs were inputted in the HDT. The main focus is on *Project Group Work*'s feasibility, capabilities, coordination, and OTTBGRW's possible side effects. *Project Group Work*'s characteristics and preconditions are essential Polymathic and holistic management of media platforms/services. This RDP proposes the *Project Group Work* which is capable of synchronizing various BTP activities and proposes to check why transformation initiatives fail in the implementation phase. The score of 8.6, which is shown in Table 1, shows that *Project Group Work*'s integration is possible, but is not a simple project. This RDP's main recommendations are:

- The PoC checks *Project Group Work*'s feasibility, which should replace traditional media platforms/services.
- BTP's implementation phase is the cause for failure, therefore there is a need an optimal IH transformation framework.
- The *Entity* must use *Patterns*, and OTTBGRW solutions.
- The RLR proved the existence of an important knowledge gap between the legacy media platforms/services and OTTP based ones.
- An ARbLP and ROE based HDT can be used to create the initial *Project Group Work* concept.
- Traditional media platforms/services cannot cope with the media's evolution.
- The *Applied Mathematical Model* based RDP CSA is possible and mature.
- OTTP based APDs, ICS support CSA is transformable and mature.
- *Project Group Work*'s main blocks CSA is transformable and mature.
- *EA Concept*'s integration CSA is complex and can cause failure.
- *Project Group Work* coordinates *Patterns* integration to deliver EA blueprints, and BBs.
- *Project Group Work* uses methodologies like SAFe, to coordinate all BTP's activities.
- The PoC proved the research feasibility and delivered the recommendations on how to integrate the *Project Group Work*.

An IH transformation framework supports the BTP and *Project Group Work* and delivers a set of technical and managerial recommendations.

## REFERENCES


Ahmad, A., Floris, A., & Atzori, L. (2016). *QoE-centric Service Delivery: A Collaborative Approach among OTTs and ISPs. Computer Networks. DIEE.* University of Cagliari., doi:10.1016/j.comnet.2016.09.022

Akamai. (2020). *Protecting the Bank of OTT-White Paper*. Akamai.

ATC. (2023). *IMPACT OF OTTPS ON YOUNG MINDS*. ATC Online. https://atconline.biz/impact-of-ott-platforms-on-young-minds/

Atos. (2019). *Tailoring cybersecurity for the Broadcast Media industry*. Atos.

AWS. (2022). *Service OTT: la voie de l'innovation vidéo basée sur le cloud-Création d'une plateforme pour réussir sur le marché du streaming et de la diffusion directe au client (D2C)*. AWS.

Banerjee, D. (2022). *The Best Video Format for OTTPs: Decoded*. Muvi One. https://www.muvi.com/blogs/the-best-video-format-for-ott-platforms-decoded

Botturi, L., & Belfer, K. (2014). *PEDAGOGICAL PATTERNS FOR ONLINE LEARNING*. ResearchGate.

Brightcove (2023). *OTT Syndication API*. Brightcove. https://apis.support.brightcove.com/ott-syndication/index.html

Calvello, M. (2023). *How OTTPs Are Changing the Content Game*. G2. https://www.g2.com/articles/ott-platforms

Castanheira, C., & Fernandes, P. (2019). *Future Operations Support Services for Digital Operations Transformation*. OSS FOR DIGITAL OPERATIONS TRANSFORMATION.

Deloitte. (2024). *Digital transformation for telecom operators*. Deloitte. https://www2.deloitte.com/ch/fr/pages/technology-media-and-telecommunications/articles/digital-transformation-for-telecom-operators.html

Dogan, S., Çalgici, P., Arditi, D., & Gunaydin, H. (2015). *Critical success factors of partnering in the building design process. METU JFA 2015/2*. Department of Architecture, İzmir Institute of Technology.

Easterbrook, S., Singer, J., Storey, M., & Damian, D. (2008). *Guide to Advanced Empirical Software Engineering-Selecting Empirical Methods for Software Engineering Research* (F. Shull, Ed.). Springer.

FastEdge. (2023). *How does educational OTT work*. FastEdge. https://www.fastedge.ai/how-does-educational-ott-work/#:~:text=Educational%20OTT%20can%20be%20thought,through%20web%2Dbased%20discussion%20boards

Giachetti, R. (2012). *A Flexible Approach to Realize an Enterprise Architecture*. Department of Systems Engineering, Naval Postgraduate School, Monterey, CA USA. Presented: New Challenges in Systems Engineering and Architecting. Conference on Systems Engineering Research (CSER). St. Louis, MO. 10.1016/j.procs.2012.01.031

Han, J., Park, D., Hua, M., & Childs, P. (2021). Is group work beneficial for producing creative designs in STEM design education? *International Journal of Technology and Design Education*. doi:10.1007/s10798-021-09709-y

Hine, G. S. C. (2013). The importance of action research in teacher education programs. In *Design, develop, evaluate: The core of the learning. environment. Proceedings of the 22nd Annual Teaching Learning Forum, 7-8 February 2013*. Murdoch University. http://ctl.curtin.edu.au/professional_development/conferences/tlf/tlf2013/refereed/hine.html

Hinkelmann, K. (2016). *Modelling in Enterprise Architecture. University of applied sciences northwest Switzerland*. Business School.

Hosiaisluoma (2022). *Holistic Enterprise Development*. Hosiaisluoma. https://www.hosiaisluoma.fi/blog/archimate-examples

Hussain, O., Dillon, Th., Chang, E., & Hussain, F. (2010). Transactional risk-based decision making system in e-business interactions. [CRl Publishing ltd Computer Systems. Science & Engineering.]. *International Journal of Computer Systems Science & Engineering*, *2010*(1), 15–28.

IAB. (2016). *Over The Top (OTT) Video: An Overview*. IAB., iab.com/ott

InfoSys. (2023). *CASE STUDY: Digital streaming service provider implements OTTP*. InfoSys. https://www.infosys.com/industries/media-entertainment/case-studies/ott-platform.html

Janinović, D. (2022). *DIGITAL TRANSFORMATION STRATEGY OF RADIO AND TELEVISION OF MONTENEGRO FOR THE PERIOD 2023-2027*. OSCE Missions.

Jonkers, H., Band, I., & Quartel, D. (2012a). *ArchiSurance Case Study*. The Open Group.

Kersten, M. (2021). *Introducing SAFe® 5*. Your Operating System for Business Agility. Scaled Agile Inc.

Kher, A. (2022). *Case Study: Incorporating all the OTT streaming into one platform*. UX Planet. https://uxplanet.org/case-study-incorporating-all-the-ott-streaming-into-one-platform-d3e5b1fbdd59

Kim, J., & Lennon, Sh. (2017). *Descriptive Content Analysis on E-Service Research. International Journal of Service Science, Management, Engineering, and Technology (IJSSMET)*. IGI Global., doi:10.4018/IJSSMET.2017010102

Kochar, M. (2022). *How Is Data Science Powering OTTP to Multi-Dollar Business*. Data Science. https://blog.learnbay.co/how-is-data-science-powering-ott-platform-to-multi-dollar-business

Koshy, J. (2022). How technology & transformation are putting the T's together in OTT. *iTWire*.

KPMG. (2017). *The 'Digital First' journey-How OTTPs can remain 'on-demand ready*. KPMG.

Levy, E. (2022). *Streaming Data Architecture in 2022: Components and Examples*. Upsolver. https://www.upsolver.com/blog/streaming-data-architecture-key-components

Llewellyn, R. (2023). *Digital Transformation Use Cases in 10Well-Known Industries*. Privacy.

Ming, Y., & Nikolaos, B. (2009). Analysis, Modeling And Simulation Of Network Traces For Video Transmission Over Ip Networks. Jacksonville State University, AL, USA

Mishra, A. (2022). *How OTTPs Help in Education*. PrepInsta.

Mohan, A. (2017). *Digital Transformation: How is OTT Redefining Media*. CASBAA, the Association for digital multichannel TV.

Mullan, L. (2020). *Massive Interactive: How over-the-top contentservices are shaking up teleco and mediaindustries*. IT Procurement. Technology Magazine. https://technologymagazine.com/it-procurement/massive-interactive-how-over-top-content-services-are-shaking-teleco-and-media-industries

Nasir, M. (2021). *5 Reasons Why OTTPs Will Be The New Education Normal For Blended Learning*. IPSB Technology. Kuala Lumpur. https://www.ipsb.com.my/reasons-ott-platform-blended-learning/

NEC (2015). *Best Practices Guide for Over-the-Top (OTT) Deployment*. UNIVERGE Cloud Services. NEC.

OMG. (2022). *DECISION MODEL AND NOTATION (DMN)*. OMG. https://www.omg.org/dmn/

Pappas, Ch. (2022). *Cloud Based e-Learning Platforms: 10 Advantages eLearning Professionals Need To Know*. Docebo. https://www.docebo.com/learning-network/blog/cloud-based-elearning-platforms/

Patil, R. (2022). Streaming Boom Accelerates Digital Transformation Of Network Operators. *Forbes*. https://www.forbes.com/sites/forbestechcouncil/2023/02/01/streaming-boom-accelerates-digital-transformation-of-network-operators/?sh=7885cd34554e

Pisonia, G., Gijlersc, H., Ha Nguyend, T., & Chene, H. (2021). Collaboration patterns in students' teams working on business cases. *CEUR Workshop Proceedings (CEUR-WS.org)*. CEUR.

Polderman, J., & Willems, J. (1998). *Introduction to Mathematical Systems Theory: A Behavioral Approach. (Texts in Applied Mathematics)* (2nd ed.). Springer Verlag. doi:10.1007/978-1-4757-2953-5

Postle, G. (2003). Online Teaching and Learning in Higher Education: A Case Study. Department of Education, Science and Training. University of Southern Queensland, USQ. Canberra. Australia.

Putri, N., & Yusof, S. M. (2009). Critical success factors for implementing quality engineering tools and techniques in malaysian's and indonesian's automotive industries: An Exploratory Study. *Journal Proceedings of the International MultiConference of Engineers and Computer Scientists.*, *2*, 18–20.

Sankaralingam, K., Ferris, M., Nowatzki, T., Estan, C., Wood, D., & Vaish, N. (2013). *Optimization and Mathematical Modeling in Computer Architecture*. Morgan & Claypool Publishers.

Scaled Agile. (2022). *Succeeding with AI in SAFe*. Scaled Agile, Inc. https://www.scaledagileframework.com/succeeding-with-ai-in-safe/

Sinha, A. (2021). *Case study: Reimagining the OTT experience*. Bootcamp. https://bootcamp.uxdesign.cc/reimagining-the-ott-experience-ui-ux-case-study-6cc42c18b2ca

Spencer, M. (2016). Agile EA – the MVP – enterprise architect-Building the Business Value Chain. *Enterprise Architect*. https://enterprisearchitect.wordpress.com/2016/04/11/agile-ea-the-mvp/1/2

Starinieri, M. (2018). *Atos in Media Pathways to digital transformation*. Atos.

STEM. (2023). *How Does Educational OTT Work?* STEM. https://www.stem.org.uk/resources/community/collection/511682/how-does-educational-ott-work

Stojanov, I. (2015). *Scaling agile using scaled agile framework. Master*. Eindhoven University of Technology. Netherlands.

Stork, Ch., Nwana, H., Esselaar, S., & Koyabe, M. (2020). *Over The Top (OTT) Applications & the Internet Value Chain Recommendations to Regulators, Policy Makers and Tax Authorities*. Commonwealth Telecommunications Organisation.

Syynimaa, N. (2015). *Enterprise Architecture Adoption Method for Higher Education Institutions*. [Doctoral Thesis, Informatics Research Centre Henley Business School University of Reading].

TCG. (2021). *Transforming OTT Business with Big Data & Analytics. A TCG Digital Whitepaper*. TCG Digital.

TeleStream. (2023). *What is OTT?* TeleStream. https://www.telestream.net/video/solutions/what-is-ott.htm

Thune, T. (2011). Success Factors in Higher Education–Industry Collaboration: A case study of collaboration in the engineering field. *Tertiary Education and Management*, *17*(1), 31–50. doi:10.1080/13583883.2011.552627

Trad, A. (2015a). *A Transformation Framework Proposal for Managers in Business Innovation and Business Transformation Projects-Intelligent aBB architecture*. Centeris.

Trad, A. (2015b). *A Transformation Framework Proposal for Managers in Business Innovation and Business Transformation Projects-An ICS's atomic architecture vision*. Centeris.

Trad, A. (2022d). *Educational Transformation Project's Remote Group Work (ETPRGW). Shaping Online Spaces Through Online Humanities Curricula*. IGI-Global.

Trad, A. (2023b). *Business Transformation Projects: The Role of Psychology-Based Resistance (RPbR). Book: Considerations on Education for Economic, Social, and Environmental Sustainability*. IGI Global.

Trad, A. (2023d). *Organizational and Digital Transformation Projects-A Mathematical Model for Building Blocks based Organizational Unbundling Process*. IGI Global.

Trad, A. (2023e). *Academic and Educational Transformation Projects-The Role Team-Based Learning in Polymathics (RTBLP)*. IGI Global.

Trad, A., & Kalpić, D. (2014b). *The Selection, and Training Framework selection and training framework (STF) for Manager's in Business Innovation Transformation Projects–Educational Recommendations*. Journal EDEN.

Trad, A., & Kalpić, D. (2020a). *Using Applied Mathematical Models for Business Transformation. IGI Complete Author Book*. IGI Global. doi:10.4018/978-1-7998-1009-4

Trad, A., & Kalpić, D. (2022a). Business Transformation Projects based on a Holistic Enterprise Architecture Pattern (HEAP)-The Basics. IGI. USA.

Trad, A., & Kalpić, D. (2022b). Business Transformation Projects based on a Holistic Enterprise Architecture Pattern (HEAP)-The Implementation. IGI. USA.

Trad, A. & Kalpić, D. (2022c). *Business Transformation Project's Holistic Agile Management (BTPHAM) CBER 2022*. Research Gate.

Trad, A., & Kalpić, D. (2023a). *Educational Transformation Project's Remote Group Work (ETPRGW). Book: Shaping Online Spaces Through Online Humanities Curricula*. IGI Global.

UI. (2023). *Two Keys to Marketing Success: Case Studies on OTT Media*. UI. https://fia.ui.ac.id/en/dua-kunci-sukses-marketing-studi-kasus-pada-ott-media/

Wong, S. (2015). *Over The Top TV (OTT), Big Data and The Future Of Predictive Buyer Analytics*. DXC.

Zandia, F., & Tavana, M. (2011). *A fuzzy group multi-criteria enterprise architecture framework selection model. Management Information Systems, Lindback Distinguished Chair of Information Systems, La Salle University, Philadelphia, PA 19141*. Elsevier.

## APPENDIX

### Glossary Defintions

ACS: Applied Case Study
AHMM: Applied Holistic MM
AHMM4OTT: (or Applied Mathematical Model) AHMM for OTT's Integration
AI: Artificial Intelligence
APD: APplication Domain
AR: Action Research
ARbLP: Based Learning Process
*BTMM:* BTP Transformation Mathematical Model
BTP: (or *Project*) Business Transformation Project
BTPOTTBGRW: (or *Project Group Work*) A BTP can be supported by the OTTBGRW.
DT: Digital Transformation
DS: Data-Science
EA: Enterprise Architecture
EERM: Empirical Engineering Research Model
HDT: Heuristics Decision Tree
IHEAC: (or EA Concept) EA IH Concept
GRW: GRoup Work.
ICS: Information and Communication Systems
IH: In-House
IHI: In-House Implemented
KMS: Knowledge Management System
LP: Learning Process
MM: Mathematical Model
RLR: Researched Literature Review
OCP: (or *Pattern*) OTT based Collaboration Pattern
OTT: Over-the-Top
OTTBGRW OTT: based Platforms and Group Work.
OTTP: OTT Platform
OTTPbE: OTTP based Education
OTTV: OTT Video-on-demand
PHMM: Polymathic-Holistic Mathematical Mode
PIGRW: Polymathic OTT based Interdisciplinary GRW
PoC: Proof of Concept
QoE: Quality of Experience
RDP: Research and Development Project
RSA: Reference SA
SA: Streaming Architectures
*TRADf*: Transformation Research Architecture Development framework
UoTW: Unit of Transformational Work

# Chapter 14
# OTT and Live Sports – Indian Premier League 2023 and its Impact on the Future of Sports Broadcasting in India

**Josraj Augustine Arakkal**
https://orcid.org/0000-0002-4153-2762
*Symbiosis Institute of Media and Communication, Symbiosis International University, India*

**Preetha Menon**
https://orcid.org/0000-0002-5054-7536
*Flame University, India*

## ABSTRACT

*The battle between JioCinema and Star Sports for live sports viewership reached its peak with Indian Premier League (IPL) 23. In stage one, this chapter uses the case study method to present the story of IPL 23 viewership in the light of free subscriptions. Data was collected through secondary sources like newspaper articles, online newsfeeds, syndicated reports, and expert comments over a period of six months. This data was then analysed on the basis of audience size, advertiser dynamics, and advertising revenues for both JioCinema and Star Sports. Focusing on free subscriptions as a tactic and the sustainability of such promotions on the long-term standing of an over-the-top platform (OTT), an OTT platform performance model (PPM) is proposed in stage two of this chapter. The model offers a five-pronged approach to long term sustainability of an OTT platform.*

## INTRODUCTION

Sports viewership through over the top (OTT) platforms is growing globally (Kwak et al., 2023). The Indian market for live telecast of sports is also sizable and growing. Such acceleration in growth has been driven by the evolving technology in the distribution of content (Bhattacharyya et al., 2022). The viewership has evolved from government owned channels on Doordarshan to private players like Star and

DOI: 10.4018/979-8-3693-3526-0.ch014

Sony in the linear or traditional television format. With the advent of OTT, the audience suddenly could decide on watching a live cricket match between a linear TV option like Star Sports and an OTT digital streaming platform like Disney Hotstar; or between Sony Six and Sony Live for a wrestling bout. The ease and convenience of watching OTT in terms of when, where and how came into play. The phenomenon of connected TV brought in a new layer to the way the audience consumes content. This chapter brings in the unique case of free live sports telecasts of a very premium sports property, namely, the edition no. 16 of Indian Premier League (IPL) 2023 in the age of subscription based (SVOD) models. In this chapter, we present a conceptual framework for its existence and sustenance in this dynamic marketplace.

The recent FICCI-EY Media & Entertainment Industry report for 2022 threw up some interesting data. While about 700 million consumers watch sports on traditional TV (either free or pay), over 150 million consumers view sports through the OTT platforms (free or bundled with data). (*Windows of Opportunity India's Media & Entertainment Sector-Maximizing across Segments*, 2023). Interestingly enough, the OTT audience base is already about 20% of the larger TV viewership base for sports.

In the past, cable or satellite TV were the default selectors or transmission agents of sports content, especially while live streaming. However, in just a few years OTT has taken over this market by storm. OTT in its nature of delivering content directly to the consumer bypassed cable, live-streaming and satellite TV distribution platforms, in terms of the number of users by the year 2019 (Bhattacharyya et al., 2022). New media democratized the media landscape by giving its consumers power and tools to shape their own content pathways (Chipp & Chakravorty, 2016). There is no doubt that sports consumers preferred live streaming over delayed views. However, new media like OTT offered advantages for every kind of sports consumer.

## FACTORS LEADING TO THE ADOPTION AND CONTINUED USE OF OTT PLATFORMS

Several factors led to the adoption and use of OTT platforms. One of the greatest catalysts overarching all other reasons of OTT adoption was Covid-19. The pandemic was pivotal in lifestyle changes like work-from-home, complete closure of other entertainment outlets like malls and theatres and a general movement of city dwellers to their home towns and villages. Admirable internet connectivity and a lack of other forms of entertainment were the prime reasons for why customers turned toward OTT platforms during the pandemic, for all their entertainment needs. In this segment, we explain specific factors that have led to the adoption of OTT platforms at length. This is followed by the reasons for continued use of the platform by customers.

*Convenience:* OTT offered the convenience of watching movies and series from the comfort of one's homes (Yoon & Chun 2023). Since the primary viewing device also changed from television to mobile phones, the convenience also came in the form of mobility. Customers could now view their favourite shows on-the-go. The flexibility of watching shows and movies at one's own pace also led to its popularity (Saini and Gupta, 2022). It made new terms like 'binge watching' commonplace amongst viewers. Ease of accessibility, also termed as playability is another key feature of OTT that has led to its adoption (Shah and Mehta, 2022; Yoon & Chun 2023).

*Perceived usefulness:* The next important factor for the adoption of OTT is its perceived usefulness. Several studies have applied the technology acceptance model (TAM) to decipher adoption behaviour of OTT consumers (Jeong et al., 2022; Singh et al., 2023). These studies discuss the technological aspects of

OTT and its popularity amongst consumers due to a positive perception of usefulness. Variety of content is another reason for why consumers consider the OTT as useful (Saini and Gupta, 2022). Typically, any viewer may have two or more genres of preferred content and the OTT platforms are developed to visibly display the genres and the content under them in a user-friendly fashion.

*Quality of content:* For television viewers, OTT was a dream come true in terms of quality of content. With each OTT platform developing cutting edge original content, the quality of content improved drastically (Saini and Gupta, 2022). It also provided legitimate 24/7 access to global cinema which was otherwise only available on pirated platforms or during international film festivals. Higher budget production and global actors, also led to an improvement in its quality.

*Social influence:* Despite its obvious usefulness and ease of use, one of the factors that really lead to conversion was social influence (Shah and Mehta, 2022). Gen Z viewers played an important role in convincing the buyers (their parents) in switching from television to OTT platforms. The need to stay in touch with new content and the fear of losing out were other offshoots of social influence visible amongst OTT subscribers. Customers were hugely influenced by online reviews while picking platforms and shows (Sharma & Lulandala, 2023) and the power of the community lead to adoption of OTT for many.

*Affordability:* The rise or fall of an OTT platform subscription is also directly impacted by its affordability. As in the case of any new service, customers are wary of the price they may have to pay to access OTT. Providing competitive low prices was a popular strategy adopted by most OTT platforms, driving customers to subscribe to annual plans of such platforms (Sharma & Lulandala, 2023). Thus, affordability is an important factor for OTT adoption.

*Ad free content:* "free of commercials" was another important factor for choosing OTT over regular television (Xu et al., 2023). Evidently, these values have been shaping consumer choices across program genres including sports. However, this might change soon with the advent of AVOD options on OTT platforms.

*Risks:* There are also some deterrents to the adoption of OTT platforms. All of these can be grouped under the header of risks. Risk could be of several kinds and customers' fear and vulnerability to each of these risks were the reasons for their hesitation in adoption OTT. Some of these risks are security risk, cost risk and technical risk (Yoon & Chun 2023). Viewers feared that since OTT could be accessed only through the internet much of their private viewing data would be captured by the platform exposing them to security risks. Passage through internet was also the reason for the risk of technical glitches. Viewers feared that their viewing experience would be limited to the access and speed of the internet leading to a poor viewing experience.

Apart from the above factors that were pivotal in the adoption of OTT platforms, it is also crucial to understand the reasons for continued use of such platforms. Below are listed some of the most popular reasons for customers to continue with their OTT subscriptions.

*Perceived value:* One of the top reasons for continued usage of OTT is the perceived value attached in continuing to stay with the platform. The factors grouped under perceived value are perceived ease of use, perceived customization benefits, mobility (Dhiman et al., 2022) In some senses this resonates with some of the important reasons for adoption as well. Thus, we can easily say that price drops and affordability can get customers in, however to keep them OTT platforms must provide them with value that is embedded in quality. Continued usage of an OTT platform also came from dedication (Hyun and Ku, 2023) which is developed due to personalization and content diversity. OTT platforms with the best options to create personal profiles and preferences topped the chart for continued use.

*Search experience:* Another reason for why customers stay with an OTT platform is its search experience (Jaiswal and Singh, 2022). Despite available content, if one has a poor search experience, the chances of customers dropping off the platform becomes high. This new finding in OTT research is useful to platform providers, trying to keep customers from switching to cheaper or newer OTT platforms.

*Screen watching time:* The hours spent in front of a screen is also directly proportion to chances of their continued usage of an OTT platform (Jaiswal and Singh, 2022). The more invested consumers are in a particular platform higher the chances that they will continue to stay with the platform. Increased screen watching time is also indicative of their ability to find shows and content of their choice on the platform and their affinity towards that platform.

Resistance to switching: Lastly, consumer's inherent inertia to switching platforms also increases their continued usage of an OTT platform (Hyun and Ku, 2023). Through their research, Hyun and Ku demonstrate a significant impact of consumer's resistance to switching platforms as a reason for them to stay within their comfort zone of existing platforms. This could be because they are satisfied with the platform or despite it, since the real reason for not switching is a personality trait rather than a platform feature/quality.

## Jio Cinema and IPL 2023: The New OTT Giant on the Block

The OTT industry has been contemplating several strategies for monetization and growth. Subscription based video on demand (SVOD), Advertisement based video on demand (AVOD) and Transactional based video on demand (TVOD) are some options available to OTT platforms to monetize the viewership. In the light of this, a free telecast of IPL 2023 was both disruptive as well as innovative in its offering.

The IPL telecast of 2023 was available free on the OTT platform - JioCinema. JioCinema is an Indian OTT media streaming service that is owned by Viacom18. Viacom 18 is a joint venture between Reliance Industries and Paramount Global. Before JioCinema was launched there were established players like Netflix and Amazon Prime ruling the market (Nagaraj et al., 2021). Each of these platforms had created a niche of themselves. Netflix is known for global shows and original content while Prime is preferred for their regional content in India. In the given scenario, a new kid on the block would find it challenging to find its footing. However, this was not the case for JioCinema. They decided to take the market by storm with their big bang announcement of airing IPL 2023 free of cost on their live streaming platform JioCinema.

IPL has been an iconic event for the T20 format cricket enthusiasts. In its 16$^{th}$ season in 2023, the satellite television telecast rights and digital streaming rights for IPL were allotted to two separate players (for the first time), namely Star Sports (Star India) and Jio Cinema (Viacom 18) respectively. This needs to be seen in the perspective of both the rights coming at around the same cost – Rs. 23,575 crores for the satellite television rights and Rs. 23, 758 crores for the digital rights (please cite). These rights won through a bidding process with BCCI (Board of Control for Cricket in India) cover a 5-year period from 2023 to 2027. Despite shelling out the huge monies, JioCinema chose to put it out to the audience without any subscription charges, while Star TV as a pay TV service came at a certain cost to the consumer. This is indicative of deep pockets and an aggression in the market that is out to wipe out competition. In the previous five years, Disney Star was the sole television and digital rights holder for the IPL and both involved a subscription fee. Rightly so, JioCinema has therefore proven to be a giant in the share of eyeballs for IPL 2023.

## OBJECTIVES OF THIS STUDY

The four research questions for this study are stated below. These four questions can be viewed as three overarching objectives. How did JioCinema's pricing offer affect the viewership dynamics between the two platforms JioCinema and Star Sports.

1. How did JioCinema's pricing offer affect the viewership between the two platforms JioCinema and Star Sports?
2. How did JioCinema's pricing offer affect the revenues garnered by the two players?
3. How sustainable is the "free" model for OTT platforms? What is the best possible strategy for long term viewership?
4. When does platform adoption lead to continued viewership amongst Gen Z customers?

The purpose of the study is therefore: to explore and understand the following:

- To understand the role of free subscriptions of premium sports properties and how it would impact the various stake-holders
- To examine the importance of other factors such as technology and experience to viewership base of live sports telecasts
- To develop a strategy for viewership retention and expansion on OTT platforms

## METHODOLOGY

This study employs a multi-method approach to answering the research questions and fulfilling the objectives stated above. In stage 1, we score through secondary data available in newspaper articles, online newsfeeds, syndicated reports and expert comments to build a nuanced understanding of the viewership and advertising dynamics between the two players – JioCinema and Star Sports. Then in stage 2, we progress to develop a conceptual model to predict the factors important to retain a viewership base for OTT platforms. Finally, we conduct a primary study amongst Gen Z viewers of IPL 23 to validate the model proposed in this chapter.

### Stage One

Stage 1 uses a case study approach, nested in the context of IPL 2023 and the play-out of strategies between OTT players, JioCinema and Star Sports. As defined by Creswell (2014), "case studies are qualitative design in which the researcher explores in depth a program, event, activity, process or one or more individuals". The case study approach is apt in the context of IPL 2023 as there is very little published research papers to gather real-time data. This research follows a descriptive method of case study. In a descriptive case study, a phenomenon (in this case IPL 23) is enumerated extensively with real-world evidence (Yin, 2014). The base data was gathered from published secondary data around the months the IPL 23 was broadcasted. It covered a period of roughly six months from the time of the announcement of distribution rights to the time of post-IPL analysis. The research set out to answer the questions around the viewership on both platforms and the revenue they generated through advertising.

It is set in the context of JioCinema offering free subscription to its viewers while Star Sports was a paid subscription service.

## Stage Two

In this stage, the authors develop a conceptual model titled the OTT Platform Performance Model focusing on the key factors important for an OTT platform to sustain competition and grow. It provides a formula to succeed, based on the learnings from the case study, building a holistic model for sustaining in the market.

## Findings

The findings of this study are presented in chronological order of the stages.

## Stage One Findings

### The Industry Narrative: Looking Back, Looking Ahead

The IPL edition of 2023 was not just a battle between the10 teams vying for the title. There was an off-the-field battle being fought between the two broadcasters – JioCinema streaming on OTT against the Star Sports bouquet of satellite television channels. It was a battle royale both for the audience eyeballs and the advertising revenue. It also needs to be stated that Disney Star had enjoyed both the rights on their OTT, namely Disney Hotstar and satellite sports channels under Star TV until the previous season. The breaking news in all this was Jio's decision to offer the premium cricketing event running into 53 days over 74 matches as a freebie!

### Audience Size

There is no doubting the crowd pulling ability of this extravaganza called IPL and the numbers it did indeed generate. A report in Economic Times indicated the respective viewership numbers for Jio Cinema and Star Sports at 44.9 crore and 50.5 crore. (Das, 2023). This still indicated Star to have a lead exceeding 5.5 crore viewers. These were however numbers put out by the respective platforms and not by an independent audit. Figure 1 is a creative representation of the numbers stated above. Disney Star effectively became the first cricket tournament to register over half a billion viewers according to the same report. "From last ball nail-biters to high-scoring thrillers, from the addition of the impact player rule to the emergence of future stars, from the last spot on the playoffs being decided in the last league stage match to the tournament winner being decided on the final ball, this edition of IPL has been special in many ways. Countless memories will be etched in cricket history including the fact that is the first time that half a billion viewers tuned in for a Cricket tournament," was the sentiment shared by Sanjog Gupta, Head-Sports at Disney Star.

The lead in audience exposure through Star Sports can be explained by the gap between smartphone and television penetration. The smartphone penetration is around 460 million as pointed out by the

*Figure 1.*

industry reports like FICCI-EY and Affle, which is about 50% of the television penetration in India. Further, India's muted fast-internet reach along with data cost would have come in the way for OTT. In the context of the above limitations, JioCinema did make a significant impact. JioCinema was expected to sweep the audience off their feet. The numbers support this confidence, with approximately 46% market share in viewership for IPL 23 in the first year of its launch.

## Sponsors and Advertisers

Just like the viewers, the advertisers too were caught up between the television and digital broadcasters. The industry experts were clearly expecting a swing towards digital (Moniz, Financial Express, 2023). A report in Business Today indicated a fall in the numbers of advertisers by 42% on Star Sports (Lidhoo, 2023). Startups like the edtechs were a major fallout from the previous year's advertisers. Brands like CRED, Paytm, Swiggy, Acko, BYJU'S, which until the previous year were huge on TV, gave it a miss in 2023. The number of advertisers had dropped from 81 to 47. At the same time, there were several advertisers like Amazon, Samsung, Indeed, TVS, Castrol, Noise, Puma, etc., who were exclusive to JioCinema.

The same article in Business Today, a leading magazine in India, also had interesting quotes from the industry experts. An important trend is narrowcasting (targeted ads) instead of broadcasting, brand expert Manish Porwal explains. "Overall, digital and TV audiences have separated. Because of digital, which allows for brands to pick and choose its audience, it has allowed different kinds of brands to get in. The ability of a client to place his ad in his required slot has gone up and hence narrowcasting has gone up," says Porwal, MD of Alchemist India. Karan Taurani, Senior VP-Research Analyst at Elara Capital., says that digital is way more acceptable of local advertisers and niche advertising verticals, which are not that relevant for TV. "Traditional verticals are more skewed towards TV. They will have 70 per cent of their ads coming on TV," according to Taurani.

Jio with its wide accessibility and sharper targeting had managed to shake up the market. The unique advertisers with them were a testimony to the growing power of OTT.

## Advertising Revenue

This is the litmus test to determine the success of any media property or event. Unfortunately, in the public domain, no reliable data has emerged on the same. According to Red Seer Strategy Consultants, the revenue across both the broadcasters combined was Rs. 4,700 crore. (Annex. G) However, they did not provide the split. A report in Financial Express stated that Star Sports was selling the ad at around Rs. 15 lakh per 10 seconds while the rates on connected TV was 6 lakh per 10 seconds (Moniz, Financial Express, 2023). Ad rates for the digital mode were running at Rs. 200- 250 CPM (cost per thousand impressions) and at Rs. 300 or higher for targeted advertising, said the same report based on the inputs received from senior media buying executives. However, there was no clarity on the overall advertising monies generated by the two players. Another report in Financial Express did estimate the IPL revenue on advertising including the Women IPL as well at Rs. 6000 crore (Moniz, Financial Express, 2023). There they did provide a split, indicating Rs. 3,500-4,000 crore for the television channels and Rs. 2,000-2,500 crore going into digital. This translates roughly into 60-70% of the revenue accounted for by television.

Jio Cinema as a media business unit falls under Viacom 18 Sports and they have been reportedly bold and aggressive in their statements on their strengths. Speaking at a media round table, Anil Jayaraj, CEO of Viacom18 Sports, stated that the company plans to eliminate all barriers to consumption across affordability, language and access by streaming the IPL on its JioCinema platform for the first time with free access for viewers. In the same interview he stated that the ambition was to provide high quality, premium long-form content on digital platforms (Moniz, Financial Express, 2023). This did not necessarily mean that it would be free of cost for life. There was an indication that the free lunches may end soon and JioCinema would also move in to a subscription-based model or at least a freemium model which combined free with premium. This means that some of the content would still be free while some others would come at a premium.

## Long Term Sustainability of the ‘Free’ JioCinema Model

Reliance as the parent company of JioCinema is known for their deep pockets and aggressive push in every market segment they get into. Their approach was no different in their foray into the telecast of live sports with IPL. At the same time, it is not clearly a sustainable model. But the rationale seems to create enough disruption and take ownership of the market. Ahead of the 16th season of the Indian Premier League (IPL), Viacom18 Sports outlined its objective of becoming “the largest aggregator of audiences in India” (Moniz, Financial Express, 2023). They seem to be sporty enough to burn some money in the beginning for a long-term dominance of the media and entertainment sector. They seem to have created enough damage for Disney Star to consider offloading the satellite television channels. The more recent reports indicate their wanting to find a buyer for the entertainment platforms (Reuters, 2023). Interestingly Disney Star is now planning to telecast Asia Cup and World Cup (both big ticket limited overs cricket tournaments) free on mobile (Srivastava, 2023). Jio is already creating ripples in the market and the way the competition is pitching itself. Have the Disney Star fallen into a trap?

How important is pricing? What are the other factors at play? It becomes critical to evaluate the success of an OTT platform beyond the role of pricing. We shall look at some of the other variables that go into deciding the popularity of an OTT platform.

In answer to these questions and as a way to look forward, we propose an OTT Platform Popularity Model (PPM) in the next section. The model looks beyond free subscriptions to a holistic model for sustainable existence. The five-pronged approach proposed in this model offers OTT platforms a formula to succeed in the current markets, cluttered with players vying for viewership and advertising budgets.

## Stage Two Findings

### OTT Platform Performance Model: A Holistic Understanding of the Important Factors

The OTT Platform Performance Model is a forward-looking model for OTT platforms that wish to sustain in this competitive market. It goes beyond short term tactics like price promotions and free subscriptions to provide a holistic understanding of the important factors for OTT platforms to cultivate a loyal audience base.

### Programming Content

The quality of the programmes would be the key in generating traction. Although all OTT platforms host a plethora of genres and content types, it is important to build on any one strength to gain audience mindshare. Netflix owns the space in terms of originals. Amazon Prime owes its popularity for its movies and regional content. Sony Liv is winning the battle in terms of regional programming. Disney Hotstar has got diluted recently in their popularity with it losing IPL and HBO rights to Jio.

### Brand Personality

Each of the OTT players have been trying to differentiate and find a niche for themselves. Netflix with its premium stands out as an elitist offering. Zee Five, meanwhile, is seen as too massy. Amazon Prime has a rich library of Bollywood movies. Sony Liv is clearly making an effort on its niche genre offerings. While OTT platforms focus on content, they should also focus on create recognizable distinct brand personas for themselves. These brand personas will be detrimental in developing long term strategies for the brand and also help in day-to-day operational choices made by the brand.

### Pricing

The categorization as we have seen earlier is between AVOD and SVOD on the pricing strategy. Even Netflix which is a SVOD might be going the AVOD route. Figure 2 represents the data from FICCI EY 2023 report on the Media & Entertainment industry. It details all the segments with respect to their pricing models and the size of these segments. The AVOD is clearly a larger base at 77 million homes as against a SVOD base of 43 million (FICCI EY, 2023)

## Audience Profile

Who is the core viewer of the channel? The advertiser is buying into the audience and not just the platform. While there is limited amount of third-party data on the viewership aspect around OTT, it would be very crucial for brands wanting to associate with the platforms. Many brands would seek a certain skew in the audience profile be it Gen Z, millennials or so on. Thus the choice of the audience and their

*Figure 2.*

profile is a strategic input while creating the brand persona for the OTT platform. For instance, an OTT platform with a largely Gen Z audience must build a brand that will appeal to Gen Z.

## Advertising Potential

The AVOD business enables the platform to go easy with their pressure on subscription. It explains why a Disney Hotstar is more reasonably priced on subscription as compared to Netflix. This advertising rate is typically a function of the audience base and its profile.

Together, this 5-pronged OTT Platform Performance Model provides a formula that is ideal for the long-term sustainability of the OTT platform. However, a word of caution is that while the model is ideal for the current market conditions, this may change dynamically over time. It is therefore advised to understand the current circumstances as detailed in stage 1 of this chapter into cognizance while absorbing the OTT PPM model. The backdrop of IPL 23 and the market disruption strategy of providing free subscriptions to live sports by JioCinema presented a unique case at hand. This in turn highlights the need for a strategy that is beyond free subscriptions. The OTT PPM is a solution to that end.

# CONCLUSION

The impact of the aggressive pricing strategy on the penetration and popularity of the platform and how it might affect the various stake-holders in the long run forms a critical area of investigation of this study. This chapter also explores the possible migration of audience and advertisers from one platform to another. While affordability and accessibility lead to a higher adoption of an OTT platform, its far-reaching implications on the Indian Media & Entertainment Industry specifically with regard to OTT business models and its sustenance are the main focus of this study. Would this be the last nail in the coffin and mark the beginning of the end for traditional linear television in the long term?"

The OTT P model proposed in this chapter, will shed light on the seminal factors that lead to the success of an OTT platform, with special emphasis on live sports telecasts. Furthermore, it provides pointers for sustained leadership of the OTT platform, despite the plethora of choices for the viewer. It will provide direction to OTT platform owners, to build a robust strategy for leadership and growth. It also provides direction to nations other than India that are yet to experience such a phenomenon. Additionally, it adds to the knowledge base of strategic pricing and promotion of emerging technologies such as OTT, paving the way for future research based on the conceptual model.

## REFERENCES


Bhattacharyya, S. S., Goswami, S., Mehta, R., & Nayak, B. (2022). Examining the factors influencing adoption of over the top (OTT) services among Indian consumers. *Journal of Science and Technology Policy Management*, *13*(3), 652–682. doi:10.1108/JSTPM-09-2020-0135

Chipp, K. F., & Chakravorty, D. (2016). Producer push to consumer pull: Who curates new media content? Developing strategies for new media environments. *Journal of Product and Brand Management*, *25*(4), 373–386. doi:10.1108/JPBM-06-2015-0918

Creswell, J. W. (2014). *Research design: Qualitative, quantitative, and mixed method approaches* (4th ed.). SAGE Publications.

Dhiman, N., Singh, A., & Sarmah, R. (2022). How continuous intentions towards over the top platform are framed? Stimulus–organism–response model perspective. *Vision (Basel)*, 09722629221104202.

FICCI EY. (2023). *Windows of opportunity: India's media & entertainment sector - Maximising across segments*. FICCI EY.

Jaiswal, S., & Singh, A. (2022). OTT Services and Intention to Continuous Subscription: A Moderated-Moderated Mediation Analysis. *PURUSHARTHA-A journal of Management. Ethics and Spirituality*, *15*(2), 104–125. doi:10.21844/16202115206

Jeong, D., Lee, M., & Park, E. (2022). Antecedents of Consumer Adoption of Over-the-Top Services in South Korea. *International Journal on Media Management*, *24*(3), 121–136. doi:10.1080/14241277.2022.2124996

Kwak, K. T., Lee, S. Y., Ham, M., & Lee, S. W. (2021). The effects of internet proliferation on search engine and over-the-top service markets. *Telecommunications Policy*, *45*(8), 102146. doi:10.1016/j.telpol.2021.102146

Lidhoo, P. (2023, May 28). Why advertisers are caught up between the tv and digital broadcasters. *Business Today*. https://www.businesstoday.in/magazine/the-buzz/story/ipl-2023-why-advertisers-are-caught-up-between-the-tv-and-digital-broadcasters-380778-2023-05-10

Moniz, C. (2023, February 14). 9000 Crore in advertisement spend riding on cricket 2023. *Financial Express*. https://www.financialexpress.com/business/brandwagon-9000-crore-in-advertisement-spend-riding-on-cricket-in-2023-2980844/

Moniz, C. (2023, February 4). To swing in favor of digital. *Financial Express*. https://www.financialexpress.com/business/brandwagon-ipl-2023-to-swing-in-favour-of-digital-viacom-2971263/

Moniz, C. (2023, March 31). Bandwagon tv advertisers stare at slow pitch as Ipl begins. *Financial Express*. https://www.financialexpress.com/business/brandwagon-tv-advertisers-stare-at-slow-pitch-as-ipl-begins-today-3027757/

Reuters. (2023, July 12). Disney considers selling or partnering its India business source. *Financial Express*. https://www.financialexpress.com/business/brandwagon-disney-considers-selling-or-partnering-its-india-business-source-says-3166592/

Saini, C. P., & Gupta, N. (2020). Interrelated factors driving the purchase of over-the-top television subscription services: A study using exploratory factor analysis and the decision-making trial and evaluation laboratory method. *Applied Marketing Analytics*, *6*(1), 73–84.

Shah, S., & Mehta, N. (2023). Over-the-top (OTT) streaming services: Studying users' behaviour through the UTAUT model. *Management and Labour Studies*, *48*(4), 531–547. doi:10.1177/0258042X221137438

Sharma, K., & Lulandala, E. E. (2023). OTT platforms resilience to COVID-19 – a study of business strategies and consumer media consumption in India. *The International Journal of Organizational Analysis*, *31*(1), 63–90. doi:10.1108/IJOA-06-2021-2816

Singh, N., Arora, S., & Kapur, B. (2022). Trends in over the top (OTT) research: A bibliometric analysis. *VINE Journal of Information and Knowledge Management Systems*, *52*(3), 411–425. doi:10.1108/VJIKMS-12-2021-0316

Srivastava, A. (2023, August 2). *Best Media Info*. Best Media Info. https://bestmediainfo.com/2023/08/as-asia-cup-world-cup-goes-free-on-mobile-hotstar-plays-on-volume-slashes-mobile-ad-rates

Xu, M., Li, B., Scott, O. K. M., & Wang, J. J. (2023). New platform and new excitement? Exploring young educated sport customers' perceptions of watching live sports on OTT services. *International Journal of Sports Marketing & Sponsorship*, *24*(4), 682–689. doi:10.1108/IJSMS-07-2022-0140

Yoon, J. H., & Kim, H. K. (2023). Why do Consumers Continue to Use OTT Services? *Electronic Commerce Research and Applications*, *60*, 101285. doi:10.1016/j.elerap.2023.101285

# Chapter 15
# Digital Streaming Content:
## Changing Social Perceptions

**Vidya Deshpande**
*Bennett University, India*

**Nithin Kalorth**
https://orcid.org/0000-0001-7969-7637
*Mahindra University, India*

## ABSTRACT

*The way we consume video content has undergone a radical transformation. Gone are the days of VCRs and brick-and-mortar video lending stores. Today, digital streaming platforms using the internet have become a global business. The seeds of streaming were sown in the 1990s when audio compression technologies like MP3 came into being and services like RealPlayer and Microsoft's Windows Media Player allowed users to listen to music of their choice. This laid the groundwork for video streaming. The 2000s saw the launch of digital video streaming, changing the entertainment landscape forever. YouTube, founded in 2005, paved the way for user-generated content. Netflix, initially a DVD rental service, switched to an unlimited viewing streaming service in 2007, based on a monthly model. These platforms faced challenges like bandwidth and piracy, but fueled by technological advancements and an insatiable consumer demand for on-demand entertainment, they thrived.*

## DIGITAL STREAMING CONTENT: CHANGING SOCIAL PERCEPTIONS

The way we consume video content has undergone a radical transformation. Gone are the days of VCRs and brick-and-mortar video lending stores. Today, digital streaming platforms using the internet have become a global business. (Zhao, 2022) The seeds of streaming were sown in the 1990s when audio compression technologies like MP3 came into being and services like RealPlayer and Microsoft's Windows Media Player allowed users to listen to music of their choice. This laid the groundwork for video streaming.

DOI: 10.4018/979-8-3693-3526-0.ch015

The 2000s saw the launch of digital video streaming, changing the entertainment landscape forever. YouTube, founded in 2005, paved the way for user-generated content. Netflix, initially a DVD rental service, switched to an unlimited viewing streaming service in 2007, based on a monthly model. These platforms faced challenges like bandwidth and piracy, but fuelled by technological advancements and an insatiable consumer demand for on-demand entertainment, they thrived.

Since 2010s, there has been an explosion of streaming services, each vying for a share of the growing market. Hulu entered the fray in 2008, catering to TV shows, while Amazon Prime Video and HBO Max launched in 2014 and 2020, respectively, offering original content and established library titles. These platforms to invest heavily in original content. Shows like "House of Cards" (Netflix), "The Handmaid's Tale" (Hulu), and "Game of Thrones" (HBO) redefined viewing, blurring the lines between traditional TV broadcasting and streaming.This "streaming war" has led to concerns about content fatigue and platform fatigue, as users struggle to keep up with the sheer volume of offerings.

The first OTT platform in India, BigFlix by Reliance was launched in 2008[1]. But it wasn't until Netflix launched in 2016 (FICCI & EY India, 2020) that the hunger for fresh content grew. OTT platforms offer a mix of films, series, documentaries, and other formats served to the audience on a "view-what-you-want, when-you-want" menu.

This has led to discernible socio-cultural impacts on society. Cultivation theory holds that the more time a person spends watching television (of all kinds) the more he or she will adopt the predominant outlook of the world that is expressed on the medium. (McQuail 2010). Early cultivation research findings (Gerbner and Gross, 1976) showed that the more television people viewed, the more likely they were to exaggerate the incidence of crime in the real world and the personal risks they run.

In today's world the theory can be tested not only with relation to TV but also content on OTT & social media platforms. And the cause and effect may still be the same as it was five decades ago.

Thirteen years after her death, Princess Diana is in the headlines once again. Thanks to the *Netflix* series, *The Crown, Season 4*, old wounds in the British royalty were re-opened. Season 4, aired in November 2020, retold the fairy tale "marriage of the century" of Prince Charles and Lady Diana Spencer. The series spun the fairy tale even further by distorting some facts and showing Prince Charles and his then mistress Camilla (now wife) in poor light compared to Princess Diana.

For the British public, memories of the Princess Diana era had dimmed and there was acceptance of Prince Charles and Camilla, especially after they got married. "However, the latest season of *"The Crown"* reignited people's animosity toward the couple and created an entirely new generation of Princess Diana fans," reported *The Business Insider.*[2]

The Instagram handles of royal family and Clarence House (Official accounts of Charles and Camilla) were flooded with comments in support of Diana. "Many of the comments were simply "Diana" with the heart emoji, though some wrote messages that more explicitly critiqued Camilla," *The Business Insider* reported. "Diana is the real MVP," one user wrote, while another wrote "Don't like her" in a reference to Camilla.

In fact, a week later, PR agents handling the royal social media accounts, shut down the comments option after trolls started getting vitriolic. A classic case of viewers believing TV's fiction instead of historic facts, the basis of Gerbner's Cultivation Theory.

Cultivation theory suggests that television, as the prevalent medium in society, can foster certain perceptions about the nature of society by repeatedly portraying society in a certain manner. (Gerbner, 2002) Extrapolating on the same theory, Gerbner et al, contend, "The facts. (which are evidently learned quite well) are likely to become the basis for a broader world view, thus making television a significant source of general values, ideologies, and perspectives as well as specific assumptions, beliefs, and images.

A hypothesis that is so well explained by the "end of the world" psychosis that swept the UK in 2020. On September 13, 2020, *BBC One* viewers watched wildlife filmmaker David Attenborough's documentary, *Extinction: The Facts*. And on September 14 morning Twitter timelines were flooded with the message of fear.[3] One viewer, according to entertainment news website, *DigitalSpy,* tweeted: *"#ExtinctionTheFacts* a painful, terrifying watch. Hard to draw any positive conclusions or hope from this catalogue of human failure. *@BBCOne #ClimateChange* Is it all too late?" Another tweeted: "A very tough watch. So many harrowing images. Who on earth would want to bring a child into this world with what is potentially waiting for them? *#ExtinctionTheFacts."*

*The Daily Mirror (UK)* said, "David Attenborough took *BBC One* viewers on an emotional roller-coaster in his latest moving documentary as he discussed the increasing risk of global pandemics like coronavirus and the possible extinction of over a million animals[4]"

Attenborough's documentary did warn viewers, but his intention was to send a message to humanity to act. In his swansong, *Life on our Planet*, released a week later on the same channel, *BBC One* and *Netflix*, he ends with a similar message: "I do truly believe that, together, we can create a better future. I might not be here to see it, but if we make the right decisions at this critical moment, we can safeguard our planet's ecosystems, its extraordinary biodiversity and all its inhabitants.

"What happens next is up to every one of us."

Gerbner says that "high frequency viewers of television are more susceptible to media messages and the belief that they are real and valid." [5] In the case of Attenborough's documentary, the viewers obviously missed the message for the woods. Later, many celebrities urged viewers to watch the documentary as a wake-up call, instead of being frightened by it. The then British PM, Boris Johnson's wife Carrie Symonds, in a video message said, "Everyone should watch this. It is massively difficult to watch - I cried - but it is vital. We cannot ignore it. It's time to act."

Cultivation Theory itself has undergone several changes since it was first proposed. But it has survived over time and proved durable. Explaining the theory in her paper, Mosharafa says: Cultivation theory tackles the long-term effects of television on viewers. The theory proposes that the danger of television lies in its ability to shape not a particular viewpoint about one specific issue but in its ability to shape people's moral values and general beliefs about the world.[6]

Public reaction to both *The Crown* series on *Netflix* and Attenborough's *Life on the Planet* further reinforce what Gerbner said in the 70s. When it comes to beliefs like death, researchers found that television programmes are filling the knowledge gap individuals have about death (Sharf, Barbara & Freimuth, Vicki. 1993). The same can be said about other forms of information: historical facts, scientific facts, etc as we have seen with two recent episodes on how people believed that the 6th mass extinction was impending following the Attenborough documentary and the belief that Diana did no wrong and was the perfect princess or a victim, as portrayed by *The Crown.* Both were proved wrong, but many viewers continue to believe so.

Media theorist Denis McQuail argues[7] that television *can* teach about social reality and that the relationship between viewing and social reality may be reciprocal: television viewing causes a social reality to be constructed in a certain way, but this construction of social reality may also direct viewing behaviour.

And it's the same with content on OTT platforms.

Digital streaming platforms or OTT (Over The Top) platforms have had significant impact on viewers. The surfeit of availability as well as the watch-when-you-want kind of menu offered, has changed the way content is consumed. Advances in internet services and better video compression technologies and cloud storage have helped the streaming business fulfil the insatiable need for on-demand entertainment. "Streaming media is considered to be an updated generation of media forms and new inventions in the digital and internet era. By challenging the traditional distribution methods of television programs and films, streaming media uses a customized and personalized strategy, applying recommendation systems to actively communicate directly with audiences and sell media viewing services and marketing brands." (Tang & Wei, 2023)

There has been an explosion of streaming services, with 200+ streaming platforms around the world, including 47 in India (with an emphasis on regional language platforms). The Ormax OTT Audience Report 2023, shows total audience for streaming platforms at 481.1 million at the end of 2023. Streaming apps allow viewers to watching multiple episodes at one go, instead of the traditional appointment viewing, i.e. watching an episode at a given hour on a particular day of the week. Till the beginning of the 2000s, viewers had to wait for a certain day of the week and time slot to watch their favourite TV serial. Streaming platforms, with their to view ready content, has changed that.

Binge watching is a recent trend in which people watch TV for hours without a break. Oxford Dictionary describes binge watching as "watching multiple episodes of a TV programme in rapid succession, typically by DVD or digital streaming."

*"Whereas previously people would watch hours of television per day in relatively synchronous manners, those hours are now divided increasingly asynchronously, allowing for much more viewer choice as to what they watch and when they watch it." (Conlin, 2016).*

According to *Forbes* magazine: "The shift from TV-style appointment viewing to streaming has been significant, especially for younger people. Over 50% of streaming viewers are under 35; viewers ages 60 and up said they spend only 14% of their entertainment time with streaming services, but those ages 15–29 said they spend 22% of their entertainment time with streaming." (Fatemi, 2022)

One company that capitalized on the concept of binge-watching was Netflix. In 2103, it conducted a survey among 1,500 TV streamers (online U.S. adults who stream TV shows at least once a week) to study viewing patterns. The company found that binge watching is a widespread behaviour among this group, with 61% binge watching regularly. They defined binge watching as streaming between 2-6 episodes of the same series in one sitting. (Netflix, 2013). "While many TV fans can admit to lost weekends devoted to binge watching, it isn't necessarily always the no-sleep, up-all-night fest as previously illustrated. In fact, many viewers who binge-watch see it as the new normal with no guilty feelings." (Pena, 2015)

In the new media environment of content surplus, viewers can control their choices. They decide what they want to watch and when, instead of being dictated by the programming of a channel.

Recent trends show that young people wait for streaming platforms to upload all episodes in one go and then watch them in one or two sittings. "The bigger fear among younger people is the "fear of missing out' (FoMO). FoMO seems to be driving many of the changes that are evident in modern media consumption; part of the enjoyment of watching TV is being able to talk about it with friends, and being part of an overarching, cultural conversation. This is particularly evident in serial dramas, where plot points and twists are likely to be prematurely revealed the longer a person waits to watch current episodes." (Conlin, 2016)

The instant gratification and viewer control over TV episodes is the biggest advantage. While some may see that as a disadvantage, many of today's viewers don't have the patience to wait for episodes to be available one at a time. "Binge-watching accelerates the consumption of a television series, making some shows more enjoyable -- yet other particular shows may be better received when given enough time to digest." (Pena, 2015).

Independent brand strategist and author, Ambi Parameswaran, in an article in *Business Standard,* says "We are now in the throes of the OTT revolution. The pandemic has turned even the humblest among us into binge-watching OTT addicts." (Parameswaran, 2022) He looks at appointment viewing with a new lens. He feels people watch a show as soon as it drops on OTT to gain social currency with their social group. "This the first-day-first-show craze is not something that will go away just because the show is available for longer. What OTT, however, is going to do is extend the life of the content," he says.

Studies done by doctors say binge-watching can have both positive and negative effects on the health of young viewers. "Binge-watching can be both harmonious and highly entertaining as well as an obsessive and compensatory behaviour. People are driven by multiple motivations for binge-watching, including those of a social and cognitive nature, by transportation into the narrative, or willingness to pass the time, but also to escape daily life problems, to regulate negative emotions, and to gain instant gratification." (Starosta & Izydorczyk, 2020)

An in-depth understanding of binge-watching may be needed to further understand the demand for content, marketing and distribution on streaming platforms.

## REFERENCES

Conlin, L. (2016). Time-shifting vs. Appointment viewing: The role of fear of missing out within TV consumption behaviors. *Communicatio Socialis*, 151–164.

Fatemi, F. (2022, November 14). How TV Viewing Habits have changed. *Forbes*. https://www.forbes.com/sites/falonfatemi/2022/11/14/how-tv-viewing-habits-have-changed/?sh=77428cac4888

Netflix. (2013, December 13). Netflix declares binge watching the new normal. *PR Newswire*. https://www.prnewswire.com/news-releases/netflix-declares-binge-watching-is-the-new-normal-235713431.html

Parameswaran, A. (2022, January 21). Appointment Viewing will live on. *Business Standard*.

Pena, L. L. (2015). Breaking Binge: Exploring The Effects Of Binge Watching On Television Viewer Reception. *Dissertations - ALL. 283*. https://surface.syr.edu/etd/283, 9.

Starosta, J., & Izydorczyk, B. (2020, June 22). Understanding the Phenomenon of Binge-Watching-A Systematic Review. *International Journal of Environmental Research and Public Health*, *17*(12), 17. doi:10.3390/ijerph17124469 PMID:32580289

Tang, W., & Wei, M. (2023). Streaming media business strategies andaudience-centered practices: A comparativestudy of Netflix and Tencent Video. *Online Media Glob. Commun.*, *2*(1), 3–24. doi:10.1515/omgc-2022-0061

## ENDNOTES

1 https://www.india.com/business/reliance-entertainment-launches-bigflix-indias-first-global-multi-language-hd-movie-platform-2074662/ .

2 https://www.businessinsider.in/thelife/news/fans-of-the-crown-are-trolling-prince-charles-and-camilla-on-their-royal-social-media-accounts/articleshow/79416770.cms. Published on Nov 26, 2020.

3 https://www.digitalspy.com/tv/a34004227/bbc-one-sir-david-attenborough-new-documentary-extinction-the-facts/ Published on Sept 13, 2020.

4 https://www.mirror.co.uk/tv/tv-news/sir-david-attenborough-terrifies-fans-22693064 Published on Sept 14,2020

5 Mass Communication Theory: Foundations, Ferment, and Future, Seventh Edition/ Stanley J. Baran and Dennis K. Davis/ Cengage Learning

6 Mosharafa, Eman. (2015). All you Need to Know About: The Cultivation Theory.

7 McQuail's Mass Communication Theories, McQuail, Denis, 6th Edition, Sage Publications

# Compilation of References


'Qala' movie review: Anvitaa Dutt's mother-daughter tale is poignant and admirable. (2022, December 2). The Hindu. https://www.thehindu.com/entertainment/movies/qala-movie-review-anvitaa-dutts-mother-daughter-tale-is-poignant-and-admirable/article66215167.ece

Pandey, G. (2008, October 10). Subaltern Citizens and their Histories. *Interventions*, *10*(3), 271–284. doi:10.1080/13698010802444587

A.G.M, J. & Dijk, V. (2006). Digital divide research, achievements andshortcomings. *Poetics*, *34*(5), 221–235.

Aaker, D. A. (2010). *Building strong brands*. Simon and Schuster.

Abidin, C. (2016). Visibility labour: Engaging with influencers' fashion brands and #OOTD advertorial campaigns on Instagram. *Media International Australia, Incorporating Culture & Policy*, *161*(1), 86–100. doi:10.1177/1329878X16665177

Abreu, J., Nogueira, J., Becker, V., & Cardoso, B. (2017). Survey of Catch-up TV and other time-shift services: A comprehensive analysis and taxonomy of linear and nonlinear television. *Telecommunication Systems*, *64*(1), 57–74. doi:10.1007/s11235-016-0157-3

ACW. (2022). *Africa Code Week*. ACW. https://coderina.org/africa-code-week/ [Accessed 12 Jun 2023].

Adgully. (2018, November 13). *ZEE5 launches its first Bengali original web series 'Kaali'*. Addgully. https://www.adgully.com: https://www.adgully.com/zee5-launches-its-first-bengali-original-web-series-kaali-81623.html

Adgully. (2018, November 13). *ZEE5 launches its first Bengali original web series 'Kaali'*. Adgully. https://www.adgully.com: https://www.adgully.com/zee5-launches-its-first-bengali-original-web-series-kaali-81623.html

Adgully. (2023, May 17). *Hoichoi launches its new Content arm hoichoi studios*. adgully. https://www.adgully.com/hoichoi-launches-its-new-content-arm-hoichoi-studios-131865.html

Afilipoaie, A., Iordache, C., & Raats, T. (2021). The 'Netflix Original' and what it means for the production of European television content. *Critical Studies in Television*, *16*(3), 304–325. doi:10.1177/17496020211023318

Agarwal, M. (2020, March 27). *Mapping The Market: India's Regional OTT Video Streaming Landscape*. Inc.42. https://inc42.com: https://inc42.com/features/mapping-the-market-indias-regional-ott-video-streaming-landscape/

Agarwal, S., & Das, M. (2022). Indian Web Series - Revolutionising the Portrayal of Indian Women in Mainstream Media. *International Journal of Creative and Innovative Research in all. Studies*, *4*(8), 95–100. http://www.ijciras.com/PublishedPaper/IJCIRAS1802.pdf

Agbiji, O. M., & Swart, I. (2015). Religion and social transformation in Africa: A critical and appreciative perspective. *Scriptura*, *114*, 1–20. doi:10.7833/114-0-1115

Ahmad, A., Floris, A., & Atzori, L. (2016). *QoE-centric Service Delivery: A Collaborative Approach among OTTs and ISPs. Computer Networks. DIEE.* University of Cagliari., doi:10.1016/j.comnet.2016.09.022

Aishwarrya,C. (2020). *Enhanced User-Experience – A key differentiator for OTT platforms in 2020?* Media Samosa. https://mediasamosa.com/2020/11/11/enhanced-user-experience-ott-strategy/

Akamai. (2020). *Protecting the Bank of OTT-White Paper.* Akamai.

Ali, A.. (2021). Impact of Smartphone: A Review on Negative Effects on Students. *PalArch's Journal of Archaeology of Egypt / Egyptology, 18*(4), 5710-5718. Retrieved from https://www.archives.palarch.nl/index.php/jae/article/view/7164

Alix. (2020, June 08). Does Gender Exist in Isolation? *Medium.* https://gender.medium.com/29-does-gender-exist-in-isolation-8688c972bf3

Alpana Kakkar, K. N. (2019, may 15). Analysis of Product Placement in Web Series and Its Influence on Consumer Buying Behavior. *Journal of Business.* https://journalofbusiness.org/index.php/GJMBR/article/view/2755

Amarkant, M. (2020, August 24). *Zee5 Churails Review: 10 Hard-Hitting Scenes that Portray the Myriad Shades of Womanhood | #Mainchurailhoon.* DiaryOfAnInsaneWriter. https://diaryofaninsanewriter.com/2020/08/24/churailspowerfulscenes/

An Unruly Modernity: The City in Bombay Cinema. (n.d.). The Book Review, Monthly Review of Important Books. https://www.thebookreviewindia.org/an-unruly-modernity-the-city-in-bombay-cinema/

Anand, S. N. (2021). We let 'Joji's' plot grow organically: Dileesh Pothan. *The Hindu.* https://www.thehindu.com/entertainment/movies/dileesh-pothan-speaks-about-his-new-filmjoji-and-the-inspiration-from-shakespeares-macbeth/article34264470.ece

Anghelcev, G., Sar, S., Martin, J. D., & Moultrie, J. L. (2021). Binge-watching serial video content: Exploring the subjective phenomenology of the binge-watching experience. *Mass Communication & Society, 24*(1), 130–154. doi:10.1080/15205436.2020.1811346

Ang, I. (1990). Culture and Communication: Toward an Ethnographic Critique of Media Consumption in the Transnational Media System. *European Journal of Communication, 5*(5), 239–266. doi:10.1177/0267323190005002006

Ang, I. (1996). *Living Room Wars: Rethinking Media Audiences for a Postmodern World.* Routledge. doi:10.4324/9780203289549

Ang, I. (2001). Desperately guarding border: media globalization, "cultural imperialism", and the rise of "Asia". In Y. Soucho (Ed.), *House of Glass: culture, modernity, and the state in Southeast Asia* (pp. 27–45). Seng Lee Press.

Ani. (2020, August 5). *LetsOTT the innovative Digital Streaming (OTT) Search Engine.* BW Businessworld. https://www.businessworld.in/article/LetsOTT-the-innovative-Digital-Streaming-OTT-Search-Engine/05-08-2020-305314/

Anjum, A. (2021, November 16). Kusha Kapila: Anything that exists will be trolled. *Outlook India.* https://www.outlookindia.com/. https://www.outlookindia.com/website/story/entertainment-news-kusha-kapila-anything-that-exists-will-be-trolled/401191

Antonio, A. & Tuffley, D. (2014). The Gender Digital Divide in Developing Countries. *Future Internet, 6*(4), 673-687.

Arigo, D., Pagoto, S., Carter-Harris, L., Lillie, S. E., & Nebeker, C. (2018). Using social media for health research: Methodological and ethical considerations for recruitment and intervention delivery. *Digital Health, 4*, 2055207618771757. doi:10.1177/2055207618771757 PMID:29942634

Arya. (2017). Imtiaz Ali as an Auteur. Amity Journal of Media & Communication Studies, Vol.7(1), 50–53.

Arya, K. K., & Verma, M. (2020). Contemporary Bollywood's Tryst with Bourgeois Romances of Hindi Cinema. Journal of Content. *Community and Communication, 13*, 308–315.

Arya, V., Verma, H., Sethi, D., & Agarwal, R. (2019). Brand authenticity and brand attachment: How online communities built on social networking vehicles moderate the consumers' brand attachment. *IIM Kozhikode Society & Management Review, 8*(2), 87–103. doi:10.1177/2277975219825508

Asemah, E. (2011). *Selected Mass Media Themes*. Jos University Press.

Asemah, E. S. (2012). Media imperialism: The cultural implication for Nigeria. In M. Mboho & H. Batta (Eds.), *Communication approaches to development: Essays in Honour of Professor Des Wilson*. BSM Resources Limited.

ATC. (2023). *IMPACT OF OTTPS ON YOUNG MINDS*. ATC Online. https://atconline.biz/impact-of-ott-platforms-on-young-minds/

Atos. (2019). *Tailoring cybersecurity for the Broadcast Media industry*. Atos.

Aufderheide, P. (2000). *The Daily Planet*. University of Minnesota Press.

AWS. (2022). *Service OTT: la voie de l'innovation vidéo basée sur le cloud-Création d'une plateforme pour réussir sur le marché du streaming et de la diffusion directe au client (D2C)*. AWS.

Ayanso, A. &. C. D. &. L. K. (2010). The Digital Divide: Global and Regional ICT Leaders and Followers. *Information Technology for Development, 16*(04), 04-319.

Azhar, M. (2013, August 27). *Gay Pakistan: Where sex is available and relationships are difficult*. BBC News. https://www.bbc.com/news/23811826

Baby, J. (2021). *The Great Indian Kitchen*. Mankind Cinemas.

Badlani, G. (2022, July 25). OTT censorship – through the lens of freedom of speech and expression. *Lucknow Law Review*. https://www.lucknowlawreview.org/post/ott-censorship-through-the-lens-of-freedom-of-speech-and-expression

Balbi, G., & Magaudda, P. (2018). *A history of digital media: An intermedia and global perspective*. Routledge., doi:10.4324/9781315209630

Banerjee, D. (2022). *The Best Video Format for OTTPs: Decoded*. Muvi One. https://www.muvi.com/blogs/the-best-video-format-for-ott-platforms-decoded

Banerjee, S., & Moniz, C. (2018, July 16). *The Rs. 5,000 Cr Opportunity for OTT Platforms in India*. Impact on Net. https://www.impactonnet.com: https://www.impactonnet.com/cover-story/the-rs-5000-cr-opportunity-for-ott-platforms-in-india-6074.html

Baran, S. (2002). *Introduction to Mass Communication, Media and Culture* (2nd ed.). McGraw Hill Higher Education.

Barker, C. (1999). *Television, Globalization and Cultural Identities*. Open University Press.

Baruh, L. (2009). Publicized intimacies on reality television: An analysis of voyeuristic content and its contribution to the appeal of reality programming. *Journal of Broadcasting & Electronic Media, 53*(2), 190–210. doi:10.1080/08838150902907678

Basu, R. (2020) *Television and OTT*. PWC. https://www.pwc.in/industries/entertainment-and-media/television-and-ott.html.

Basuroy, T. (2023, March 20). *Number of unique visitors of video over-the-top (OTT) platform in India in April 2022, by platform*. Statista. https://www.statista.com/statistics/1373116/india-video-ott-unique-visitors-by-platform/

BBC. (2023, July 20). *Netflix ends password sharing in "big market" India*. BBC News. https://www.bbc.com/news/world-asia-india-66253656

Beck, D., Hellmueller, L. C., & Aeschbacher, N. (2012). Factual entertainment and reality TV. *Communication Research Trends, 31*(2), 4–27.

Bennett, B. (2012). Digital Natives. In: Z. Yan, ed. Encyclopedia of Cyber Behavior. Research Gate.

Benshoff, H. (2015). *Film and television analysis: An introduction to methods, theories, and approaches*. Routledge. doi:10.4324/9780203129968

Berger, J., & Milkman, K. L. (2012). What makes online content viral? *JMR, Journal of Marketing Research, 49*(2), 192–205. doi:10.1509/jmr.10.0353

Berry, S. (2005). Marketing and promotion. In R. E. Pearson & P. Simpson (Eds.), *The Critical Dictionary of Film and Television Theory* (pp. 371–375). Routledge.

Bharadwaj, A., El Sawy, O. A., Pavlou, P. A., & Venkatraman, N. V. (2013). Digital business strategy: Toward a next generation of insights. *Management Information Systems Quarterly, 37*(2), 471–482. doi:10.25300/MISQ/2013/37:2.3

Bharatvaraj, S. S. G., & Subramani, R. R. (2018). Manifestation of Subaltern Narratives in Cinema. *American International Journal of Research in Humanities, Arts and Social Sciences. 25*(1), 05-08. http://iasir.net/AIJRHASSpapers/AIJRHASS19-102.pdf

Bhatt, A. (2023, April 11). *"Professor of Tomfoolery" is Kenny Sebastian's most personal show yet*. 5X Fest. https://www.5xfest.com/5xpress/professor-of-tomfoolery-is-kenny-sebastians-most-personal-show-yet

Bhattacharya, M. (2020). Growth of OTT in India the Covid Factor. E-cineIndia. (ISSN)2582-2500.

Bhattacharyya, M. (2020, June). Growth of OTT in India: The Covid Factor. *E-Cine India*, 1-11. https://www.researchgate.net/publication/342520995_21A-Article-Mausumi-Bhattacharyya-Growth-of-OTT-in-India-The-Covid-Factor

Bhattacharyya, R., Kohli, S., & More, R. (2020). Change In Consumption Habits Of Entertainment During Lockdown And The Role Of OTT Platforms. *Elementary Education Online, 19*(4), 4686–4695. https://bibliomed.org/?mno=98027

Bhattacharyya, S. S., Goswami, S., Mehta, R., & Nayak, B. (2022). Examining the factors influencing adoption of over the top (OTT) services among Indian consumers. *Journal of Science and Technology Policy Management, 13*(3), 652–682. doi:10.1108/JSTPM-09-2020-0135

Bijolia, D. (2023, May 16). *Dahaad's Nuanced Portrayal of Casteism and Patriarchy Goes Where Few Web Series Dare*. Homegrown. https://homegrown.co.in/homegrown-creators/dahaads-nuanced-portrayal-of-casteism-and-patriarchy-goes-where-few-web-series-dare?fbclid=IwAR0Q5pnNt0vVbuY70uKNBt0SCnoPMCgelJkszISppQYBQzl9mfbz6RlP5D0

Bilal, M. (2023, May 25). *The true story of Dahaad and Cyanide Mohan, the serial killer who inspired Vijay Varma's character*. DailyO. https://www.dailyo.in/entertainment/the-true-story-of-dahaad-and-cyanide-mohan-the-serial-killer-who-inspired-vijay-varmas-character-39833

Bleier, A., & Eisenbeiss, M. (2015). Personalized online advertising effectiveness: The interplay of what, when, and where. *Marketing Science, 34*(5), 669–688. doi:10.1287/mksc.2015.0930

Bohra, R. (2021, 30 November). Impact of OTT on traditional mode of entertainment. *The Times of India*. https://timesofindia.indiatimes.com/readersblog/raghavi/impact-of-ott-on-traditional-mode-of-entertainment-39292/

Booth, G. D. (1995). Traditional Content and Narrative Structure in the Hindi Commercial Cinema. *Asian Folklore Studies, 54*(2), 169–190. https://www.jstor.org/stable/1178940. doi:10.2307/1178940

Bordwell, D. (1985). *Narration in the Fiction Film*. University of Wisconsin Press.

Bordwell, D. (2008). *Poetics of Cinema*. Routledge.

Bordwell, D., & Thompson, K. (2010). *Film Art: An Introduction*. McGraw-Hill Companies,Incorporated.

Botturi, L., & Belfer, K. (2014). *PEDAGOGICAL PATTERNS FOR ONLINE LEARNING*. ResearchGate.

Bouquillion, P., & Ithurbide, C. (2022). Audiovisual industry and digital platforms in India: A contribution from the political economy of communication. *Global Media and Communication*, *18*(3), 345–364. doi:10.1177/17427665221125548

Boyle, R. (2020). A new Golden Age? In R. Steen, J. Novick, & H. Richards (Eds.), *Routledge Handbook of Sports Journalism* (pp. 355–358). Routledge., doi:10.4324/9781315616704-33

Brightcove (2023). *OTT Syndication API*. Brightcove. https://apis.support.brightcove.com/ott-syndication/index.html

Brody, R. (2021, June 1). "Joji," reviewed: The first major film of the COVID-19 pandemic. *The New Yorker*. https://www.newyorker.com/culture/the-front-row/joji-reviewed-the-first-major-film-of-the-covid-19-pandemic

Bryant, H. (2020). What 'The Last Dance' reveals about Michael Jordan's legacy. https://www.espn.com/nba/story/_/id/29365615/what-last-dance-reveals-michael-jordan-legacy

Buchholz, K. (2020b, September 23). *Infographic: International Brands Rise to the top of Indian streaming market*. Statista Daily Data. https://www.statista.com/chart/23009/market-share-streaming--ott--platforms-in-india/

Buckingham, D. (1987). *Public Secrets*. BFI.

Calvello, M. (2023). *How OTTPs Are Changing the Content Game*. G2. https://www.g2.com/articles/ott-platforms

Cambridge Dictionary. (2023). *Voyeurism*. Cambridge Press. https://dictionary.cambridge.org/dictionary/english/voyeurism

Capgemini. (2020). Impact of the Covid-19 Outbreak on Media & Entertainment. Capgemini. https://www.capgemini.com/wp-content/uploads/2020/06/Covid-19-Impact-on-Media-Entertainment-industry_VF.pdf

Carter, L. (2014). The cultural imperialism model in international communication.

Cashmore, E. (2006). *Celebrity culture*. Routledge. doi:10.4324/9780203962916

Castanheira, C., & Fernandes, P. (2019). *Future Operations Support Services for Digital Operations Transformation*. OSS FOR DIGITAL OPERATIONS TRANSFORMATION.

Chakrabarty, D. (1998). Minority Histories, Subaltern Pasts. *Economic and Political Weekly*, *33*(9), 473–479. https://www.jstor.org/stable/4406471

Chakraborty, Y. (2021, January 13). *Is the OTT boom passing Bengal by?* Get Bengal. https://www.getbengal.com: https://www.getbengal.com/details/is-the-ott-boom-passing-bengal-by

Chakraborty, P. (2022). Postcolonial Theory and the Challenge of Caste. *Interventions*, *25*(2), 157–173. doi:10.1080/1369801X.2022.2029531

Chand, S. (2021). *Sales Promotion Tools: Consumer-Oriented and Trade-Oriented Sales Promotion*. Your Article Library. https://www.yourarticlelibrary.com/sales/sales-promotion-tools-consumer-oriented-and-trade-oriented-sales-promotion/32305 [Accessed 22 Jun 2023].

Chan-Olmsted, S. M., & Park, J. T. (2000). From On-Air to Online World: Examining the Content and Structures of Broadcast TV Stations' Web Sites. *Journalism & Mass Communication Quarterly*, *77*(2), 321–339. doi:10.1177/107769900007700207

Chatterjee, P. (2023, May 05). Ranajit Guha, the Unconventional Historian. *The Wire*. https://thewire.in/history/ranajit-guha-the-unconventional-historian

Chattopadhyay, B. (2019, April 3). *Astey Ladies review: New Hoichoi series is a ton of fun in first half, but frustrating in second*. First Post. https://www.firstpost.com: https://www.firstpost.com/entertainment/astey-ladies-review-new-hoichoi-series-is-a-ton-of-fun-in-first-half-but-frustrating-in-second-6380621.html

Chen, L. (2016). *Talking about health and health related issues: an inquiry into the social media use of Chinese celebrity physicians and their fans*. [PhD Thesis, University of Iowa].

Cheng, L., & Kim, S. (2016). The relationship between brand-program congruency and purchase intention. *International Journal of Applied Sports Sciences : IJASS*, *28*(2), 123–136. doi:10.24985/ijass.2016.28.2.123

Che, X., & Ip, B. (2017). *Social networks in China*. Chandos Publishing.

Chibike, J. J. (2021). *Photography in a human rights crisis: Readers' perspectives and views of BBC's representation of the 2018 post-election violence in Zimbabwe*. [Unpublished Master's thesis, National University of Science and Technology].

Chibike, J. J., & Nkomo, T. (2022). Public relations education and cultural sensitivity in the digital epoch: A case of National University of Science and Technology's Journalism and Media Studies Department. *Journal of African Education*, *3*(3), 55–78. doi:10.31920/2633-2930/2022/v3n3a3

Chipp, K. F., & Chakravorty, D. (2016). Producer push to consumer pull: Who curates new media content? Developing strategies for new media environments. *Journal of Product and Brand Management*, *25*(4), 373–386. doi:10.1108/JPBM-06-2015-0918

Chiumbu, S. (2004). Redefining the National Agenda, Media and Identity-Challenges of Building a New Zimbabwe. In H. Melber (Ed.), *Media, Public Discourse and Political Contestation in Zimbabwe*. Nordiska AfrikaInstitutet.

Cho, H. (2023). Nostalgia in sport and leisure. *Current Opinion in Psychology*, *49*, 101551. Advance online publication. doi:10.1016/j.copsyc.2022.101551 PMID:36702009

Cho, H., Pyun, D. Y., & Lim, S. H. (2019). Enhancing the stability of the leisure nostalgia scale: Assessment of the measurement invariance between two different leisure populations. *Leisure Studies*, *39*(2), 294–303. doi:10.1080/02614367.2019.1653356

Chung, J. M. (2022). Introduction to Metaverse and Video Streaming Technology and Services. In *Emerging Metaverse XR and Video Multimedia Technologies: Modern Streaming and Multimedia Systems and Applications* (pp. 1–60). Apress.

Chung, S., & Cho, H. (2014). Parasocial relationship via reality TV and social media: its implications for celebrity endorsement. In *Proceedings of the ACM International Conference on Interactive Experiences for TV and Online Video* (pp. 47–54). ACM. 10.1145/2602299.2602306

Colakoglu, B.E. (2009). Perceptional difference of international news: Western media influence on non-Western media. *China Media Research,* 55-63

Conlin, L. (2016). Time-shifting vs. Appointment viewing: The role of fear of missing out within TV consumption behaviors. *Communicatio Socialis*, 151–164.

Creswell, J. W. (2014). *Research design: Qualitative, quantitative, and mixed method approaches* (4th ed.). SAGE Publications.

Cunningham, S., & Craig, D. (2019). *Social media entertainment: The new intersection of Hollywood and Silicon Valley*. NYU Press.

Cutting, J. E. (2016). Narrative theory and the dynamics of popular movies. *Psychonomic Bulletin & Review, 23*(6), 1713–1743. doi:10.3758/s13423-016-1051-4 PMID:27142769

Cyrill, M. (2023, April 11). Exploring India's Thriving OTT Video Segment: Trends, Growth, and Investor Interest. *India Briefing*. https://www.india-briefing.com: https://www.india-briefing.com/news/india-ott-video-segment-growth-trends-investments-27681.html/#:~:text=In%202022%2C%20market%20estimates%20showed,Hindi%2C%20Tamil%2C%20and%20Telegu

Dahiya, S., & Nijhawan, G. S. (2020). Role of Covid as a Catalyst in Increasing Adoptions of OTT in India: A Study of Evolving Consumer Consumption Pattern and Future Business Scope. *Journal of Content, Community & Communication., 12*(6), 298–311. doi:10.31620/JCCC.12.20/28

Das Gupta, C. (2008). *Seeing is Believing: Selected Writings on Cinema*. Penguin Viking.

Deloitte. (2024). *Digital transformation for telecom operators*. Deloitte. https://www2.deloitte.com/ch/fr/pages/technology-media-and-telecommunications/articles/digital-transformation-for-telecom-operators.html

Deuze, M., Witschge, T., & Brunow, D. (2016). Prefiguring the Post-Demise of Print: Digital Life After Newspapers. *Journal of Media Innovations, 3*(1), 109–127.

Dey, A. (2021, December 15). Family entertainment, regional content: OTT players craft strategy for 2022. *Fortune India*. https://www.fortuneindia.com: https://www.fortuneindia.com/enterprise/family-entertainment-regional-content-ott-players-craft-strategy-for-2022/106377

Dhaktode, N. (2023). 'Can Subaltern Speak?': Discourse Without Considering Caste an Exploitative Institution. *Contemporary Voice of Dalit, 0*(0), 2455328X2211455. doi:10.1177/2455328X221145597

Dhar, D. (2021, November 09). *Redefining The 'Mainstream': Queer Representation and Content on OTT Platforms*. Feminism in India. https://feminisminindia.com/2021/11/09/redefining-the-mainstream-queer-representation-and-content-on-ott-platforms/

Dhiman, B. (2023) Key Issues and New Challenges in New Media Technology in 2023: A Critical Review. *Journal of Media & Management. SRC/JMM-184.*

Dhiman, N., Singh, A., & Sarmah, R. (2022). How continuous intentions towards over the top platform are framed? Stimulus–organism–response model perspective. *Vision (Basel)*, 09722629221104202.

Diggs-Brown, B. (2011). *Strategic public relations: An audience-focused approach*. Cengage Learning.

DiMaggio, P. H. E. C. C. &. S. S. (2004). Digital inequality: From unequal access to differentiated use. Social Inequality, 355-400.

DISD/DPIDG. (2021). *Leveraging digital technologies for social inclusion*. UN. https://www.un.org/development/desa/dspd/2021/02/digital-technologies-for-social-inclusion/

Djafarova, E., & Rushworth, C. (2017). Exploring the credibility of online celebrities' Instagram profiles in influencing the purchase decisions of young female users. *Computers in Human Behavior, 68*, 1–7. doi:10.1016/j.chb.2016.11.009

Dogan, S., Çalgici, P., Arditi, D., & Gunaydin, H. (2015). *Critical success factors of partnering in the building design process. METU JFA 2015/2*. Department of Architecture, İzmir Institute of Technology.

Drennan, M., Baranovsky, Y., & Baranovsky, V. (2018). *Scriptwriting for web series: Writing for the Digital age*. Routledge. doi:10.4324/9781351237857

Driessens, O. (2013). Celebrity capital: Redefining celebrity using field theory. *Theory and Society, 42*(5), 543–560. doi:10.1007/s11186-013-9202-3

Duggal, S. (2023, April 13). *The insane journey of creator Prajakta Koli*. Entrepreneur. https://www.entrepreneur.com/en-in/entrepreneurs/the-insane-journey-of-creator-prajakta-koli/449636

Dutta, M. B. (1999). Taming the victim: Rape in soap opera. *The Journal of Popular Film and Television, 27*(1), 34–39. doi:10.1080/01956059909602795

Dwyer, R. (1998). The Hindi Romantic Cinema. *Journal of South Asian Studies*.21(1). Retrieved from: https://www.tandfonline.com/loi/cas20

Dwyer, R. (2000). The erotics of the wet sari in Hindi films. *South Asia: Journal of South Asian Studies*, 23(2), 143-160, oi:10.1080/00856400008723418

Dwyer, Rachel (2012). Fire and Rain, The Tramp and The Trickster: Romance and the Family in the early films of Raj Kapoor. *The South Asianist.* 2(3), 9–32. Retrieved from: www.southasianist.ed.ac.uk

Dwyer, R. (2010). Bollywood's India: Hindi Cinema as a Guide to Modern India. *Asian Affairs, XLI*(III), 381–398. https://www.tandfonline.com/loi/raaf20. doi:10.1080/03068374.2010.508231

Dwyer, R. (2014). *Picture Abhi Baaki Hai: Bollywood as a Guide to Modern India*. Hachette India.

Dyer, R. (2004). *Heavenly bodies: Film stars and society*. Routledge.

Easterbrook, S., Singer, J., Storey, M., & Damian, D. (2008). *Guide to Advanced Empirical Software Engineering-Selecting Empirical Methods for Software Engineering Research* (F. Shull, Ed.). Springer.

Ebbrecht, T. (2007). Docudramatizing history on TV: German and British docudrama and historical event television in the memorial year 2005. *European Journal of Cultural Studies, 10*(1), 35–53. doi:10.1177/1367549407072969

Eisenhardt, K. M., & Martin, J. A. (2000). Dynamic capabilities: What are they? *Strategic Management Journal, 21*(10-11), 1105–1121. doi:10.1002/1097-0266(200010/11)21:10/11<1105::AID-SMJ133>3.0.CO;2-E

Eishita. (2022, July 14). *How OTT Platforms Expanded Representation In India & Helped Break Taboos*. ScoopWhoop. https://www.scoopwhoop.com/entertainment/how-ott-platforms-expanded-representation-in-india/

Endavo. (2019). *Everything You Need to Know About OTT.* Endavo. https://www.endavomedia.com/what-is-ott/ [Accessed 14 Jun 2022].

Espinoza, D. & Reed, D. (2018). Wireless technologies and policies for connecting rural areas in emerging countries: a case study in rural Peru. *Digital Policy, Regulation and Governance, 20*(5), 479-511.

ESPN Press Room. (2020). *ESPN and Netflix set new April 19 premiere date for highly anticipated documentary series "The Last Dance"*. ESPN. https://espnpressroom.com/us/press-releases/2020/03/espn-and-netflix-set-new-april-19-premiere-date-for-highly-anticipated-documentary-series-the-last-dance/

ESPN. (2023). *The Last Dance: The untold story of Michael Jordan's Chicago Bulls*. ESPN. https://www.espn.com/nba/story/_/id/28973557/the-last-dance-untold-story-michael-jordan-chicago-bulls

European Commission. (2022). *Digital inclusion.* EC. https://digital-strategy.ec.europa.eu/en/policies/digital-inclusion#:~:text=Digital%20inclusion%20is%20an%20EU,with%20the%20COVID%2D19%20pandemic. [Accessed 15 Jun 2023].

Evens, T., & Donders, K. (2018). *Platform power and policy in transforming television markets*. Palgrave Macmillan. doi:10.1007/978-3-319-74246-5

Farooqui, J. (2022, March 30). *OTT platforms in India spent Rs 3700 crore on original content in 2021*. Exchange4Media. https://www.exchange4media.com: https://www.exchange4media.com/digital-news/ott-platforms-in-india-spent-rs-3700-crore-on-original-content-in-2021-119384.html

Farooqui, J. 2021. *Regional OTTs on the rise as native audiences demand local language content*. [Online] Available at: https://www.exchange4media.com/digital-news/regional-platformsdriving-the-second-ott-wave-in-india-114698.html [Accessed 01 Jul 2023].

Farooqui, M. (2021). OTTs set to launch more shows in 2021, invest more in regional content. *Moneycontrol*. https://www.moneycontrol.com/news/trends/entertainment/otts-set-to-launch-more-shows-in-2021-invest-more-in-regional-content-6326001.html, Accessed on July10, 2021

FastEdge. (2023). *How does educational OTT work*. FastEdge. https://www.fastedge.ai/how-does-educational-ott-work/#:~:text=Educational%20OTT%20can%20be%20thought,through%20web%2Dbased%20discussion%20boards

Fatemi, F. (2022, November 14). How TV Viewing Habits have changed. *Forbes*. https://www.forbes.com/sites/falonfatemi/2022/11/14/how-tv-viewing-habits-have-changed/?sh=77428cac4888

Fetchko, M. J., Roy, D. P., & Clow, K. E. (2018). *Sports marketing*. Routledge. doi:10.4324/9781315175614

FICCI EY. (2023). *Windows of opportunity: India's media & entertainment sector - Maximising across segments*. FICCI EY.

FICCI FRAMES 2023 - Industry Reports. (n.d.). FICCI FRAMES 2023 - Industry Reports. https://www.ficci-frames.com/reports.html

FICCI. (2022). *Tuning into consumer: Indian M&E rebounds with a customer-centric approach. Ernst & Young Associates LLP*. FICCI. https://www.ey.com/en_in/media-entertainment/tuning-into-consumer-indian-m-and-e-rebounds-with-a-customer-centric-approach

Financial Express. (2020). Now streaming, more than ever: Netflix, Amazon, others need to up their game. *Financial Express*. https://www.financialexpress.com/industry/technology/ now-streaming-more-than-ever-netflix-amazon-others-need-to-up-their-game/1918370/

Fitzgerald, S. (2019). Over-the-top video services in India: Media imperialism after globalisation. *Media Industries Journal*, *6*(2), 89–115. doi:10.3998/mij.15031809.0006.206

Flayelle, M., Canale, N., Vögele, C., Karila, L., Maurage, P., & Billieux, J. (2019). Assessing binge-watching behaviors: Development and validation of the "Watching TV Series Motives" and "Binge-watching Engagement and Symptoms" questionnaires. *Computers in Human Behavior*, *90*, 26–36. doi:10.1016/j.chb.2018.08.022

Fracchia, J. (2004). Subaltern Studies and Collective Memories in Piana degli Albanesi: Methodological Reflections on a Historiographical Encounter. *Asian Journal of Social Science*, *32*(2), 247–273. https://www.jstor.org/stable/23654588. doi:10.1163/1568531041705112

Frater, P. (2023). *Netflix to Increase Asia Content Spending by 15% in 2023 as Revenue Growth Accelerates in the Region – Study*. Varriety. https://variety.com/2023/tv/news/netflix-content-spending-asia-pacific-1235543665/

Freberg, K., Graham, K., McGaughey, K., & Freberg, L. A. (2011). Who are the social media influencers? A study of public perceptions of personality. *Public Relations Review*, *37*(1), 90–92. doi:10.1016/j.pubrev.2010.11.001

Friedman, T. (1992). The World of The World of Coca-Cola. *Communication Research, 19*(5), 642–662. doi:10.1177/009365092019005005

Fuchs, J. (2022). *Freemium Pricing: How to Leverage It & 4 Popular Models.* Hubspot. https://blog.hubspot.com/sales/freemium-pricing-strategy

Fuchs, C. (2010). New Imperialism: Information and media imperialism? *Global Media and Communication, 6*(1), 33–60. doi:10.1177/1742766510362018

Fung, A. Y., & Chik, G. H. (2021). When East Asian media industries are faced with digitalization: Transformation and survival strategies. In *The Routledge companion to media industries* (pp. 312–323). Routledge. doi:10.4324/9780429275340-31

Gaikwad, S. (2019). How Indian cinema evolved into being one of the most powerful modes of entertainment. *Businesstoday.* https://www.businesstoday.in/opinion/columns/how-indian-cinema-evolved-intobeing-one-of-the-most-powerful-modes-of-entertainment/story/307454.html

Gamble, C. T. (2023). *Bridging the Digital Divide: The Imperative of Digital Inclusion.* LinkedIn. https://www.linkedin.com/pulse/bridging-digital-divide-imperative-inclusion-tournis-gamble/

Ganti, T. (2013). *Bollywood: A Guidebook to Popular Hindi Cinema.* Routledge. doi:10.4324/9780203834411

Gantz, W. (1981). An exploration of viewing motives and behaviors associated with television sports. *Journal of Broadcasting, 25*(3), 263–275. doi:10.1080/08838158109386450

Gautam, P., Upadhyay, A. K., & Deepmala, S. (2023). Digital Literacy among Educators. Noida, UP, India, 11th International Library Information Professionals summit (l-LIPS) on new and Innovative Libraries in Digital Era: Services and Practice, (pp. 413-418). Noida.

Gensler, S., Völckner, F., Liu-Thompkins, Y., & Wiertz, C. (2013). Managing brands in the social media environment. *Journal of Interactive Marketing, 27*(4), 242–256. doi:10.1016/j.intmar.2013.09.004

Geraghty, C. (2005). The study of the soap opera. In J. Wasko (Ed.), *Companion to Television* (pp. 308–323). Blackwell Publishing. doi:10.1002/9780470997130.ch17

Geraldine de Bastion, S. M. (2020). Data and the global South:Key issues for inclusive digital development. In: D. Heinrich-Böll-Stiftung Washington, ed. Data and the global South:Key issues for inclusive digital development. Washington, USA: Heinrich-Böll-Stiftung Washington, DC.

Ghosh, S. (2023, June 28). Raima Sen: OTT contents now celebrate and amplify the voices of women from all walks of life. *The Times of India.* https://timesofindia.indiatimes.com/entertainment/bengali/movies/news/raima-sen-ott-contents-now-celebrate-and-amplify-the-voices-of-women-from-all-walks-of-life/articleshow/101341971.cms?from=mdr

Ghosh, S. (2023, June 28). Raima Sen: OTT contents now celebrate and amplify the voices of women from all walks of life. *Times of India.* https://timesofindia.indiatimes.com: https://timesofindia.indiatimes.com/entertainment/bengali/movies/news/raima-sen-ott-contents-now-celebrate-and-amplify-the-voices-of-women-from-all-walks-of-life/articleshow/101341971.cms?from=mdr

Giachetti, R. (2012). *A Flexible Approach to Realize an Enterprise Architecture.* Department of Systems Engineering, Naval Postgraduate School, Monterey, CA USA. Presented: New Challenges in Systems Engineering and Architecting. Conference on Systems Engineering Research (CSER). St. Louis, MO. 10.1016/j.procs.2012.01.031

Giles, D. C. (2000). *Illusions of immortality: A psychology of fame and celebrity.* Macmillan. doi:10.1007/978-1-137-09650-0

Goldsmith, B. (2013). “SportTV”: The legacies and power of television. In B. Hutchins & D. Rowe (Eds.), *Digital media sport: Technology, power and culture in the network society* (pp. 52–65). Routledge. doi:10.4324/9780203382851

Goonatilake, S. (2005). Cultural Imperialism: A Short History, Future, and a Postscript from the Present. In B. Hamm & R. Smandych (Eds.), *Cultural Imperialism: Essays on the Political Economy of Cultural Domination* (pp. 33–51). Broadview Press.

Gopalkrishnan, S. (2019). Marginalised in the New Wave Tamil Film: Subaltern Aspirations in three films by Bala, Kumararaja and Mysskin. *Rupkatha Journal on Interdisciplinary Studies in Humanities,11*(2),1-9. https://rupkatha.com/V11/n3/v11n303.pdf

Goyal, S. (2022, January 15). Is The Darkness On OTT Forced Or Organic? *Outlook India.* https://www.outlookindia.com: https://www.outlookindia.com/website/story/entertainment-news-is-the-darkness-on-ott-forced-or-organic/409809

Grainger, A., Newman, J. I., & Andrews, D. L. (2006). Sport, the media, and the construction of race. In A. A. Raney & J. Bryant (Eds.), *Handbook of Sports and Media* (pp. 482–505). Lawrence Erlbaum Associates. doi:10.4324/9780203873670.ch27

Gupta, C. D. (2008). *Seeing is believing: Selected writings on cinema.* Penguin India.

Gupta, G., & Singharia, K. (2021). Consumption of OTT media streaming in COVID-19 lockdown: Insights from PLS analysis. *Vision (Basel), 25*(1), 36–46. doi:10.1177/0972262921989118

GVR. (2023). *GVR Report coverOver The Top Devices And Services Market Size, Share & Trends Report.* GVR. https://www.grandviewresearch.com/industry-analysis/over-the-top-ott-devices-services-market

Hadebe, S. (2005). Reflections on the proposed Ndebele/Shona-Shona/Ndebele dictionary. *Lexikos, 15,* 265–275.

Hahm, J., Kang, T. A., & Matsuoka, H. (2021). Understanding the relationship between past experience of a sports mega-event and current spectatorship: The mediating role of nostalgia. *Sustainability (Basel), 13*(6), 3504. doi:10.3390/su13063504

Hall, J., O’Mahony, B., & Vieceli, J. (2010). An empirical model of attendance factors at major sporting events. *International Journal of Hospitality Management, 29*(2), 328–334. doi:10.1016/j.ijhm.2009.10.011

Hall, S. (1980). Encoding/decoding. In S. Hall, D. Hobson, A. Lowe, & P. Willis (Eds.), *Culture, media, language.* Hutchinson.

Hall, S. (1992). The question of cultural identity. In S. Hall, D. Held, & T. McGrew (Eds.), *Modernity and its futures.* Polity Press.

Hamdar, Y. (2021). *Digital Transformation in the Global South: Enhancing Citizens’ Trust.* Data Appliance. https://datapopalliance.org/digital-transformation-in-the-global-south-enhancing-citizens-trust/

Han, J., Park, D., Hua, M., & Childs, P. (2021). Is group work beneficial for producing creative designs in STEM design education? *International Journal of Technology and Design Education.* doi:10.1007/s10798-021-09709-y

Hanna, R., Rohm, A., & Crittenden, V. L. (2011). We’re all connected: The power of the social media ecosystem. *Business Horizons, 54*(3), 265–273. doi:10.1016/j.bushor.2011.01.007

Hardiman, D. (2023, October 31). Ranjit Guha: A Tribute. *South Asian Studies, 0*(0), 1–6. doi:10.1080/02666030.2023.2270837

Hargittai, E., & Litt, E. (2011). The tweet smell of celebrity success: Explaining variation in Twitter adoption among a diverse group of young adults. *New Media & Society, 13*(5), 824–842. doi:10.1177/1461444811405805

Hasan, S. (2005). India and Pakistan: Common Identity and Conflict. *Refugee Survey Quarterly*, *24*(4), 74–80. https://www.jstor.org/stable/45054038. doi:10.1093/rsq/hdi085

Hayward, J. (1997). *Consuming Pleasures Active Audiences and Serial Fictions from Dickens to Soap Opera*. University Press of Kentucky.

Heeks, R. (2022). Digital inequality beyond the digital divide: conceptualizing adverse digital incorporation in the global South. *Information Technology for Development, 28*(04), 688-704.

Hesmondhalgh, D. (2002). *The Cultural Industries*. Sage.

Himabindu Reddy, S. J. A. J. V. W. (2022). A Critical Review of Global Digital Divide and the Role of Technology in Healthcare. *Cureus, 14*(09).

Hindu Business Line. (2020, December 16). Paid ott subscriptions up by nearly 60%: CII-BCG Report. *The Hindu BusinessLine*. https://www.thehindubusinessline.com/info-tech/paid-ott-subscriptions-up-by-nearly-60-cii-bcg-report/article33346493.ece

Hine, G. S. C. (2013). The importance of action research in teacher education programs. In *Design, develop, evaluate: The core of the learning. environment. Proceedings of the 22nd Annual Teaching Learning Forum, 7-8 February 2013*. Murdoch University. http://ctl.curtin.edu.au/professional_development/conferences/tlf/tlf2013/refereed/hine.html

Hinkelmann, K. (2016). *Modelling in Enterprise Architecture. University of applied sciences northwest Switzerland*. Business School.

Hipke, M., & Hachtmann, F. (2014). Game changer: A case study of social-media strategy in Big Ten athletic departments. *International Journal of Sport Communication*, *7*(4), 516–532. doi:10.1123/IJSC.2014-0022

Hogan, P. (2009). Hindi Cinema as a Challenge to Film Theory and Criticism. *Projections*, 3(2), 5-9. Retrieved from: https://doi.org/ doi:10.3167/proj.2009.030201

Hogan, P. C. (2009). *Understanding Indian Movies: Culture, Cognition, and Cinematic Imagination*. University of Texas Press.

Hosiaisluoma (2022). *Holistic Enterprise Development*. Hosiaisluoma. https://www.hosiaisluoma.fi/blog/archimate-examples

Hosseinichimeh, N., & Rahmani, A. (2020). A computational approach to managing and delivering video contents on OTT platforms. *Computers & Industrial Engineering*, *140*, 106262.

Huang, Z., & Benyoucef, M. (2013). From e-commerce to social commerce: A close look at design features. *Electronic Commerce Research and Applications*, *12*(4), 246–259. doi:10.1016/j.elerap.2012.12.003

Hugo Munsterberg's The Photoplay. (2021, February 4). YouTube. https://www.youtube.com/watch?v=vd32nkyUjgM

Hussain, D. M. (2014). SOCIAL IMPLICATIONS OF DIGITAL DIVIDE IN GLOBAL CONTEXT: AN APPRAISAL. *ZENITH International Journal of Multidisciplinary Research, 4*(2), 271-281.

Hussain, O., Dillon, Th., Chang, E., & Hussain, F. (2010). Transactional risk-based decision making system in e-business interactions. [CRl Publishing ltd Computer Systems. Science & Engineering.]. *International Journal of Computer Systems Science & Engineering*, *2010*(1), 15–28.

Hussein, M., & Elsayed, H. (2021). Over the Top (OTT) Platforms: A New Era in the Television Industry. *The Journal of Business and Retail Management Research*, *16*(1), 1–12.

Hutchins, B., Li, B., & Rowe, D. (2019). Over-the-top sport: Live streaming services, changing coverage rights markets and the growth of media sport portals. *Media Culture & Society, 41*(7), 975–994. doi:10.1177/0163443719857623

IAB. (2016). *Over The Top (OTT) Video: An Overview*. IAB., iab.com/ott

IEEE. (2019). *What Is Digital Inclusion? The Global Effort to Bring Everyone Online.* IEEE. https://ctu.ieee.org/what-is-digital-inclusion-the-global-effort-to-bring-everyone-online/

IEEE. (2022). *Solutions to the Digital Divide: Moving Toward a More Equitable Future.* IEEE. https://ctu.ieee.org/solutions-to-the-digital-divide-moving-toward-a-more-equitable-future/

InfoSys. (2023). *CASE STUDY: Digital streaming service provider implements OTTP*. InfoSys. https://www.infosys.com/industries/media-entertainment/case-studies/ott-platform.html

Intelligence, M. (2023). *OTT MARKET SIZE & SHARE ANALYSIS - GROWTH TRENDS & FORECASTS (2023 - 2028).* Mordor Intelligence. https://www.mordorintelligence.com/industry-reports/over-the-top-market

Iordache, C., Raats, T., & Mombaerts, S. (2023). The Netflix original documentary, explained: Global investment patterns in documentary films and series. *Studies in Documentary Film, 17*(2), 151–171. doi:10.1080/17503280.2022.2109099

IT for Change. (2013). *ICTs for empowerment and social transformation - A note prepared by IT for Change for ActionAid International.* IT for Change. https://itforchange.net/index.php/icts-for-empowerment-and-social-transformation-a-note-prepared-by-it-for-change-for-actionaid-0 [Accessed 03 Jul 2023].

ITU. (2020). *International Telecommunication Union.* ITU. https://dig.watch/actor/international-telecommunication-union [Accessed 21 Jun 2023].

ITU. (2022). *Digital inclusion of all.* https://www.itu.int/en/mediacentre/backgrounders/Pages/digital-inclusion-of-all.aspx#:~:text=Overview,and%20services%2C%20equally%20and%20equitably. [Accessed 04 JUn 2023].

Iversen, M. H. (2015). Employing a rhetorical approach to the practice of audience research on political communication. In Kramp, L., Carpentier, N., Hepp, A. Nieminen, H and Olsson, T. (Eds). *Journalism, Representation and the Public Sphere*. Lumiere.

Jaiswal, S., & Singh, A. (2022). OTT Services and Intention to Continuous Subscription: A Moderated-Moderated Mediation Analysis. *PURUSHARTHA-A journal of Management. Ethics and Spirituality, 15*(2), 104–125. doi:10.21844/16202115206

Janinović, D. (2022). *DIGITAL TRANSFORMATION STRATEGY OF RADIO AND TELEVISION OF MONTENEGRO FOR THE PERIOD 2023-2027.* OSCE Missions.

Jansen, B. J., & Spink, A. (2006). How are we searching the World Wide Web? A comparison of nine search engine transaction logs. *Information Processing & Management, 42*(1), 248–263. doi:10.1016/j.ipm.2004.10.007

Jenkins, H. (2017). *Convergence culture: where old and new media collide. Is the Universe a Hologram?* MIT Press. doi:10.7551/mitpress/9780262036016.003.0012

Jeong, D., Lee, M., & Park, E. (2022). Antecedents of Consumer Adoption of Over-the-Top Services in South Korea. *International Journal on Media Management, 24*(3), 121–136. doi:10.1080/14241277.2022.2124996

Jha, L. (2021, September 24). Bengali OTT service Hoichoi announces new line-up as watch-time doubles. *Live Mint.* https://www.livemint.com: https://www.livemint.com/industry/media/bengali-ott-service-hoichoi-announces-new-line-up-as-watch-time-doubles-11632484289520.html

Jha, L. (2022, August 2). Local OTT platforms battle rising costs of production. *Live Mint.* https://www.livemint.com: https://www.livemint.com/industry/media/local-ott-platforms-battle-rising-costs-of-production-11659381966885.html

Jha, L. (2022, March 28). Regional Languages to Account for 54% of OTT Content by 2024. *Live Mint*. https://www.livemint.com: https://www.livemint.com/industry/media/regional-languages-to-account-for-54-of-ott-content-by-2024-11648457191989.html

Jonkers, H., Band, I., & Quartel, D. (2012a). *ArchiSurance Case Study*. The Open Group.

Jose, R. J. (2020). Factors influencing the shift from traditional TV to OTT platforms in India. *International Journal of Advanced Science and Technology*, *29*(7s), 4044–4051. http://sersc.org/journals/index.php/IJAST/article/view/22888

Joshi, S. (2015). Impact of Over The Top Services on Telecom Service Providers. *Indian Journal of Science and Technology*, *8*(S4), 145–160. doi:10.17485/ijst/2015/v8iS4/62238

Juel, H. (2006). Defining documentary film. *POV A Danish Journal of Film Studies, 22*, 5–15.

Junzhi, D. (2021). The development of e-commerce in China during the COVID-19 pandemic on the example of the textile industry. *BRICS Journal of Economics*, *2*(3), 54–69. doi:10.38050/2712-7508-2021-3-3

Kakar, S. (1981). The Ties that Bind: Family Relationships in the Mythology of Hindi Cinema. *India International Centre Quarterly*. 8:1, 11-21. Retrieved from- https://www.jstor.org/stable/23001932

Kamble, C., & Biswal, S. K. (2023, August 23). The Changing Role of Marginalised Women in Hindi Cinema: A Shift from Cinematic Negotiations to Empowerment. *Media Watch*, *14*(3), 386–402. doi:10.1177/09760911231191266

Kaplan, A. M., & Haenlein, M. (2010). Users of the world, unite! The challenges and opportunities of Social Media. *Business Horizons*, *53*(1), 59–68. doi:10.1016/j.bushor.2009.09.003

Kashmir, R. (2022, March 7). How the demand for local language content increased in the last 5 years. *Rising Kashmir*. http://risingkashmir.com: http://risingkashmir.com/how-the-demand-for-local-language-content-increased-in-the-last-5-years#:~:text=In%20recent%20years%2C%20we%20have,and%20receptive%20to%20new%20experiences

Kashyap, S. (2019, January 25). To win over Indian Audiences, OTT platforms need to understand the sensibilities of Bharat. *YourStory*. https://yourstory.com: https://yourstory.com/2019/01/content-language-india-ott-bharat

Katz, E., Gurevitch, M., & Haas, H. (1973). On the use of the mass media for important things. *American Sociological Review*, *38*(2), 164–181. doi:10.2307/2094393

Kaur, S. P., & Sharma, V. "A transition from Male Mas- culinity to Female Androgyny in Hindi Cinema." Media Watch, vol. 8, no. 2, 2017, pp. 287-94. https://mediawatchjournal.in/325mwj/may17-12.pdf

Kaushal, S. (2020). *OTT platforms in India See subscriber bases surge UP to 80% amid coronavirus lockdown*. Forbes. https://www.forbes.com/sites/swetakaushal/2020/05/26/good-news-amid-lockdownott-platforms-register-60-80-surge-in-subscription-base/#77d0fab7413a

Kay, K. (2018). 'New' India and the NUTTAHS. *New Indian Nuttahs*, (Sept), 1–20. doi:10.1007/978-3-319-97867-3_1

Keane, M., Yecies, B., & Flew, T. (Eds.). (2018). *Willing collaborators: Foreign partners in Chinese media*. Rowman & Littlefield.

Keane, M., Yu, H., Zhao, E. J., & Leong, S. (2020). *Chinas Digital Presence in the Asia-Pacific: Culture, Technology and Platforms*. Anthem Press. doi:10.2307/j.ctv20pxxzt

Keller, K. L. (1993). Conceptualizing, measuring, and managing customer-based brand equity. *Journal of Marketing*, *57*(1), 1–22. doi:10.1177/002224299305700101

Kerr, G., Delle Fave, A., & Vella-Brodrick, D. (2017). The art and science of a sustainable performing career in music. In *The science of wellbeing* (pp. 201–227). Springer.

Kersten, M. (2021). *Introducing SAFe® 5.* Your Operating System for Business Agility. Scaled Agile Inc.

Khalid, M. Z., & Ahmed, A. (2014). Entertainment –education strategies for social change: Opportunities and emerging trends. *Review of Journalism and Mass Communication*, *2*(1), 69–89.

Khamis, S., Ang, L., & Welling, R. (2017). Self-branding, 'micro-celebrity' and the rise of Social Media Influencers. *Celebrity Studies*, *8*(2), 191–208. doi:10.1080/19392397.2016.1218292

Kher, A. (2022). *Case Study: Incorporating all the OTT streaming into one platform.* UX Planet. https://uxplanet.org/case-study-incorporating-all-the-ott-streaming-into-one-platform-d3e5b1fbdd59

Khurana, A. (2023, May 11). Dahaad Season 1 Review: Sonakshi Sinha and Vijay Varma both roar in this Cop Drama. *Entertainment Times.* https://timesofindia.indiatimes.com/web-series/reviews/hindi/dahaad/season-1/seasonreview/100101778.cms

Kietzmann, J. H., Hermkens, K., McCarthy, I. P., & Silvestre, B. S. (2011). Social media? Get serious! Understanding the functional building blocks of social media. *Business Horizons*, *54*(3), 241–251. doi:10.1016/j.bushor.2011.01.005

Kim, J., & Lennon, Sh. (2017). *Descriptive Content Analysis on E-Service Research. International Journal of Service Science, Management, Engineering, and Technology (IJSSMET).* IGI Global., doi:10.4018/IJSSMET.2017010102

Kjeldsen, J. E. (2016). Studying Rhetorical Audiences–a Call for Qualitative Reception Studies in Argumentation and Rhetoric. *Informal Logic*, *36*(2), 136–158. doi:10.22329/il.v36i2.4672

Kochar, M. (2022). *How Is Data Science Powering OTTP to Multi-Dollar Business.* Data Science. https://blog.learnbay.co/how-is-data-science-powering-ott-platform-to-multi-dollar-business

Kohli, D. (2019). *Looking into the future – Ott trends.* YourStory. https://yourstory.com/2019/02/looking-into-future-top-ott-trends-2019

Kohli-Khandekar, V. (2023, July 25). *Farzi, Citadel top OTT charts in first half of 2023: Ormax Media report.* www.business-standard.com. https://www.business-standard.com/entertainment/farzi-citadel-top-ott-charts-in-first-half-of-2023-ormax-media-report-123072500905_1.html

Kohli-Khandekar, V. (2022). *The Indian Media Business.* SAGE Publications.

Koshy, J. (2022). How technology & transformation are putting the T's together in OTT. *iTWire.*

Koskinen, K. (2019). Digital Platforms in the Global South:Foundations and Research Agenda. In: *Information and Communication Technologies for Development. Strengthening Southern-Driven Cooperation as a Catalyst for ICT4D.* Switzerland: Springer Nature Switzerland.

Koskinen, K., Bonina, C., & Eaton, B. (2018). Digital Platforms in the Global South: Foundations and Research Agenda. *Development Implications of Digital Economies*, *1*(1), 1–28.

KPMG. (2017). The 'digital first' journey- how OTT platforms can remain 'on-demand' ready. *KPMG Report.* KPMG. https://assets.kpmg/content/dam/kpmg/in/pdf/2017/10/The-Digital-First-journey.pdf

KPMG. (2017). *The 'Digital First' journey-How OTTPs can remain 'on-demand ready.* KPMG.

Krick, K., Boswell, J. S. & Poretsky, S. (2021). *Bridging the digital divide for an inclusive digital economy.* Research Gate.

Kroonenberg, S. (2023). Gramsci's Writing Body. On Embodiment and Subaltern Knowledge. *Interventions*, *0*(0), 1–16. doi:10.1080/1369801X.2023.2191861

Kucharska, W., Confente, I., & Brunetti, F. (2020). The power of personal brand authenticity and identification: Top celebrity players' contribution to loyalty toward football. *Journal of Product and Brand Management*, *29*(6), 815–830. doi:10.1108/JPBM-02-2019-2241

Kulkarni, M. (2019). Digital accessibility: Challenges and opportunities. *IIMB Management Review, 31*(01), 91-98.

Kumar, A. (2023, May 31). 2 counts on which Amazon Prime's Dahaad won at showing caste—and 1 where it utterly failed. *The Print*. https://theprint.in/opinion/2-counts-on-which-amazon-primes-dahaad-won-at-showing-caste-and-1-where-it-utterly-failed/1603687/

Kumar, P. (2020, August 14). 'Churails' review: Justice League of women, by women, for women. *The Hindu*. https://www.thehindu.com/entertainment/reviews/churails-review-justice-league-of-women-by-women-for-women/article32351662.ece

Kumar, S. A. (2023, April 4). The Multilingual Era Has Arrived And Here's Why It's A Blessing For Regional Stories. *India Times*. https://www.indiatimes.com: https://www.indiatimes.com/entertainment/binge/the-multilingual-era-has-arrived-and-heres-why-its-a-blessing-for-regional-stories-598030.html

Kumar, V. (2014). *Making "Freemium" Work*. HBR. https://hbr.org/2014/05/making-freemium-work [Accessed 26 Jun 2023].

Kumar, R. (2013). The making of the National Film Archive of India: Notes from the Archive of the Archive. *The Moving Image : the Journal of the Association of Moving Image Archivists*, *13*(1), 98–128. doi:10.5749/movingimage.13.1.0098

Kwak, K. T., Lee, S. Y., Ham, M., & Lee, S. W. (2021). The effects of internet proliferation on search engine and over-the-top service markets. *Telecommunications Policy*, *45*(8), 102146. doi:10.1016/j.telpol.2021.102146

Labrecque, L. I., Markos, E., & Milne, G. R. (2011). Online personal branding: Processes, challenges, and implications. *Journal of Interactive Marketing*, *25*(1), 37–50. doi:10.1016/j.intmar.2010.09.002

Labrecque, L. I., vor dem Esche, J., Mathwick, C., Novak, T. P., & Hofacker, C. F. (2013). Consumer power: Evolution in the digital age. *Journal of Interactive Marketing*, *27*(4), 257–269. doi:10.1016/j.intmar.2013.09.002

Lal, N. (2023, March 17). 'Hours of observing women in khaki led to realistic portrayal'. *Times of India*.

Lal, V. (2001). Subaltern Studies and Its Critics: Debates over Indian History [Review of A Subaltern Studies Reader 1986-1995; Nationalism, Terrorism, Communalism: Essays in Modern Indian History; Writing Social History; The Furies of Indian Communalism: Religion, Modernity and Secularization, by R. Guha, P. Heehs, S. Sarkar, & A. Vanaik]. *History and Theory*, *40*(1), 135–148. https://www.jstor.org/stable/2678074. doi:10.1111/0018-2656.00158

Laskar, M. (2023). Examining the emergence of digital society and the digital divide in India: A comparative evaluation between urban and rural areas. *Frontiers in Sociology, 12*.

Laxmii to Dil Bechara, big Bollywood films released on OTT in 2020. (2020, December 22). India Today. https://www.indiatoday.in/movies/bollywood/story/laxmii-to-dil-bechara-big-bollywood-films-released-on-ott-in-2020-1751978-2020-12-22

Lee, C. (1980). *Economic Dependency and cultural imperialism: Theoretical perspectives Media imperialism reconsidered: The homogenizing of television culture*. Sage.

Lee, J. E., & Watkins, B. (2016). YouTube vloggers' influence on consumer luxury brand perceptions and intentions. *Journal of Business Research*, *69*(12), 5753–5760. doi:10.1016/j.jbusres.2016.04.171

Lee, J., & Kang, J. (2022). Sport brand authenticity: Scale development and validation. *Korean Journal of Sport Science, 33*(4), 659–673. doi:10.24985/kjss.2022.33.4.659

Leslie-Walker, A., & Mulvenna, C. (2022). The Football Association's Women's Super League and female soccer fans: Fan engagement and the importance of supporter clubs. *Soccer and Society, 23*(3), 314–327. doi:10.1080/14660970.2022.2037218

Levy, A. (2018). *Why Netflix Loves to Highlight Its Local Content.* Fool.com. https://www.fool.com/investing/2018/04/24/why-netflix-loves-to-highlight-its-local-content.aspx.

Levy, E. (2022). *Streaming Data Architecture in 2022: Components and Examples.* Upsolver. https://www.upsolver.com/blog/streaming-data-architecture-key-components

Liaqat, Q. (2023, March 20). The Notion of the 'Subaltern' and the Drone Victim Subjectivities in Pakistani Anglophone Fiction. *Interventions, 0*(0), 1–19. doi:10.1080/1369801X.2023.2190906

Lidhoo, P. (2023, May 28). Why advertisers are caught up between the tv and digital broadcasters. *Business Today.* https://www.businesstoday.in/magazine/the-buzz/story/ipl-2023-why-advertisers-are-caught-up-between-the-tv-and-digital-broadcasters-380778-2023-05-10

Lim, J. S., Hwang, Y. C., Kim, S., & Biocca, F. A. (2015). How social media engagement leads to sports channel loyalty: Mediating roles of social presence and channel commitment. *Computers in Human Behavior, 46*, 158–167. doi:10.1016/j.chb.2015.01.013

Lin, L. (2022). *Convergent Chinese television industries: An ethnography of Chinese production cultures.* Springer Naturc. doi:10.1007/978-3-030-91756-2

Lin, Y. W. M. Z. H. Z. R. S. Z.-J. M. (2016). *Content Promotion for Online Content Platforms with the Diffusion Effect., 21*(Jun), 49.

Llewellyn, R. (2023). *Digital Transformation Use Cases in 10Well-Known Industries.* Privacy.

Lotz, A. D. (2017). *Portals: A treatise on Internet-distributed television.* Michigan Publishing, University of Michigan Library.

Lutgendorf, P. "Is There an Indian Way of Filmmaking?" *International Journal of Hindu Studies*, vol. 10, no. 3, 2006, pp. 227–56. *JSTOR*, https://www.jstor.org/stable/20106974. Accessed 24 Aug. 2023.

Lutgendorf, P. (2005). Sex in the Snow: The Himalayas as Erotic Topos in Popular Hindi Cinema. *Himalaya: Journal of the Association for Nepal and Himalayan Studies, 25*(1-2), 29–37.

Mabweazara, H., & Strelitz, L. (2009). Investigating the popularity of the Zimbabwean tabloid newspaper UMthunywa: A reception study of Bulawayo, *African. Journalism Studies, 30*(2), 113–133.

Macwan, S. S. (2023, August 07). Ranajit Guha, Subaltern Studies, and the Indian English Novel. *Economic and Political Weekly, 58*(31). https://www.epw.in/journal/2023/31/postscript/ranajit-guha-subaltern-studies-and-indian-english.html

Mahalakshmi, N., & Gopalan, K. (2021, January 1). *OTT players are splurging on content. What's their game?* Outlook Business. https://www.outlookbusiness.com: https://www.outlookbusiness.com/strategy-4/feature-17/ott-players-are-splurging-on-content-whats-their-game-6146

Malewar, S., & Bajaj, S. (2020). Acceptance of OTT video streaming platforms in India during COVID-19: Extending UTAUT2 with content availability. *Journal of Content. Community and Communication, 12*(6), 89–106. doi:10.31620/JCCC.12.20/09

Malhotra, S., & Alagh, T. (2004). Dreaming the Nation: Domestic Dramas in Hindi Films Post 1990. *South Asian Popular Culture, 2*(1), 19–37. doi:10.1080/1474668042000210492

Mallon, F. E. (1994). The Promise and Dilemma of Subaltern Studies: Perspectives from Latin American History. *The American Historical Review, 99*(5), 1491–1515. doi:10.2307/2168386

MAMBROL. N. (2018, January 24). Key Theories of Mikhail Bakhtin. Key Theories of Mikhail Bakhtin – Literary Theory and Criticism. https://literariness.org/2018/01/24/key-theories-of-mikhail-bakhtin/

Mancini, G. A. (2013). Can the effects of brand user imagery congruence be important if the brand is not as well known? *Journal of International Management Studies, 13*(4), 85–90. doi:10.18374/JIMS-13-4.10

Mankekar, P. (2000). National texts and gendered lives: An ethnography of television viewers in a North Indian city. *American Ethnologist, 20*(3), 299–322.

Mansouri, N. (2023). *Money, magic, and machines: International Telecommunication Union and liberalisation of telecommunications networks and services*. London Review of International Law. doi:10.1093/lril/lrad011

Manuel, P. (2014). The regional North Indian popular music industry in 2014: From cassette culture to cyberculture. *Popular Music, 33*(3), 389–412. doi:10.1017/S0261143014000592

Marwan, K. (2008). *Hybridity: The Cultural Logic of Globalization*. Temple University Press.

Marwick, A. E. (2015). Instafame: Luxury selfies in the attention economy. *Public Culture, 27*(1 75), 137-160.

Marwick, A. E., & boyd. (2011). To see and be seen: Celebrity practice on Twitter. *Convergence (London), 17*(2), 139–158. doi:10.1177/1354856510394539

Marzuki, S. Z., Zreik, M., Buyong, S. Z., Goyipnazarov, S., Mirzaliev, S., & Bakhramovna, I. A. (2023). An Importance Satisfaction Analysis of Hosting Events in Hangzhou, China: A Pilot Study. *Advances in Social Sciences Research Journal, 10*(6.2), 107–117. doi:10.14738/assrj.106.2.14996

Massimo Ragnedda, B. M. 2018. Digital Inclusion Empowering People Through Information and Communication Technologies. In: Digital Inclusion. An International Comparative Analyses. s.l.:Lexington Book.

Mazumdar, R. (2007). *Bombay Cinema: An Archive of the City*. University of Minnesota Press.

McDonald, M. (1995). *Representing Women*. Arnold.

McDonald, P., & Smith-Rowsey, D. (2016). *The Netflix effect: Technology and entertainment in the 21st century*. Bloomsbury Publishing USA.

McLuhan, M. (1964). *The extensions of man*.

McQuail, D. (2010). The Uses and Gratification Approach. In D. McQuail (Ed.), *McQuail's Mass Communication Theory* (6th ed.). Sage.

McQuail, D., & Deuze, M. (2020). *McQuail's Media and Mass Communication Theory*. SAGE Publications.

Meenakshi Sundareshwar Review. Meenakshi Sundareshwar treads off the beaten path. (2021, November 5). Meenakshi Sundareshwar Review: Meenakshi Sundareshwar Treads off the Beaten Path. https://timesofindia.indiatimes.com/web-series/reviews/hindi/meenakshi-sundareshwar/ottmoviereview/87282604.cms

Meenakshi Sundareshwar review: A languid film let down by its unbelievable conflict. (2021, November 5). The Indian Express. https://indianexpress.com/article/entertainment/movie-review/meenakshi-sundareshwar-movie-review-a-languid-film-let-down-by-its-unbelievable-conflict-7609080/

Mehta, S. (2022, March 17). End of the stereotypical Bollywood 'heroine'? How OTT is empowering women-led narratives. *Khaleej Times*. https://www.khaleejtimes.com/arts-and-culture/end-of-the-stereotypical-bollywood-heroine-how-ott-is-empowering-women-led-narratives

Menon, D. (2022). Purchase and continuation intentions of over-the-top (OTT) video streaming platform subscriptions: a uses and gratification theory perspective. *Telematics and Informatics Reports*, *5*, 01–15.

Menon, D. (2022). Purchase and continuation intentions of over-the-top (OTT) video streaming platform subscriptions: A uses and gratification theory perspective. *Telematics and Informatics Reports*, *5*, 100006. doi:10.1016/j.teler.2022.100006

Merriam, S. (2002). *Qualitative research in practice: Examples for discussion and analysis*. Jossey-Bass.

Miller, R., Liu, K., & Ball, A. F. (2020). Critical Counter-Narrative as Transformative Methodology for Educational Equity. *Review of Research in Education*, *44*(1), 269–300. doi:10.3102/0091732X20908501

Ming, Y., & Nikolaos, B. (2009). Analysis, Modeling And Simulation Of Network Traces For Video Transmission Over Ip Networks. Jacksonville State University, AL, USA

Mishra, M. S., & Deewan, G. (2019). Representation of Women in Post Colonial Hindi Movies. *Think India Journal*. *22*(41), 127-140. https://thinkindiaquarterly.org/index.php/think-india/article/view/19270

Mishra, A. (2022). *How OTTPs Help in Education*. PrepInsta.

Mishra, V. (2002). *Bollywood Cinema: Temples of Desire*. Routledge.

Mlambo, A. (2013). Becoming Zimbabwe or becoming Zimbabwean: Identity, nationalism and state-making. *Africa Spectrum*, *48*(1), 49–70. doi:10.1177/000203971304800103

Mohammed, Z. (2020). *Halal Love Story*. Papaya Films.

Mohan, A. (2017). *Digital Transformation: How is OTT Redefining Media*. CASBAA, the Association for digital multichannel TV.

Molala, T. S., & Makhubele, J. C. (2021). The Connection Between Digital Divide And Social Exclusion: Implications For Social Work. *Humanities & Social Sciences Reviews*, *9*(04), 194–201. doi:10.18510/hssr.2021.9427

Moniz, C. (2023, February 14). 9000 Crore in advertisement spend riding on cricket 2023. *Financial Express*. https://www.financialexpress.com/business/brandwagon-9000-crore-in-advertisement-spend-riding-on-cricket-in-2023-2980844/

Moniz, C. (2023, February 4). To swing in favor of digital. *Financial Express*. https://www.financialexpress.com/business/brandwagon-ipl-2023-to-swing-in-favour-of-digital-viacom-2971263/

Moniz, C. (2023, March 31). Bandwagon tv advertisers stare at slow pitch as Ipl begins. *Financial Express*. https://www.financialexpress.com/business/brandwagon-tv-advertisers-stare-at-slow-pitch-as-ipl-begins-today-3027757/

Montero, J., & Finger, M. (2021). *The rise of the new network industries: Regulating digital platforms*. Routledge. doi:10.4324/9781003141327

Morganosky, M. A., & Cude, B. J. (2000). Consumer response to online grocery shopping. *International Journal of Retail & Distribution Management*, *28*(1), 17–26. doi:10.1108/09590550010306737

Morris, R. C. (Ed.). (2010). *Can the Subaltern Speak?: Reflections on the History of an Idea*. Columbia University Press., https://www.jstor.org/stable/10.7312/morr14384

Mpofu, S. (2014). *Public and Diasporic Online Media in the Discursive Construction of National Identity: A Case of Zimbabwe*. [Unpublished PhD thesis, University of the Witwatersrand].

Mpofu, S. (2021). Social Media and COVID-19: Taking Humour During Pandemics Seriously. *Digital Humour in the Covid-19 Pandemic: Perspectives from the Global South*, 1-16.

Mubarki, M. A. (2018, December 06). Looking beyond Post-Colonial Modernity: Subaltern Masculinity and the Mumbai Cinema. *South Asia*, *41*(4), 723–743. doi:10.1080/00856401.2018.1527430

Mudaliar, A. (2023, February 13). The evolution of content in India's Ott Ecosystem. *Times of India Blog*. https://timesofindia.indiatimes.com/blogs/voices/the-evolution-of-content-in-indias-ott-ecosystem/

Mullan, L. (2020). *Massive Interactive: How over-the-top contentservices are shaking up teleco and mediaindustries*. IT Procurement. Technology Magazine. https://technologymagazine.com/it-procurement/massive-interactive-how-over-top-content-services-are-shaking-teleco-and-media-industries

Mushtaq, W. (2020, November 02). *Churails: Why a feminist detective show was banned in Pakistan*. BBC News. https://www.bbc.com/news/world-asia-54629439

Nabi, R. L., Biely, E. N., Morgan, S. J., & Stitt, C. R. (2003). Reality-based television programming and the psychology of its appeal. *Media Psychology*, *5*(4), 303–330. doi:10.1207/S1532785XMEP0504_01

Naik, A. (2023b, March 2). Regulation of OTT Platforms: Challenges and solutions. *WION*. https://www.wionews.com/opinions-blogs/regulation-of-ott-platforms-challenges-and-solutions-567573

Nandy. Ashis. (1981). The Popular Hindi Film: Ideology and First Principles.*India International Centre Quarterly*. Vol. 8(1), 89-96. Retrieved from:https://www.jstor.org/stable/23001938

Napoli, P. M. N. (2019). Media as platforms: Definition, dynamics, and drivers. *Journal of Communication*, *69*(2), 277–301.

Narayanan, M. (2019). *Kumbhlangi Nights*. Fahadh Faasil and Friends.

Narayanan, M. (2020). *C U Soon*. Fahadh Faasil and Friends.

Nasir, M. (2021). *5 Reasons Why OTTPs Will Be The New Education Normal For Blended Learning*. IPSB Technology. Kuala Lumpur. https://www.ipsb.com.my/reasons-ott-platform-blended-learning/

Nath, P. J. (2021, August 26). *Streaming a new reality: Malayalam movies in the Ott World*. BusinessLine. https://www.thehindubusinessline.com/blink/cover/ott-lifeline-a-moment-of-reckoning-for-malayalam-cinema/article36111055.ece#:~:text=C%20U%20Soon%2C%20Joji%2C%20Sara's%2C,knees%20by%20a%20rampaging%20virus

Nayak, S. C., & Biswal, S. K. (2021). OTT Media streaming in COVID-19 lockdown: The Indian experience. *Media Watch*, *12*(3), 440–450. doi:10.15655/mw_2021_v12i3_166019

Ncube, B. J. (2018). *Radio Mthwakazi and the construction of Ndebele Ethnic Identity*. PhD thesis. University of Johannesburg.

Ndlovu, M. (2017). *Site of remembrance and forgetting: New media reconstructions of distinct Ndebele collective memory and history in the context of hegemonic Zimbabwean nationalism*. [Unpublished PhD thesis, Stellenbosch University].

Ndlovu-Gatsheni, S. (2009). *Do 'Zimbabweans' exist? Trajectories of nationalism, national identity formation and crisis in a postcolonial state*. Peter Lang.

NEC (2015). *Best Practices Guide for Over-the-Top (OTT) Deployment*. UNIVERGE Cloud Services. NEC.

Netflix. (2013, December 13). Netflix declares binge watching the new normal. *PR Newswire*. https://www.prnewswire.com/news-releases/netflix-declares-binge-watching-is-the-new-normal-235713431.html

*Netflix's Monika Shergill on understanding audiences, stories and the storytelling journey*. (2021, May 6). Afaqs! https://www.afaqs.com/news/media/netflixs-monika-shergill-on-understanding-audiences-stories-and-the-storytelling-journey

Newman, T., Peck, J., Harris, C., & Wilhide, B. (2013). *Social media in sport marketing*. Holcomb Hathaway Publishers., doi:10.4324/9781315212975

Nguyen, A. (2022). Digital Inclusion: Social Inclusion in the Digital Age. In *Handbook of Social Inclusion, Research & Practices in health and social care* (pp. 1–13). Springer Nature Switzerland.

Nguyen, A. 2023. Social Inclusion in the Digital Age. In: V. H. V. College of Health Sciences, ed. Handbook of Social Inclusion. Springer Nature Switzerland.

Nielsen.com. (2022). *Sports fandom is increasing, powered by new digital platforms, global report finds*. Nielsen. https://www.nielsen.com/news-center/2022/sports-fandom-is-increasing-powered-by-new-digital-platforms-global-report-finds/

Nigam, A. (2022). Online gaming and OTT consumption: An exploratory study of generation Z. *Journal of Promotion Management*, *28*(4), 420–442. doi:10.1080/10496491.2021.2008576

Niyogi, A. (2023, February 2). ZEE5 series Roktokorobi is a taut mystery thriller around family secrets and mental health. *Telegraph India*. https://www.telegraphindia.com: https://www.telegraphindia.com/entertainment/zee5-series-roktokorobi-is-a-taut-mystery-thriller-around-family-secrets-and-mental-health/cid/1915890

Noiriel, G. (1994). Foucault and History: The Lessons of a Disillusion. *The Journal of Modern History*, *66*(3), 547–568. https://www.jstor.org/stable/2124485. doi:10.1086/244886

Novak, T. P., Hoffman, D. L., & Yung, Y. (2000). Measuring the Customer Experience in Online Environments: A Structural Modeling Approach. *Marketing Science*, *19*(1), 22–42. doi:10.1287/mksc.19.1.22.15184

NPR. (2020). *'The Last Dance' tries to cut through all the Bulls*. NPR. https://www.npr.org/2020/04/30/848229079/the-last-dance-tries-to-cut-through-all-the-bulls

O'Connor, B., & Boyle, R. (1993). Dallas with balls: Televized sport, soap opera and male and female pleasures. *Leisure Studies*, *12*(2), 107–119. doi:10.1080/02614369300390101

OECD. (2018). *BRIDGING THE DIGITAL GENDER DIVIDE: INCLUDE, UPSKILL, INNOVATE*. Australian government.

OET. (2022). *Advancing Digital Equity for All Barriers & Strategies*. OET. https://tech.ed.gov/advancing-digital-equity-for-all/barriers-and-strategies/

OMG. (2022). *DECISION MODEL AND NOTATION (DMN)*. OMG. https://www.omg.org/dmn/

Osama, E. (2001). *Media imperialism in the global era*. Middle East News Online.

Ott, B. L., & Mack, R. L. (2020). *Critical media studies: An introduction*. John Wiley & Sons.

Otto, B. K. (2007). *Fools Are Everywhere: The Court Jester around the World*. University of Chicago Press.

P T, Y. (2020, October 18). *Dileesh Pothan and reframing the rural landscape in Malayalam cinema • reading room co*. Reading Room Co. https://readingroomco.com/2020/10/18/dileesh-pothan-and-the-reframing-of-the-rural-landscape-in-malayalam-cinema/

P, N. S., & Thomas, P. C. (2022). Understanding the Factors Lead to the Adoption of Content Localization Strategy by OTTP in India: A Review. *International Journal of Scientific Research and Management*, *10*(05), 19–23. doi:10.18535/ijsrm/v10i5.com01

Paige, T. (2005). Golden age is now for sports documentaries: Genre gains popularity among programmers. *Multichannel News*, *26*(33), 16.

Panjiar, T., Chaudhuri, R., & Waghre I, P. (2023, June 12). *A Public Brief on Demand on Fair Share*. Internet Freedom Foundation. https://content.internetfreedom.in/api/files/divco3ywedt9rpe/sce88gbwixr5ro4/fair_share_analysis_brief_1_zq0MuI6hrX.pdf/?ref=static.internetfreedom.in

Papacharissi, Z., & Mendelson, A. L. (2007). An exploratory study of reality appeal: Uses and gratifications of reality TV shows. *Journal of Broadcasting & Electronic Media*, *51*(2), 355–370. doi:10.1080/08838150701307152

Pappas, Ch. (2022). *Cloud Based e-Learning Platforms: 10 Advantages eLearning Professionals Need To Know*. Docebo. https://www.docebo.com/learning-network/blog/cloud-based-elearning-platforms/

Parameswaran, A. (2022, January 21). Appointment Viewing will live on. *Business Standard*.

Patel, K. (2015). Social Media in the Indian Context: -New flavor of the Season. *Global Journal Of Multidisciplinary Studies*, *4*(6), 2680-270.

Patel, K. (2017). An Analysis of Impact of Personal Communication Technology on Public Policy Making Process In India. in*ternational Journal of Emerging Trends in Information & Knowledge Management*, *1*(1). Retrieved 06 12, 2020.

Patel, M. K., Khadia, R., & Awasya, G. (2020). A study: OTT viewership in "lockdown" and viewer's dynamic watching experience. *International Journal on Transformations of Media. Journalism & Mass Communication*, *5*(2), 10–22.

Patil, P. (2023, May 17). *Dahaad: A Crime Thriller Exploring the Hard-Hitting and Grim Realities of Modern Day India*. Feminism in India. https://feminisminindia.com/2023/05/17/dahaad-a-crime-thriller-exploring-the-hard-hitting-and-grim-realities-of-modern-day-india/

Patil, R. (2022). Streaming Boom Accelerates Digital Transformation Of Network Operators. *Forbes*. https://www.forbes.com/sites/forbestechcouncil/2023/02/01/streaming-boom-accelerates-digital-transformation-of-network-operators/?sh=7885cd34554e

Patra, P. (2023, June 06). Anatomy of a villain: Why Vijay Varma's Anand in Dahaad is just a few steps away from greatness. *The Telegraph Online*. https://www.telegraphindia.com/entertainment/anatomy-of-a-villain-why-vijay-varmas-anand-in-dahaad-is-just-a-few-steps-away-from-greatness/cid/1942774

Pattanaik, D. (2010). *Jaya*. Penguin Books.

Paul, S. (2017). A new public sphere? English-language stand-up comedy in India. *Contemporary South Asia*, *25*(2), 121–135. doi:10.1080/09584935.2017.1321618

Pellissery, L. J. (2018). *Ee. Ma. Yau.* OPM Cinemas.

Pellissery, L. J. (2017). *Angamaly Diaries*. Friday Film House.

Pellissery, L. J. (2019). *Jallikattu*. Opus Penta.

Pena, L. L. (2015). Breaking Binge: Exploring The Effects Of Binge Watching On Television Viewer Reception. *Dissertations - ALL. 283.* https://surface.syr.edu/etd/283, 9.

PennState. (2019). The Digital Divide. In: INFORMATION, PEOPLE, AND TECHNOLOGY. Research Gate.

*Performance Indicators Reports*. (n.d.). Telecom Regulatory Authority of India. https://www.trai.gov.in/release-publication/reports/performance-indicators-reports

Perrin, A. (2015). *Social media usage: 2005-2015*. Pew Research Center.

Perse, E. M., & Ferguson, D. A. (1993). The impact of the newer television technologies on television satisfaction. *The Journalism Quarterly*, *70*(4), 843–853. doi:10.1177/107769909307000410

Pertierra, R., &. C. J. V. (2018). Digital inclusion in the Global South: An analysis of mobile and OTT services in the Philippines. *Telematics and Informatics*, *35*(03), 540–550.

Pettersson, R. (1993). Visual information. *Educational Technology*.

Pew Research Center. (2019). *Defining generations: Where Millennials end and Generation Z begins*. Pew Research. https://www.pewresearch.org/short-reads/2019/01/17/where-millennials-end-and-generation-z-begins/

Pillai, S. (2021, October 21). *Ott revolution: A boon for filmmakers*. Forbes India. https://www.forbesindia.com/article/south-celebrity-special-2021/ott-revolution-a-boon-for-filmmakers/71057/1

Pine, B. J., & Gilmore, J. H. (2020). *The experience economy: Competing for customer time, attention, and money*. Harvard Business Review Press.

Pisonia, G., Gijlersc, H., Ha Nguyend, T., & Chene, H. (2021). Collaboration patterns in students' teams working on business cases. *CEUR Workshop Proceedings (CEUR-WS.org)*. CEUR.

Pittman, M., & Sheehan, K. (2015). Sprinting a media marathon: Uses and gratifications of binge-watching television through Netflix. *First Monday*, *20*(10). Advance online publication. doi:10.5210/fm.v20i10.6138

Piu, P. (2023, September 12). The journey of subalternity in Gayatri Spivak's work: Its sociological relevance. *The Sociological Review*, *0*(0), 1–19. doi:10.1177/00380261231194495

Poikayil, G. (2023, May 29). Plight of Cyanide Mohan's victim's family proves true Dahaad's dissection of casteism, patriarchy. *Onmanorama*. https://www.onmanorama.com/news/kerala/2023/05/29/plight-cyanide-mohan-victim-family-proves-true-dahaad-dissection-casteism-patriarchy-amazon-prime.html

Polderman, J., & Willems, J. (1998). *Introduction to Mathematical Systems Theory: A Behavioral Approach. (Texts in Applied Mathematics)* (2nd ed.). Springer Verlag. doi:10.1007/978-1-4757-2953-5

Postle, G. (2003). Online Teaching and Learning in Higher Education: A Case Study. Department of Education, Science and Training. University of Southern Queensland, USQ. Canberra. Australia.

Pothan, D. (2021). *Joji*. Bhavana Studios.

Prakash, G. (1994). Subaltern Studies as Postcolonial Criticism. *The American Historical Review*, *99*(5), 1475–1490. doi:10.2307/2168385

Prakash, G. (2012). Book Review: The Small Voice of History. *The Indian Economic and Social History Review*, *49*(1), 147–148. doi:10.1177/001946461104900105

Praveen, S. R. (2022, August 13). Explained: Tussle between Malayalam cinema exhibitors and producers over Ott releases. *The Hindu*. https://www.thehindu.com/news/national/kerala/malayalam-cinema-exhibitors-producers-clash-over-ott-releases-explained/article65764558.ece

PTI. (2021, August 15). Balaji Telefilms to invest in original content with affordable subscription for OTT platform. *New India Express*. https://www.newindianexpress.com: https://www.newindianexpress.com/business/2021/aug/15/balaji-telefilms-to-invest-in-original-content-with-affordable-subscription-for-ott-platform-2345046.html

Puthiyakath, H. H., & Goswami, M. P. (2021). Is Over the Top Video Platform the Game Changer over Traditional TV Channels in India? A Niche Analysis. *Asia Pacific Media Educator*, *31*(1), 133–150. doi:10.1177/1326365X211009639

Putri, N., & Yusof, S. M. (2009). Critical success factors for implementing quality engineering tools and techniques in malaysian's and indonesian's automotive industries: An Exploratory Study. *Journal Proceedings of the International MultiConference of Engineers and Computer Scientists.*, *2*, 18–20.

Qiang, X. (2011). Liberation technology: The battle for the Chinese internet. *Journal of Democracy*, *22*(2), 47–61. doi:10.1353/jod.2011.0020

Quiles, L. (2019, April 26). *Original Content as the Selling Point for OTT.* Intraway. https://www.intraway.com: https://www.intraway.com/blog/original-content-selling-point-ott/

Qureshi, A. (2020, August 24). Review: *Why Churails Is A Revolutionary Show For Pakistan*. Feminism in India. https://feminisminindia.com/2020/08/24/churails-feminist-review-pakistan-show/

Qureshi, N. (2020, March 26). *Original content is way forward for OTT platforms: Abhimanyu Singh of Contiloe Pictures.* Best Media Info. https://bestmediainfo.com: https://bestmediainfo.com/2020/03/original-content-is-way-forward-for-ott-platforms-abhimanyu-singh-of-contiloe-pictures

Rabel, G. (2020, November 21). 'Men Will Suffer': A Review of Churails. *The Gazelle*. https://www.thegazelle.org/issue/190/review-of-churails

Rachelle Bosua, M. G. (2021). Access to Flexible Work Arrangements for People With Disabilities: An Australian Study. In M. B. L. Queen Aigbefo (Ed.), *Anywhere Working and the Future of Work* (pp. 134–161). IGI. doi:10.4018/978-1-7998-4159-3.ch006

Radhakrishnan, R. (1994). Postmodernism and the Rest of the World. *Organization*, *1*(2), 305–340. doi:10.1177/135050849412004

Raghavan, V. (2020). In The comic element in Sanskrit literature. Dr. V. Raghavan Centre for Performing Arts.

Raghavendra, M. K., & Pillai, M. (2017). Bearing Witness: Malayalam Cinema and the Making of Keralam. In Beyond Bollywood: The Cinemas of South India (pp. 267–330). HarperCollins Publishers India.

Rahman, K. T., & Arif, M. Z. U. (2021). Impacts of Binge-Watching on Netflix during the COVID-19 pandemic. *South Asian Journal of Marketing*, *2*(1), 97–112. doi:10.1108/SAJM-05-2021-0070

Raj, S. J., & Suresh, A. K. (2021). *Postmodern transpositions of Shakespeare in Malayalam cinema–a transformative discourse of regional tragedy.*

Rajadhakshya. (2016. Indian Cinema: A Very Short Introduction (1st ed.). Oxford.

Rajdhakshya, A. (2003). The 'Bollywoodization' of the Indian Cinema: Cultural Nationalism in a Global Arena. *Inter-Asia Cultural Studies*, *4*(1), 25–39. doi:10.1080/1464937032000060195

Ramasubramanian, V. (2021). Disruption in the Indian OTT industry. *International Journal of Information Management*, *56*, 102–106.

Ramnath, N. (2020, August 11). 'Churails' review: Veils, deception and secrets in Zee5 web series set in Karachi. *Scroll.in.* https://scroll.in/reel/969985/churails-review-veils-deception-and-secrets-in-zee5-web-series-set-in-karachi

Ramnath, N. (2021, May 28). 'Maharani' review: This potboiler about Bihar politics never stops simmering.*Scroll.in.* https://scroll.in/reel/995992/maharani-review-this-potboiler-about-bihar-politics-never-stops-simmering

Ramon, X., & Rojas-Torrijos, J. L. (2023). European Public Service Media, Disability Sports and Cultural Citizenship in the Digital Age: An Analysis of Agenda Diversity in the Tokyo 2020 Paralympic Games. *Journalism and Media*, *4*(1), 289–303. doi:10.3390/journalmedia4010020

Ranger, T. (2004). Historiography, Patriotic History and the History of the Nation: The Struggle over the Past in Zimbabwe. *Journal of Southern African Studies, 30*(2), 215–234. doi:10.1080/0305707042000215338

Ranipeta, S. (2023, March 3). *Move over Netflix, Prime Video, Hotstar! Regional OTT platforms are new favourites among Indian consumers.* CNBC TV. https://www.cnbctv18.com: https://www.cnbctv18.com/business/move-over-netflix-prime-video-hotstar-regional-ott-platforms-are-new-favourites-among-indian-consumers-16095301.htm

Ravi, R. (2016). *Kammatipaadam.* Global United Media.

Ray, S. (1976). *Our films, their films.* Orient Longman.

Reams, M. (2020). *Nike and Jordan logos appear over 100 times in a single episode of The Last Dance.* AdWeek. https://www.adweek.com/brand-marketing/nike-and-jordan-logos-appeared-over-100-times-in-one-episode-of-the-last-dance/

Reddick, C. G., Enriquez, R., Harris, R. J. & Sharma, B. (2020). Determinants of broadband access and affordability: An analysis of a community survey on the digital divide. *Elsevier - PMC COVID-19 Collection.*

ReelnReel. (2022). *OTT Content Pricing Strategies for OTT Platforms.* ReelInReel. https://reelnreel.com/ott-content-pricing-strategies-for-ott-platforms/

Rein, I., Kotler, P., & Shields, B. (2006). *The elusive fan.* McGraw-Hill.

Reiss, S. (2019). *Sport, Representation and culture in the modern world, 1920-2020.* Faculty Research and Creative Activities Symposium, Northeastern Illinois University. https://neiudc.neiu.edu/cgi/viewcontent.cgi?article=1095&context=frcas

Reiss, S., & Wiltz, J. (2004). Why people watch reality TV. *Media Psychology, 6*(4), 363–378. doi:10.1207/s1532785xmep0604_3

Research and Ranking. (2021, April 3) *Rise of OTT Platforms in India* [Video]. YouTube. https://www.youtube.com/watch?v=Pn3paQBgRIg

Research, P. (2022). *ICTOver the Top (OTT) Market.* North America: Precendence Research.

Rettie, R. (2003). *Connectedness, awareness, and social presence.* Paper presented at the 6th International Presence Workshop, Aalborg, Denmark.

Reuters. (2023, July 12). Disney considers selling or partnering its India business source. *Financial Express.* https://www.financialexpress.com/business/brandwagon-disney-considers-selling-or-partnering-its-india-business-source-says-3166592/

Review, F. Kartik Aaryan is a revelation in this dark psycho-thriller. (2022, December 2). Freddy Review: Kartik Aaryan Is a Revelation in This Dark Psycho-thriller. https://timesofindia.indiatimes.com/entertainment/hindi/movie-reviews/freddy/ottmoviereview/95835669.cms

Review, T. (2023). *Kenya's Digital Future: A Journey of Connectivity.* Innovation and Prosperity. [Online]

Richardson, J. (2007). *Analysing Newspapers: An Approach from Critical Discourse Analysis.* Palgrave Macmillan. doi:10.1007/978-0-230-20968-8

Rizvi, A. (2020, August 15). Watch Churails for the catharsis, not for the politics. *Images.* https://images.dawn.com/news/1185653

Rohit Jacob Jose. (2020). Factors influencing the shift from traditional TV to OTT platforms in India. *International Journal of Advanced Science and Technology, 29*(No. 7s), 4044–4051.

Roshan. (2017). *5 benefits of Video on Demand over traditional TV*. muvi. https://www.muvi.com/blogs/5-benefits-video-demandtraditional-tv.html

Rowe, P. (2005). *East Asia modern: shaping the contemporary city*. Reaktion books.

Roy, P. (2016). The Aesthetics of Emotional Acting: An Argument for a Rasa-based Criticism of Indian Cinema and Television. (Doctoral Thesis, University of Edinburgh). Retrieved: https://era.ed.ac.uk/handle/1842/22910

Roy, P. K. (2014). *India's era of digital activism*. Aljazeera. https://www.aljazeera.com/opinions/2014/1/6/indias-era-of-digital-activism

Rubin, A. (1983). Television uses and gratifications: The interactions of viewing patterns and motivations. *Journal of Broadcasting*, *27*(1), 37–51. doi:10.1080/08838158309386471

Rubin, A. M. (2009). Uses-and-gratifications perspective on media effects. In J. Bryant & M. B. Oliver (Eds.), *Media effects: Advances in theory and research* (3rd ed.). Routledge.

Rudolf Arnheim's Formalist Film Theory. (2021, February 8). YouTube. https://www.youtube.com/watch?v=_qvrzA1b-p0

Ruhiya Seward. (2020). *Understanding digital access and use in the Global South*. IDRC.

Sahoo, S., & Das, R. (2018). Is English stand-up comedy in India intersectional. *Language in India*, *18*(9), 240–249.

Saini, C. P., & Gupta, N. (2020). Interrelated factors driving the purchase of over-the-top television subscription services: A study using exploratory factor analysis and the decision-making trial and evaluation laboratory method. *Applied Marketing Analytics*, *6*(1), 73–84.

Sankaralingam, K., Ferris, M., Nowatzki, T., Estan, C., Wood, D., & Vaish, N. (2013). *Optimization and Mathematical Modeling in Computer Architecture*. Morgan & Claypool Publishers.

Santhanam, R. (2020, April 03). Do OTT platforms provide more space for sociopolitical content? *The Hindu*. https://www.thehindu.com/opinion/op-ed/do-ott-platforms-provide-more-space-for-sociopolitical-content/article31240686.ece

SAP. (2022). *Conquering Africa's Digital Divide: SAP Africa Code Week Empowers 2.6 million Youth in 2022*. SAP.

Saraswat, S. (2023). *Revolutionizing Entertainment: The Growth of OTT Platforms in India and Around the World*. LinkedIn. https://www.linkedin.com/pulse/revolutionizing-entertainment-growth-ott-platforms-india-saraswat/

Sarkar, G. (2023, July 5). Bengali OTT players prioritise market growth over competition. *India Television*. https://www.indiantelevision.com: https://www.indiantelevision.com/iworld/over-the-top-services/bengali-ott-players-prioritise-market-growth-over-competition-180705

Sarkar, R. (2021, September 27). *Hoichoi announces 20 original shows*. Cinestaan. https://www.cinestaan.com: https://www.cinestaan.com/articles/2021/sep/27/31469/hoichoi-announces-20-original-shows

Sarkar, S., Shah, A. M., & Dutta, S. K. (2020). Bollywood in Transition: Chiragdeep's Reflection of the Film Industry. *Emerging Economies Cases Journal*, *2*(2), 79–86. doi:10.1177/2516604220968537

Sarkar, S., & Soni, M. J. (2023). Is Reality Less Real? Proliferation of Sports Documentaries in a Locked-Down World. In B. Basu, M. Desbordes, & S. Sarkar (Eds.), *Sports Management in an Uncertain Environment* (pp. 221–253). Springer. doi:10.1007/978-981-19-7010-8_10

Sarrazin, N. (2008). Celluloid love songs: Musical *modus operandi* and the dramatic aesthetics of romantic Hindi film. *Popular Music, 27*(3), 393–411. https://www.researchgate.net/publication/291909806_Celluloid_love_songs_Musical_modus_operandi_and_the_dramatic_aesthetics_of_romantic_Hindi_film_Popular_Music_2008_27_3_393_DOI101017_S0261143008102197. doi:10.1017/S0261143008102197

Scaled Agile. (2022). *Succeeding with AI in SAFe*. Scaled Agile, Inc. https://www.scaledagileframework.com/succeeding-with-ai-in-safe/

Schechner. (2001). *Rasaesthetics*. California State University, Northridge.

Schiller, H. (1976). *Cultural Domination: Sources, Context and Current Styles. Communication and Cultural Domination*. International Arts and Science Press.

Schmitt, B. (1999). *Experiential marketing: How to get customers to sense, feel, think, act, relate to your company and brands*. Free Press. doi:10.1362/026725799784870496

Schneider, A. (2009, January 1). *Hum Aapke Hain Koun. . .!: An Example of the Coding of Emotions in Contemporary Hindi Mainstream Film*. Projections; Berghahn Books. https://doi.org/ doi:10.3167/proj.2009.030205

Senft, T. M. (2013). *Microcelebrity and the branded self. A Companion to New Media Dynamics,* 346-354.

Sengupta, A. (2021, October 19). Subterranean parallels: Common threads between Satyajit Ray, Mrinal Sen and Ritwik Ghatak. *The Hindu*. https://frontline.thehindu.com/cover-story/subterranean-parallels-common-threads-between-satyajit-ray-mrinal-sen-and-ritwik-ghatak/article37018238.ece

Sethi, B. K., & Nayak, A. (2020). Speaking and Speaking Differently: Language as Resistance, Liberation and Celebration in Dalit Women's Autobiographical Narratives. *Contemporary Voice of Dalit, 12*(2), 138–148. doi:10.1177/2455328X20924882

Shah, H. (2018, April 30). *How Netflix became a $100 billion company in 20 Years*. Product Habits. https://producthabits.com/how-netflix-became-a-100-billion-company-in-20-years/

Shah, S., & Mehta, N. (2023). Over-the-top (OTT) streaming services: Studying users' behaviour through the UTAUT model. *Management and Labour Studies, 48*(4), 531–547. doi:10.1177/0258042X221137438

Sharma, K. (2020) Will movies releasing directly on OTT platforms become a norm post-Covid? Experts think otherwise. *Business insider*. https://www.businessinsider.in/advertising/media/article/will-movies-releasingdirectly-on-ott-platforms-become-a-norm-post-covid-experts-thinkotherwise/articleshow/75808931.cms

Sharma, K. S. (2020). Are OTT platforms taking over theatrical experience? *Economic Times*. https://brandequity.economictimes.indiatimes.com/news/media/are-ott-platforms-taking-over-theatrical experience/73887372

Sharma, S. (2021). *India's Zee is banking on a Pakistani show for its success*. Aljazeera. https://www.aljazeera.com/economy/2021/12/10/indias-zee-is-banking-on-a-pakistani-show-for-its-success

Sharma, V. (2022, July 19). How OTT has changed the story narrative gameplan of content creators. How OTT Has Changed the Story Narrative Gameplan of Content

Sharma, A., & Banerjee, A. (2022). Socio-Economic Determinants of Digital Divide in India. *Demography India, 51*(1), 78–92.

Sharma, K., & Lulandala, E. E. (2023). OTT platforms resilience to COVID-19 – a study of business strategies and consumer media consumption in India. *The International Journal of Organizational Analysis, 31*(1), 63–90. doi:10.1108/IJOA-06-2021-2816

Sharma, S. (2017). Transnational Publics, Nationalist Ideology, and the Women Question in Hindi Cinema. *South Asia*, *41*(1), 106–120. doi:10.1080/00856401.2017.1357094

Shastri, P. (2021, March 11). comedy was Ott Remedy for covid blues. *The Times of India*. https://timesofindia.indiatimes.com/city/ahmedabad/comedy-remedy-for-covid-blues-ott-platform-survey-reveals-fav-genre-during-lockdown/articleshow/81440605.cms

Shehzad, A. (2017). *India and Pakistan: A Land of Opportunity for Smartphones.* Mystory. https://yourstory.com/mystory/729539f5c4-india-and-pakistan-a

Shen, H. (2016). China and global internet governance: Toward an alternative analytical framework. *Chinese Journal of Communication*, *9*(3), 304–324. doi:10.1080/17544750.2016.1206028

Shrikrishna, A. (2020). Asim Abbasi's 'Churails', a Pakistani web series on women divided by class privilege. *The New Indian Express*. https://www.newindianexpress.com/entertainment/hindi/2020/aug/27/asim-abbasis-churails-a-pakistani-web-series-on-women-divided-by-class-privilege--2188677.html

Sidhardhan, S. (2020, October 6). Fahadh, Mahesh Narayanan donated rs 10 lakh to fefka: Malayalam movie News - Times of India. *The Times of India*. https://timesofindia.indiatimes.com/entertainment/malayalam/movies/news/fahadh-mahesh-narayanan-donate-rs-10-lakh-to-fefka/articleshow/78507178.cms

Siegel, L. (1989). *Laughing matters: Comic tradition in India*. Motilal Banarsidass Publisher.

Singh, D. (2020) How is coronavirus impacting the streaming platforms with an increasing appetite of viewers. *Financial express*. https://www.financialexpress.com/brandwagon/how-is-coronavirus-impacting-thestreaming-platforms-with-an-increasing-appetite-of-viewers/1919916/

Singh, P. (2019). New Media as a Change Agent of Indian Television and Cinema: A Study of Over the Top Platforms. *Journal of Content, Community & Communication, 9*(5), 131.

Singh, N., Arora, S., & Kapur, B. (2022). Trends in over the top (OTT) research: A bibliometric analysis. *VINE Journal of Information and Knowledge Management Systems*, *52*(3), 411–425. doi:10.1108/VJIKMS-12-2021-0316

Singh, P., & Sharma, P. (2023). Entertaining the Royalty. A forgotten tradition of the Vidushaka in Ancient Indian History. *Historia i Świat*, *12*, 237–254. doi:10.34739/his.2023.12.14

Sinha, A. (2021). *Case study: Reimagining the OTT experience*. Bootcamp. https://bootcamp.uxdesign.cc/reimagining-the-ott-experience-ui-ux-case-study-6cc42c18b2ca

Sins, R. K. (2018). The commenting persona: Reception theory and the digital rhetorical audience. *The Journal of Medical Research*, *11*(1), 55–70.

Smilansky, S. (2017). *Experiential marketing: A practical guide to interactive brand experiences*. Kogan Page Publishers.

Smith, A. N., Fischer, E., & Yongjian, C. (2012). How does brand-related user-generated content differ across YouTube, Facebook, and Twitter? *Journal of Interactive Marketing*, *26*(2), 102–113. doi:10.1016/j.intmar.2012.01.002

Sokolova, K., & Kefi, H. (2020). Instagram and YouTube bloggers promote it, why should I buy? How credibility and parasocial interaction influence purchase intentions. *Journal of Retailing and Consumer Services*, *53*, 101742. doi:10.1016/j.jretconser.2019.01.011

Spencer, M. (2016). Agile EA – the MVP – enterprise architect-Building the Business Value Chain. *Enterprise Architect.* https://enterprisearchitect.wordpress.com/2016/04/11/agile-ea-the-mvp/1/2

Spivak, G. C. (1982). The Politics of Interpretations. *Critical Inquiry*, *9*(1), 259–278. https://www.jstor.org/stable/1343284. doi:10.1086/448199

Spivak, G. C. (1985). Three Women's Texts and a Critique of Imperialism. *Critical Inquiry*, *12*(1), 243–261. https://www.jstor.org/stable/1343469. doi:10.1086/448328

Sreekumar, T. T., &. A. A. I. (2020). OTT platforms in the Global South: Challenges and opportunities for digital inclusion. *Journal of Global Information Management*, *28*(4), 50–70.

Srinivasulu, K. (2017). The Caste Question in the Naxalite Movement. *Economic and Political Weekly*, *52*(21), 47–52. https://www.jstor.org/stable/26696125

Srivastava, A. (2023, August 2). *Best Media Info*. Best Media Info. https://bestmediainfo.com/2023/08/as-asia-cup-world-cup-goes-free-on-mobile-hotstar-plays-on-volume-slashes-mobile-ad-rates

Srivastav, S., & Rai, S. (2021). Transforming the entertainment industry: Video on Demand services in India. *Media Asia*, *49*(1), 44–57. doi:10.1080/01296612.2021.1977499

Stadler, J. (2017). Empathy in Film. In H. Maibom (Ed.), *The Routledge Handbook of Philosophy of Empathy* (pp. 317–326). Routledge. doi:10.4324/9781315282015-29

Standard, B. (2022, August 17). Partho Dasgupta: Change in the strategy of OTT platforms needed to succeed in India. https://www.business-standard.com/content/specials/partho-dasgupta-change-in-the-strategy-of-ott-platforms-needed-to-succeed-in-india-122081700945_1.html

Starinieri, M. (2018). *Atos in Media Pathways to digital transformation*. Atos.

Starosta, J., & Izydorczyk, B. (2020, June 22). Understanding the Phenomenon of Binge-Watching-A Systematic Review. *International Journal of Environmental Research and Public Health*, *17*(12), 17. doi:10.3390/ijerph17124469 PMID:32580289

STEM. (2023). *How Does Educational OTT Work?* STEM. https://www.stem.org.uk/resources/community/collection/511682/how-does-educational-ott-work

Stojanov, I. (2015). *Scaling agile using scaled agile framework. Master*. Eindhoven University of Technology. Netherlands.

Stoll, J. "Ott Video Revenue Worldwide 2028." *Statista*, 17 Apr. 2023, www.statista.com/statistics/260179/over-the-top-revenue-worldwide/

Stork, Ch., Nwana, H., Esselaar, S., & Koyabe, M. (2020). *Over The Top (OTT) Applications & the Internet Value Chain Recommendations to Regulators, Policy Makers and Tax Authorities*. Commonwealth Telecommunications Organisation.

Straits. (2023). *Over-the -Top (OTT Market)*. Strait Research. https://straitsresearch.com/report/over-the-top-market

Strelitz, L. (2000). Approaches to Understanding the Relationship between Texts and Audiences. *Communication*, *26*(2), 37–51.

Su, C. (2019). *Changing dynamics of digital entertainment media in China* [Doctoral dissertation, Queensland University of Technology].

Subramanian, B. (2022, March 22). How India is 'Cutting the Cord' to drive an OTT revolution. *Business Insider*. https://www.businessinsider.in: https://www.businessinsider.in/advertising/media/article/how-india-is-cutting-the-cord-to-drive-an-ott-revolution/articleshow/90347924.cms

Su, C., & Flew, T. (2021). The rise of Baidu, Alibaba and Tencent (BAT) and their role in China's Belt and Road Initiative (BRI). *Global Media and Communication, 17*(1), 67–86. doi:10.1177/1742766520982324

Sukwadi, R., Lasiman, S. C., Inderawati, M. M. W., & Suprata, F. (2019). Niche theory within video platform competition: Traditional vs modern. *IOP Conference Series. Materials Science and Engineering, 567*(1), 012003. doi:10.1088/1757-899X/567/1/012003

Sutera, D. (2013). *Sports Fans 2.0: How Fans Are Using Social Media to Get Closer to the Game.* Scarecrow Press.

Swalehin, M. & Nasir, M. S. (2011). Digital Divide in India and China: A Comparison. *Research J. Humanities and Social Sciences, 2*(4), 181-185.

Syynimaa, N. (2015). *Enterprise Architecture Adoption Method for Higher Education Institutions.* [Doctoral Thesis, Informatics Research Centre Henley Business School University of Reading].

T, S. (n.d.). *An Overview of India's Booming OTT Market | marketfeed.* https://www.marketfeed.com/read/en/an-overview-of-indias-booming-ott-market

Tah, S. (2022, May 19). *Indian OTT platforms get box clever to become sustainable.* Voice&Data. https://www.voicendata.com/indian-ott-platforms-get-box-clever-to-become-sustainable/

Tamandehrou, A., & Khan, S. U. (2015). Assessing the importance of globalisation and new media technology in 21st century: An analytical overview. *International Journal of Multifaceted & Multilingual Studies, 1*(4).

Tang, W., & Wei, M. (2023). Streaming media business strategies andaudience-centered practices: A comparativestudy of Netflix and Tencent Video. *Online Media Glob. Commun., 2*(1), 3–24. doi:10.1515/omgc-2022-0061

Taqdees, M. (2020, September 05). *Queer Representation in Churails: A Review.* Naya Daur. https://nayadaur.tv/05-Sep-2020/queer-representation-in-churails-a-review

TCG. (2021). *Transforming OTT Business with Big Data & Analytics. A TCG Digital Whitepaper.* TCG Digital.

Teece, D. J. (2007). Explicating dynamic capabilities: The nature and microfoundations of (sustainable) enterprise performance. *Strategic Management Journal, 28*(13), 1319–1350. doi:10.1002/smj.640

Tefertiller, A., & Sheehan, K. (2019). TV in the streaming age: Motivations, behaviors, and satisfaction of post-network television. *Journal of Broadcasting & Electronic Media, 63*(4), 595616. doi:10.1080/08838151.2019.1698233

Tele.net. (2022, May). *OTT Action: Demand rises for high quality content and seamless streaming.* Tele.net. https://tele.net.in: https://tele.net.in/ott-action-demand-rises-for-high-quality-content-and-seamless-streaming/

TeleStream. (2023). *What is OTT?* TeleStream. https://www.telestream.net/video/solutions/what-is-ott.htm

Thanh, T., & Thahn, N. (2015). The Interconnection Between Interpretivist Paradigm and Qualitative Methods in Education. *American Journal of Education Science, 1*(2), 24–27.

The Subaltern in South Asian History and Society. Report of a Conference. (1983). *Economic and Political Weekly, 18*(9), 299–301. https://www.jstor.org/stable/4371898

The Telegraph. (2022, March 22). Hoichoi scores a century with its Originals. *The Telegraph.* https://www.telegraphindia.com/my-kolkata/lifestyle/hoichoi-scores-a-century-with-its-originals/cid/1857039

The Tribune. (2021). Rise and Expansion of OTT Platforms: Making Quality Content Accessible to All. *The Tribune*. https://www.tribuneindia.com/news/brand-connect/rise-and-expansion-of-ott-platforms-making-quality-content-accessible-to-all-indian-355860

The Tribune. (2021, September 19). OTT Platforms, books, movies Thrive Together. *The Tribune*. https://www.tribuneindia.com/news/amritsar/ott-platforms-books-movies-series-thrive-together-313274

The World Bank. (2023). *Digital Development*. World Bank. https://www.worldbank.org/en/topic/digitaldevelopment/overview

Thomas, P. D. (2018). Refiguring the Subaltern. *Political Theory*, *46*(6), 861–884. https://www.jstor.org/stable/26617620. doi:10.1177/0090591718762720

Thune, T. (2011). Success Factors in Higher Education–Industry Collaboration: A case study of collaboration in the engineering field. *Tertiary Education and Management*, *17*(1), 31–50. doi:10.1080/13583883.2011.552627

Times of India. (2022, December 7). Subhashree Ganguly's debut OTT project 'Indubala Bhater Hotel' to stream soon. *Times of India*. https://timesofindia.indiatimes.com: https://timesofindia.indiatimes.com/tv/news/bengali/subhashree-gangulys-debut-ott-project-indubala-bhater-hotel-to-stream-soon/articleshow/96049509.cms

Tiwary, I. (2020). Amazon Prime Video: A Platform Ecosphere. In Global transformations in media and communication research (pp. 87–106). Springer International Publishing. doi:10.1007/978-3-030-44563-8_5

Tiwary, I. (2023). Streaming and India's film-centred video culture: Linguistic and formal diversity. *International Journal of Cultural Studies*, *0*(0). Advance online publication. doi:10.1177/13678779231197696

Today, B. (n.d.). *Netflix Next*. Netflix Next. https://www.businesstoday.in/interactive/immersive/netflix-india-popular-ott-platform-in-india/

TOI. (2023, June 6). Film exhibitor's body calls for theatre shutdown in protest against '2018's OTT release: Malayalam Movie News - Times of India. *The Times of India*. https://m.timesofindia.com/entertainment/malayalam/movies/news/film-exhibitors-body-calls-for-theatre-shutdown-in-protest-against-2018s-ott-release/articleshow/100797557.cms

Toppan. (2021). *Cultural Sensitivity: How to Market Your Brand to Other Culture*. Toppan. https://toppandigital.com/us/blog-usa/cultural-sensitivity-market-brand-cultures/

Trad, A. & Kalpić, D. (2022c). *Business Transformation Project's Holistic Agile Management (BTPHAM) CBER 2022*. Research Gate.

Trad, A., & Kalpić, D. (2022a). Business Transformation Projects based on a Holistic Enterprise Architecture Pattern (HEAP)-The Basics. IGI. USA.

Trad, A., & Kalpić, D. (2022b). Business Transformation Projects based on a Holistic Enterprise Architecture Pattern (HEAP)-The Implementation. IGI. USA.

Trad, A. (2015a). *A Transformation Framework Proposal for Managers in Business Innovation and Business Transformation Projects-Intelligent aBB architecture*. Centeris.

Trad, A. (2015b). *A Transformation Framework Proposal for Managers in Business Innovation and Business Transformation Projects-An ICS's atomic architecture vision*. Centeris.

Trad, A. (2022d). *Educational Transformation Project's Remote Group Work (ETPRGW). Shaping Online Spaces Through Online Humanities Curricula*. IGI-Global.

Trad, A. (2023b). *Business Transformation Projects: The Role of Psychology-Based Resistance (RPbR). Book: Considerations on Education for Economic, Social, and Environmental Sustainability*. IGI Global.

Trad, A. (2023d). *Organizational and Digital Transformation Projects-A Mathematical Model for Building Blocks based Organizational Unbundling Process*. IGI Global.

Trad, A. (2023e). *Academic and Educational Transformation Projects-The Role Team-Based Learning in Polymathics (RTBLP)*. IGI Global.

Trad, A., & Kalpić, D. (2014b). *The Selection, and Training Framework selection and training framework (STF) for Manager's in Business Innovation Transformation Projects–Educational Recommendations*. Journal EDEN.

Trad, A., & Kalpić, D. (2020a). *Using Applied Mathematical Models for Business Transformation. IGI Complete Author Book*. IGI Global. doi:10.4018/978-1-7998-1009-4

Trad, A., & Kalpić, D. (2023a). *Educational Transformation Project's Remote Group Work (ETPRGW). Book: Shaping Online Spaces Through Online Humanities Curricula*. IGI Global.

Trail, G. (2012). *Manual for the Motivation Scale for Sport Consumption (MSSC)*. Sport Consumer Research Consultants LLC.

Trust, T. (2020). *Stories of Change*. Tata Trust.

Tryon, C. (2015). TV got better: Netflix's original programming strategies and the onset of "binge" viewing. *Media Industries Journal*, *2*(2), 104–117.

Tsai, L. L. (2023). A deeper understanding of switching intention and the perceptions of non-subscribers. *Information Technology & People*, *36*(2), 785–807. doi:10.1108/ITP-04-2021-0255

Turner, G. (2004). Understanding celebrity. *Sage (Atlanta, Ga.)*.

UI. (2023). *Two Keys to Marketing Success: Case Studies on OTT Media*. UI. https://fia.ui.ac.id/en/dua-kunci-sukses-marketing-studi-kasus-pada-ott-media/

Ulin, J. (2013). *The business of media distribution: Monetizing film, TV and video content in an online world*. CRC Press. doi:10.4324/9780240824543

UN. (2021). *Leveraging digital technologies for social inclusion*. UN. https://www.un.org/development/desa/dspd/2021/02/digital-technologies-for-social-inclusion/

Unesco, (2022). *Digital Learning Week*. UNESCO. https://www.unesco.org/en/weeks/digital-learning [Accessed 01 July 2023].

UNICEF. (2019). Digital literacy for children:exploring definitions and frameworks. UNICEF Office of Global Insight and Policy.

Upadhye, A. (2019b, April 4). Regional Content - The next big thing for digital in India. *Reputation Today*. https://reputationtoday.in/regional-content-the-next-big-thing-for-digital-in-india/

Vaidya, R. (2019, March 8). *What Kusha Kapila learnt from being famous on the internet*. Vogue India. https://www.vogue.in/content/kusha-kapila-interview-indian-feminism-instagram-influencer

Vakaao. (2020, November 01). *Churails Season 1 Episode 6 – Husn*. Vakaao. https://vakaao.com/news/churails-season-1-episode-6-husn

van Deursen, A. J. A. M., Helsper, E. J., & Eynon, R.J.A.M. (2016). Development and validation of the Internet Skills Scale (ISS). *Information Communication and Society*, *19*(6), 804–823. doi:10.1080/1369118X.2015.1078834

Van Esler, M. (2019). Market strategies of video on demand platforms. *International Journal on Media Management*, *21*(1), 41–66.

Van Heerden, M. (2019). A little kind of community: South African students dance for self, other and society. In M. Marx (Ed.), *Dance and the quality of life* (pp. 383–397). Springer Publishers.

Vasudevan, R. (2008). Shifting Codes, Dissolving Identities. *Third Text,* 10(34), 59-77. Retrieved from: http://dx.doi.org/. doi:10.1080/09528829608576595 62

Vemireddy, V. (2011). Ram Gopal Varma as an Auteur. *Journal of Creative Communications*, *6*(1-2), 195–202. doi:10.1177/0973258613499218

Vermeulen, J. (2022). To nudge or not to nudge: News recommendation as a tool to achieve online media pluralism. *Digital Journalism (Abingdon, England)*, *10*(10), 1671–1690. doi:10.1080/21670811.2022.2026796

Vidooly. (n.d.). *Industry Insights Archives*. Vidooly. https://vidooly.com/blog/category/video-industry-insights/

Virdi, J. (1999). Reverence, rape- and then revenge: Popular Hindi cinema's 'women's film'. *Screen*, *40*(1), 17–37. http://screen.oxfordjournals.org. doi:10.1093/screen/40.1.17

Vogan, T. (2014). *Keepers of the flame: NFL Films and the rise of sports media*. University of Illinois Press. doi:10.5406/illinois/9780252038389.001.0001

Vogel, H. L. (2020). *Entertainment industry economics: A guide for financial analysis*. Cambridge University Press. doi:10.1017/9781108675499

Voorveld, H. A., & van der Goot, M. (2013). Age differences in media multitasking: A diary study. *Journal of Broadcasting & Electronic Media*, *57*(3), 392–408. doi:10.1080/08838151.2013.816709

VS. G. (2022, February 8). The "homely" shift in Malayalam cinema. *The Indian Express*. https://indianexpress.com/article/entertainment/malayalam/the-homely-shift-in-malayalam-cinema-7761953/

Vyavahare, R. (2022, September 17). Strong, real, imperfect: Is OTT redefining female characters? *TOI*. https://timesofindia.indiatimes.com/entertainment/hindi/bollywood/news/strong-real-imperfect-is-ott-redefining-female-characters/articleshow/94253705.cms

Wang, J., & Yu, X. (2021). Three currents of platformization in China. *The South Atlantic Quarterly*, *120*(4), 777–794. doi:10.1215/00382876-9443336

Wang, W. Y., & Lobato, R. (2019). Chinese video streaming services in the context of global platform studies. *Chinese Journal of Communication*, *12*(3), 356–371. doi:10.1080/17544750.2019.1584119

Wann, D. L., & James, J. D. (2019). *Sport fans: The psychology and social impact of fandom* (2nd ed.). Routledge.

Warschauer, M., & Matuchniak, T. (2010). New technology and digital worlds: Analyzing evidence of equity in access, use, and outcomes. *Review of Research in Education*, *34*(1), 179–225. doi:10.3102/0091732X09349791

Wayne, M. L. (2018). Netflix, Amazon, and branded television content in subscription video on-demand portals. *Media Culture & Society*, *40*(5), 725–741. doi:10.1177/0163443717736118

Weinish, S., & Lev, A. (2021). Who leads The Last Dance? *International Review for the Sociology of Sport, 56*(3), 436–441. doi:10.1177/1012690220969345

Westwood, R. (2004). Comic relief: Subversion and catharsis in organizational comedic theatre. *Organization Studies, 25*(5), 775–795. doi:10.1177/0170840604042414

Wheeler, M. (2014). The mediatization of celebrity politics through the social media. *International journal of digital television, 5*(3), 221-235.

Wong, S. (2015). *Over The Top TV (OTT), Big Data and The Future Of Predictive Buyer Analytics*. DXC.

Won, J. U., & Kitamura, K. (2006). Motivational factors affecting sports consumption behavior of K-league and J-league spectators. *International Journal of Sport and Health Science, 4*, 233–251. doi:10.5432/ijshs.4.233

Wright, N. S. (2015). *Bollywood and Postmodernism: Popular Indian Cinema in the 21st Century*. Edinburgh University Press. doi:10.3366/edinburgh/9780748696345.001.0001

Xalxo, J. (2020, June 4). *Kenny Sebastian: "everyone warned me that comedy is a bubble."* Rolling Stone India. https://rollingstoneindia.com/kenny-sebastian-everyone-warned-me-that-comedy-is-a-bubble/

Xiao, H. F. (2019). *Youth economy, crisis, and reinvention in twenty-first-century China: Morning sun in the tiny times*. Routledge. doi:10.4324/9780429326905

Xu, M., Li, B., Scott, O. K. M., & Wang, J. J. (2023). New platform and new excitement? Exploring young educated sport customers' perceptions of watching live sports on OTT services. *International Journal of Sports Marketing & Sponsorship, 24*(4), 682–689. doi:10.1108/IJSMS-07-2022-0140

Yin, R. K. (2009). *Case Study Research, Design and Method*. Sage Publications Ltd.

Yoon, J. H., & Kim, H. K. (2023). Why do Consumers Continue to Use OTT Services? *Electronic Commerce Research and Applications, 60*, 101285. doi:10.1016/j.elerap.2023.101285

YouTube. (2014b). *STAND UP COMEDY SPECIAL PROMO - Kenny Sebastian : Journey to the Center of My Brain*. [Video]. YouTube. https://www.youtube.com/watch?v=Mh-Nh9xs4pw

YouTube. (2023). *The Making of MostlySane | Prajakta Koli on Relationships, Money & Movies*. [Video]. YouTube. https://www.youtube.com/watch?v=J8iFOoUKLHo&t=2291s

Zandia, F., & Tavana, M. (2011). *A fuzzy group multi-criteria enterprise architecture framework selection model. Management Information Systems, Lindback Distinguished Chair of Information Systems, La Salle University, Philadelphia, PA 19141*. Elsevier.

Zankar, A. (2013). *Mughal-E-Azam*. HarperCollins Publishers India.

Zhang, C. (2019). China's new regulatory regime tailored for the sharing economy: The case of Uber under Chinese local government regulation in comparison to the EU, US, and the UK. *Computer Law & Security Report, 35*(4), 462–475. doi:10.1016/j.clsr.2019.03.004

Zhang, X. (2022). An Engagement-Based Model: Chinese Online Video Streaming Services for Chinese Viewers. In *Streaming and Screen Culture in Asia-Pacific* (pp. 163–180). Springer International Publishing. doi:10.1007/978-3-031-09374-6_9

Zhao, E. J. (2019). *Digital China's informal circuits: Platforms, labour and governance*. Routledge. doi:10.4324/9781315173795

Zhao, E. J., & Keane, M. (2013). Between formal and informal: The shakeout in China's online video industry. *Media Culture & Society*, *35*(6), 724–741. doi:10.1177/0163443713491301

Zreik, M. (2023a). From Boom to Bust: A Study of China's Economy in the Wake of COVID-19 Outbreak in H1 2020. *BRICS Journal of Economics*, *4*(1), 147–171. doi:10.3897/brics-econ.4.e101050

Zreik, M. (2023b). Stirring Up Soft Power: The Role of Chinese Cuisine in China's Cultural Diplomacy. In K. Kankaew (Ed.), *Global Perspectives on Soft Power Management in Business* (pp. 292–306). IGI Global. doi:10.4018/979-8-3693-0250-7.ch015

Zuora. (2021). *OTT CONTENT PRICING STRATEGIES.* Zuora. https://www.zuora.com/guides/ott-pricing-strategies/

# About the Contributors

**Kanika Arya** is a PhD in Film Studies; post-graduation in Mass Communication and English literature, A UGC-NET qualified educator with fourteen years of teaching experience at varied colleges affiliated to University of Mumbai. Her most recent is a book which is a commentary on RK Narayan's work. Her academic oeuvre includes working as a full-time faculty with Amity University, Noida; SPDS, Thane and K.C.College; besides she has actively worked as a visiting faculty at Birla College, Kalyan; Bedekar College, Thane; Mumbai University- Thane Campus; National College, Bandra; Vidyalankar School of Information Technology, Wadala and Hinduja College, Mumbai. She has been a syllabus reframing committee member for Mass Media at University of Mumbai, a paper setter, resource person and viva voce examiner for mass media course at various colleges affiliated with University of Mumbai .She is an active blogger and an explorative writer; She is a Hindi cinema enthusiast albeit a Film Appreciation course with FTII and my PhD research labyrinth on Hindi cinema has broadened the horizons of cinematic world for me. She is an observant writer and whatever intrigues her; she collects on her blog; besides she writes film reviews and some didactic pieces.

**Nitesh Behare** is an accomplished author with a wealth of knowledge and 15 years of experience in the academic field. He currently holds the position of Associate Professor at the Balaji Institute of International Business (BIIB), Sri Balaji University Pune (SBUP). Throughout his career, Dr. Behare has made significant contributions to the academic world through the publication of numerous research papers and book chapters, establishing himself as a recognized authority in his field.

**Lokesh Chakma** is a Research Scholar, pursuing his Ph.D. degree from Visva-Bharati, Santiniketan, India. He has worked as an Assistant Professor (full-time) in Journalism and Mass Communication at the Sidho Kanho Birsha University, West Bengal, India. He represented India in a prestigious India-Greece research project supported by the US Government and was selected to present his observations at the UNESCO Global Media Literacy Week in 2021. He is also a Citizen Historian and former Field Officer and Story Scholar at the 1947 Partition Archive.

**Jennings Joy Chibike** is a Media and Communication Lecturer in the department of Languages at Lupane State University in Zimbabwe. His research interests are in indigenous language media, new media, identity politics, decoloniality, human rights and democracy and political communication

**Sayani Das** is a researcher with over two years of experience in Mass Communication and Journalism. Presently, am working as a Guest Lecturer at the Institute of Mass Communication, Film, and Television

Studies, Kolkata. This role has not only given me the chance to share my thoughts, but it has also allowed me to learn from the amazing next generation of media enthusiasts. In 2022, I had the amazing honour of receiving the Gold Medal from The University of Burdwan for my outstanding achievement in my M.A. in Mass Communication. I've always been so curious and fascinated by the profound impact that mass media has on society. It's something that really drives me! I have been fortunate to conduct extensive research, which has resulted in several publications for international conferences and book chapters. These experiences have helped me develop expertise in my field and demonstrate my commitment to it. I'm always on the lookout for exciting challenges and opportunities to learn and grow.

**Sonali Jha** is a Ph.D. student in the School of Media Arts and Studies at Ohio University. Sonali holds a bachelor's and a master's degree in English Literature. Sonali has two years of experience in the marketing industry working as a content writer. She focuses on unraveling media and social media usage inequalities, examining how it could help in creating an impact in society such as Human Trafficking. She explores the subjective nature of people's tech relationships, along with the significance of comedy in raising awareness. Sonali is passionate about interpersonal communication, media's impact on daily life, digital inequalities, and international media.

**Sadaf Khan** has extensive teaching experience of more than ten years. A core academician and a prolific researcher, she has done incredible work in the domain of English Literature, Cultural Studies, and Afghanistan Studies. Having an international teaching experience, she has been felicitated multiple times as 'Best Teacher' by various academic institutions in which she had worked. Besides, she has also served as the Joint Editor of Women's College Annual magazine of Aligarh Muslim University for the session 2007-2008.

**Saad Ullah Khan** has more than a decade long extensive experience at university level teaching and is currently working at Amity School of Communication, Amity University Rajasthan, Jaipur (India). He has authored various research papers in journals of national as well as international repute. He was awarded 'Rajat Jayanti Vigyan Sancharak Fellowship-2009' by DST, Ministry of Science and Technology for his contribution to the field of communication and its social implications. Further he was bestowed with MANF Fellowship 2011 by UGC to carry out his research work. A PhD from prestigious Aligarh Muslim University, Dr. Saad Ullah Khan has presented 90 Research Papers in various National and International Conferences. He was awarded 'Best Researcher Award' twice (2020 & 2023) and 'Best Journalism Teacher Award' in 2021.

**Sebastiano Mereu** is a Researcher for the Sports Business Research Academy, as well as Lecturer of Digital Marketing and Branding at BHMS Business & Hotel Management School and Kalaidos University of Applied Sciences in Switzerland. His main research interests include second-screening and spectator sports, uses and gratifications of watching unscripted shows, metaverse adoption, retro marketing and apparel promotion, driving social media engagement, intentions to watch women's sports, and more. He also works as a Digital Marketing and Communications Specialist for the Swiss National Museum. Dr. Mereu holds a Doctor of Business Administration degree and a Master of Science in Marketing degree from Heriot-Watt University, UK, and a Bachelor of Science (Hons.) in International Management degree from FHNW University of Applied Sciences Northwestern Switzerland.

**Archan Mitra** is an Assistant Professor and Media Center In-charge at the School of Media Film and Entertainment at Sharda University, formerly he was an Assistant Professor at School of Communication, Amity University Kolkata. He is a former Academic Associate (Communication Area) of the Indian Institute of Management Kashipur (Institute of National Importance) and pursuing his doctoral thesis from Visva-Bharati Santiniketan, West Bengal (Central University of National Importance)

**Tanushri Mukherjee** has been a teaching professional with an experience of 18 years. Dr. Tanushri Mukherjee is presently working as Associate Professor at Amity School of Communication, Amity University Rajasthan. Her area of specialization is Public Relations, Corporate Communication and Event Management. She is recipient of many awards like Educator of the Year Award by Exchange4media Group, Distinguished Teacher for Holistic Development of Students Award, Teaching Excellence Award in "PR and Communication" and Media Innovation Award. She has been Session Chair & Guest of Honor in various conferences, member of Editorial Board as well as Reviewer of eminent Journals as well as author of quality research papers.

**Nitish Ojha** is a distinguished scholar, Researcher, and mentor having rich Industrial experience as well as varied experience in Teaching and Research. He has academic experience of over 15 years with renowned universities like American University of Europe, University of Stirling UK, and others where he worked in the capacity of Assistant Professor, also worked as a Core Member of Research and Development in Indian Universities i.e., Chandigarh University, LPU, Amity University. He has worked as a software engineer in Nokia Siemens Network, India. He served in several research projects in different profiles of researchers funded by DST/MCIT/MOHFW Govt. of India having National and International importance at IIT Delhi and in AIIMS.

**Tamanna M. Shah** is an Assistant Professor of Instruction in the Department of Sociology and Anthropology at Ohio University. She completed her Ph.D. in Sociology from the University of Utah. She holds a master's degree in sociology from Kansas State University and a bachelor's degree in economics. She has an interest in comparative political sociology, gender, and race, social change, and inequality. She has authored journal articles and other publications, notably, "Adjustment to Divorce (Spouses)" for the Wiley Blackwell Encyclopaedia of Family Studies, "Chaos and Fear: Creativity and Hope in an Uncertain World" in International Sociology, and book chapters on innovation, sustainability issues, and job readiness among young people. She has also worked with the Asian Development Bank on water and sanitation policy papers.

**Suraj Sharma** is an experienced Chairperson with a demonstrated history of working in the education management industry. Skilled in Sales, Management, Marketing, Strategic Planning, and Business Development. Strong business development professional with a M.B.S focused in International Marketing from Pune University.

**Sanjna Sebastian Thoppil** is a PhD Research Scholar in the Department of English and Cultural Studies at CHRIST University, Bangalore, India. Her research areas of interest include film studies, visual culture, and semiotics. She has published and presented papers on the representation of urban spaces in cinema and the pedagogical potential of films in Indian social studies education.

**Abhishek Vaish** - faculty at IIIT Allahabad and Guest Faculty at Moscow Institute of Physics and Technology, Moscow and University of Computer Studies, Yangon, Myanmar, Sampoerna University, Jakarta and Dhofar University, Salaleh, Oman. He set up a lab of dedicated Cyber Security Lab for University of Computer Studies, Yangon, Myanmar. He has published more than 50 scientific papers in International Journals, conferences, and Book Chapters. He was the person involved behind the foundation and starting the course of 'Information Security' in the Indian Institute of Information Technology, Allahabad, India. He was a scientific advisor to Ministry of Science and Technology from Indian Govt. side to the Russian Federation. He is also an Academic Advocate for the Infosec Audit and Control Association, USA. supervised 5 PhD candidates and 100+ Masters Students, Handled various Cyber Security Projects at Wipro Ltd. The major domains of include web application testing, wireless penetration testing, and Implementation of an Information Security Management system for Bharti Airtel India, Wipro, TCS.

**Manish Verma** is an internationally renowned media educator and leader. After his formative education and doctorate from 'A' and 'A+' Indian universities, industry experience in India's top media houses, he has continued his journey towards excellence with international study and training programmes doing a post-graduate program in Media Arts and Production from the University of Technology Sydney and an Executive Program in Management and Leadership in Education (MLE) from Harvard University. Prof. Verma has under his belt the distinction of being the founder editor of the well-reputed Amity Journal of Media and Communication Studies, and the honour of organizing and hosting three editions of the prestigious International Conference on Media Studies. He has guided many doctoral research and has a number of research publications to his name. In his career spanning 22 years, and as a professor of 12-year standing, he has worked in top-ranking Indian institutions like Amity University, JICM Bhopal, and National Institute of Fashion Technology.

**Shrikant Waghulkar** has 6 years of teaching experience in the field of management education. He has been awarded with Doctorate degree in management faculty by Savitribai Phule Pune University, Pune. His academic interest includes subjects like Research methodology, Personal Management, Rural Marketing, and Digital Marketing. He is passionate social worker and leading the charitable trust aiming to the social elevation of farmer's children by helping them to receive better education.

**Serene Runping Zhu** received her Bachelor of Law from Sichuan University (China), Master of Arts from Sichuan University (China), Master of Communication Studies from Nanyang Technological University (Singapore), and Doctor of Philosophy (August 2020) from the University of Western Australia (Australia). Currently Dr. Zhu is an Associate Professor at School of Journalism and Communication, Lanzhou University, China. Dr. Zhu's research interests include foreign students' and immigrants' media usage and cultural adaptation, women's working conditions from domestic workers to the glass ceiling, representation & framing research, social media & society, AI and journalism education, interactions between humans and virtual influencers and avatars, and the media systems in Australia and China.

**Mohamad Zreik**, a Postdoctoral Fellow at Sun Yat-sen University, is a recognized scholar in International Relations, specializing in China's Arab-region foreign policy. His recent work in soft power diplomacy compares China's methods in the Middle East and East Asia. His extensive knowledge spans Middle Eastern Studies, China-Arab relations, East Asian and Asian Affairs, Eurasian geopolitics, and

Political Economy, providing him a unique viewpoint in his field. Dr. Zreik is a proud recipient of a PhD from Central China Normal University (Wuhan). He's written numerous acclaimed papers, many focusing on China's Belt and Road Initiative and its Arab-region impact. His groundbreaking research has established him as a leading expert in his field. Presently, he furthers his research on China's soft power diplomacy tactics at Sun Yat-sen University. His significant contributions make him a crucial figure in understanding contemporary international relations.

# Index

## F



## G



## H



## I



## K



## L



## M



## N



## O

## P



## R



## S



## T



## U



## V



## Y



## Z

www.ingramcontent.com/pod-product-compliance
Lightning Source LLC
LaVergne TN
LVHW060417300926
849162LV00006B/1456
*9798369335260*